Methods in Molecular Biology

Series Editor
John M. Walker
School of Life and Medical Sciences
University of Hertfordshire
Hatfield, Hertfordshire, AL10 9AB, UK

For further volumes:
http://www.springer.com/series/7651

PTEN

Methods and Protocols

Edited by

Leonardo Salmena

Department of Pharmacology and Toxicology, University of Toronto,
Princess Margaret Cancer Centre, Toronto, ON, Canada

Vuk Stambolic

University of Toronto, Princess Margaret Cancer Centre, Toronto, ON, Canada

Editors
Leonardo Salmena
Department of Pharmacology and Toxicology
University of Toronto
Princess Margaret Cancer Centre
Toronto, ON, Canada

Vuk Stambolic
University of Toronto
Princess Margaret Cancer Centre
Toronto, ON, Canada

ISSN 1064-3745 ISSN 1940-6029 (electronic)
Methods in Molecular Biology
ISBN 978-1-4939-3297-9 ISBN 978-1-4939-3299-3 (eBook)
DOI 10.1007/978-1-4939-3299-3

Library of Congress Control Number: 2015956040

Springer New York Heidelberg Dordrecht London

Cover image: Phosphorylated Akt (green) in a mouse pten-/- prostate gland.(Belongs to Leonardo Salmena)

Printed on acid-free paper

Humana Press is a brand of Springer
Springer Science+Business Media LLC New York is part of Springer Science+Business Media (www.springer.com)

Preface

The discovery of PTEN was made possible by the advancement of methodologies that identify the localization of allelic loss to specific chromosomal regions including cytogenetics, haplotype mapping, and linkage analysis. Similarly, the unique functional and biochemical features of the PTEN including dynamic and regulated subcellular localization, and intriguing phosphatase activity and unique catalytic pocket, extensive posttranslational modifications, and conformational heterogeneity resulting from its phosphorylation continue to pose great difficulties for studying PTEN function. Just as cell lines, antibodies, PTEN animal models, and enzymatic assays have allowed researchers to shed light on the fascinating biology of PTEN as we know it today, novel discoveries continue to be dependent upon the development of unique tools and novel experimental methodologies.

In this volume entitled *PTEN: Methods and Protocols*, experts in the PTEN field have contributed detailed descriptions of some of the prevailing and novel techniques that allow the study of PTEN function in disease and biology. Chapters include background information, lists of the materials and reagents, step-by-step protocols, and various notes describing pitfalls or specific intricacies of the respective experiments. These include methods to detect expression levels of PTEN in normal and diseased human specimens (Chapters 3 and 4) and methods to evaluate specific PTEN function in brain cancer (Chapter 5), *PTEN* hamartoma tumor syndrome (Chapter 6), and diabetes (Chapter 7). Early PTEN studies focused on mutation and genomic losses; more recently, advances have been made to understand that PTEN function is intricately controlled at the level of transcriptional stability and protein conformation, localization, activity, and stability. To study such features methods that utilize a novel biosensor to measure PTEN regulation (Chapter 8) in addition to techniques to measure posttranscriptional regulation of PTEN by microRNAs (Chapter 9) and ceRNAs (Chapter 10) or posttranslational regulation through ubiquitination (Chapter 11) are presented. Furthermore, this volume presents methods describing novel techniques to detect PTEN localization (Chapter 12 and 13) and previously unstudied structural features of PTEN measured through X-Ray Crystallography and Hydrogen Deuterium Exchange Mass Spectrometry (Chapter 14). Advanced techniques to investigate the function of PTEN in normal and cancer stem cells are presented in Chapter 14. The final two chapters describe methods to study PTEN function in model organisms including mice and *C. elegans.*

Lastly, we are grateful to all of the scientists and clinicians who have contributed their knowledge, time, and passion to this book. Without them this would not have been possible.

Toronto, ON, Canada

Leonardo Salmena
Vuk Stambolic

Contents

Contributors

SUZANNE J. BAKER • *Department of Developmental Neurobiology, St Jude Children's Research Hospital, Memphis, TN, USA*

JOHN E. BURKE • *Department of Biochemistry and Microbiology, University of Victoria, Victoria, BC, Canada*

MIREIA CASTILLO-MARTIN • *Department of Pathology, Icahn School of Medicine at Mount Sinai, New York, NY, USA*

IAN D. CHIN-SANG • *Department of Biology, Queen's University, Kingston, ON, Canada*

CARLOS CORDON-CARDO • *Department of Pathology, Icahn School of Medicine at Mount Sinai, New York, NY, USA*

CHARIS ENG • *Genomic Medicine Institute, Cleveland Clinic, Cleveland, OH, USA; Lerner Research Institute, Cleveland Clinic, Cleveland, OH, USA; Taussig Cancer Institute, Cleveland Clinic, Cleveland, OH, USA; CASE Comprehensive Cancer Center, Case Western Reserve University, Cleveland, OH, USA; Department of Genetics and Genome Sciences, Case Western Reserve University, Cleveland, OH, USA*

ANABEL GIL • *Centro de Investigación Príncipe Felipe, Valencia, Spain; Department of Hematology and Medical Oncology, Biomedical Research Institute INCLIVA, Valencia, Spain*

CARLOTTA GIORGI • *Department of Morphology, Surgery and Experimental Medicine, Section of Pathology, Oncology and Experimental Biology and LTTA Center, University of Ferrara, Ferrara, Italy*

OLENA GORBENKO • *Princess Margaret Cancer Centre, University Health Network, Toronto, ON, Canada*

AMIT GUPTA • *Institute of Biological Chemistry, Biophysics and Bioengineering, Heriot Watt University, Edinburgh, UK*

JING JIAO • *Department of Molecular and Medical Pharmacology, University of California at Los Angeles, Los Angeles, CA, USA*

NISHA KRIPLANI • *Institute of Biological Chemistry, Biophysics and Bioengineering, Heriot Watt University, Edinburgh, UK*

HONG LEI • *The MOE Key Laboratory of Cell Proliferation and Differentiation, School of Life Sciences, Peking-Tsinghua Center for Life Sciences, Peking University, Beijing, China*

NICHOLAS R. LESLIE • *Institute of Biological Chemistry, Biophysics and Bioengineering, Heriot Watt University, Edinburgh, UK*

EVELYNE LIMA-FERNANDES • *Inserm, U1016, Institut Cochin, Paris, France; CNRS, UMR8104, Paris, France; University of Paris Descartes, Sorbonne Paris Cité, Paris, France; Structural Genomics Consortium, University of Toronto, Toronto, ON, Canada*

JOSÉ I. LÓPEZ • *Department of Pathology, Cruces University Hospital, University of the Basque Country (UPV/EHU), Barakaldo, Spain; Biocruces Health Research Institute, Barakaldo, Bizkaia, Spain*

ANA COLLAZO LORDUY • *Department of Pathology, Icahn School of Medicine at Mount Sinai, New York, NY, USA*

Cynthia T. Luk • *Toronto General Research Institute, University Health Network, Toronto, ON, Canada; Institute of Medical Science, University of Toronto, Toronto, ON, Canada*
Helene Maccario • *Faculte de Pharmacie, Aix Marseille Universite, Marseille, UK*
Andrea Marranci • *Oncogenomics Unit, Istituto Toscano Tumori, Pisa, Italy; University of Siena, Siena, Italy*
Glenn R. Masson • *Laboratory of Molecular Biology, Medical Research Council (MRC), Cambridge, UK*
Sonia Missiroli • *Department of Morphology, Surgery and Experimental Medicine, Section of Pathology, Oncology and Experimental Biology and LTTA Center, University of Ferrara, Ferrara, Italy*
Stanislas Misticone • *Inserm, U1016, Institut Cochin, Paris, France; CNRS, UMR8104, Paris, France; University of Paris Descartes, Sorbonne Paris Cité, Paris, France*
Claudia Morganti • *Department of Morphology, Surgery and Experimental Medicine, Section of Pathology, Oncology and Experimental Biology and LTTA Center, University of Ferrara, Ferrara, Italy*
Jonathan Nakashima • *Department of Molecular and Medical Pharmacology, University of California at Los Angeles, Los Angeles, CA, USA*
Joanne Ngeow • *Genomic Medicine Institute, Cleveland Clinic, Cleveland, OH, USA; Cancer Genetics Service, Division of Medical Oncology, National Cancer Centre, Singapore, Singapore; Oncology Academic Clinical Program, Duke-NUS Graduate Medical School, Singapore, Singapore*
Pier Paolo Pandolfi • *Cancer Research Institute, Beth Israel Deaconess Cancer Center, Departments of Medicine and Pathology, Beth Israel Deaconess Medical Center, Harvard Medical School, Boston, MA, USA*
Antonella Papa • *Monash Biomedicine Discovery Institute and Department of Biochemistry and Molecular Biology, Monash University, Clayton, VIC, Australia*
Paolo Pinton • *Department of Morphology, Surgery and Experimental Medicine, Section of Pathology, Oncology and Experimental Biology and LTTA Center, University of Ferrara, Ferrara, Italy*
Laura Poliseno • *Oncogenomics Unit, Istituto Toscano Tumori, Pisa, Italy; CNR-IFC, Pisa, Italy*
Rafael Pulido • *Centro de Investigación Príncipe Felipe, Valencia, Spain; Biocruces Health Research Institute, Barakaldo, Bizkaia, Spain; IKERBASQUE, Basque Foundation for Science, Bilbao, Spain*
Marcus Ruscetti • *Department of Molecular and Medical Pharmacology, University of California at Los Angeles, Los Angeles, CA, USA*
Francesco Russo • *LISM, CNR, IIT-IFC, Pisa, Italy; Department of Computer Science, University of Pisa, Pisa, Italy*
Leonardo Salmena • *Department of Pharmacology and Toxicology, University of Toronto, Princess Margaret Cancer Centre, Toronto, ON, Canada*
Stephanie A. Schroer • *Toronto General Research Institute, University Health Network, Toronto, ON, Canada*
Suzanne Schubbert • *Department of Molecular and Medical Pharmacology, University of California at Los Angeles, Los Angeles, CA, USA*
Mark G.H. Scott • *Inserm, U1016, Institut Cochin, Paris, France; CNRS, UMR8104, Paris, France; University of Paris Descartes, Sorbonne Paris Cité, Paris, France*
Vuk Stambolic • *University of Toronto, Princess Margaret Cancer Centre, Toronto, ON, Canada*

YVONNE TAY • *Cancer Science Institute of Singapore, Centre for Translational Medicine, National University of Singapore, Singapore, Singapore; Department of Biochemistry, Yong Loo Lin School of Medicine, National University of Singapore, Singapore, Singapore*
TIN HTWE THIN • *Department of Pathology, Icahn School of Medicine at Mount Sinai, New York, NY, USA*
ANDREA TUCCOLI • *Oncogenomics Unit, Istituto Toscano Tumori, Pisa, Italy*
MARIANNA VITIELLO • *Oncogenomics Unit, Istituto Toscano Tumori, Pisa, Italy; University of Siena, Siena, Italy*
ROGER L. WILLIAMS • *Laboratory of Molecular Biology, Medical Research Council (MRC), Cambridge, UK*
MINNA WOO • *Toronto General Research Institute, University Health Network, Toronto, ON, Canada; Institute of Medical Science, University of Toronto, Toronto, ON, Canada; Division of Endocrinology, Department of Medicine, University Health Network, University of Toronto, Toronto, ON, Canada; Toronto General Research Institute, Toronto, ON, Canada*
JOHN F. WOOLLEY • *Department of Pharmacology and Toxicology, University of Toronto, ON, Canada*
HONG WU • *Department of Molecular and Medical Pharmacology, University of California at Los Angeles, Los Angeles, CA, USA; The MOE Key Laboratory of Cell Proliferation and Differentiation, School of Life Sciences, Peking-Tsinghua Center for Life Sciences, Peking University, Beijing, China*
SHUMIN WU • *The MOE Key Laboratory of Cell Proliferation and Differentiation, School of Life Sciences, Peking-Tsinghua Center for Life Sciences, Peking University, Beijing, China*
QINZHI XU • *The MOE Key Laboratory of Cell Proliferation and Differentiation, School of Life Sciences, Peking-Tsinghua Center for Life Sciences, Peking University, Beijing, China*
WENKAI YI • *The MOE Key Laboratory of Cell Proliferation and Differentiation, School of Life Sciences, Peking-Tsinghua Center for Life Sciences, Peking University, Beijing, China*
SHANQING ZHENG • *Department of Biology, Queen's University, Kingston, ON, Canada*
GUO ZHU • *Department of Pathology, University of Tennessee Health Sciences Center, Memphis, TN, USA*
HAICHUAN ZHU • *The MOE Key Laboratory of Cell Proliferation and Differentiation, School of Life Sciences, Peking-Tsinghua Center for Life Sciences, Peking University, Beijing, China*

Part I

Introductory Chapters

Chapter 1

PTEN: History of a Tumor Suppressor

Leonardo Salmena

Abstract

Starting from the discovery of "inhibitory chromosomes" by Theodor Boveri to the finding by Henry Harris that fusing a normal cell to a cancer cell reduced tumorigenic potential, the notion of tumor suppression was recognized well before any tumor-suppressor genes were discovered. Although not the first to be revealed, PTEN has been demonstrated to be one of the most frequently altered tumor suppressors in cancer. This introductory chapter provides a historical perspective on our current understanding of PTEN including some of the seminal discoveries in the tumor suppressor field, the events leading to PTEN's discovery, and an introduction to some of the most important researchers and their studies which have shed light on PTEN biology and function as we know it today.

Key words PTEN, Tumor suppressor, Cancer

1 Inhibitory Chromosomes and Cancer

Cancer was first posited to have a genetic basis in pioneering work by German geneticist Theodor Boveri at the beginning of the 20th century. In his article entitled *In Zur Frage der Entstehung Maligner Tumoren*, Boveri investigated chromosomal segregation and abnormal growth characteristics in sea-urchin eggs [1]. Despite the fact that the notion of a "gene" was not articulated until many years later, a direct line can be drawn from Boveri's near-prophetic predictions about the role of "*chromosomen*" in cancer and the discovery of countless features of cancer biology. This includes the concept that "a tumor cell that proliferated without restraint would be generated if '*teilungshemmende chromosomen*' were eliminated" which points to the existence of inhibitory chromosomes, or as we know them today: tumor-suppressor genes [1, 2].

Precisely as predicated by Boveri in 1914, tumor-suppressor genes are now recognized as the safeguards of the cell, which suppress damaged cells from developing into full-blown tumors. Physiologically, a tumor-suppressor gene's function is almost singularly focused on the preservation of the organism. However,

Leonardo Salmena and Vuk Stambolic (eds.), *PTEN: Methods and Protocols*, Methods in Molecular Biology, vol. 1388, DOI 10.1007/978-1-4939-3299-3_1, © Springer Science+Business Media New York 2016

at the cellular level, tumor-suppressor genes have pleiotropic roles including the maintenance, the repair of, or elimination of cells with anything less than an intact and unaltered genome. For instance, upon the occurrence of a spontaneous DNA damage event, tumor-suppressor genes are poised to "sense" the damage, directly or indirectly, and initiate an appropriate course of action prior to cell division. Tumor suppressors often respond by slowing down cell cycling which will allow other tumor suppressors time to repair the DNA, and in other cases where the damage is beyond repair, tumor suppressors initiate signaling to eliminate the damaged cell. It is increasingly apparent that tumor suppression functions as an exquisitely organized network of tumor-suppressor proteins. Critically, when even one of these genes is lost due to gross chromosomal deletion, or incapacitated due to mutation, a an essential cellular defense is disabled. Such an event can increase the chance of generating cells with uncontrollable growth characteristics, and ultimately leading to cancer.

2 PTEN and the Rise of the Tumor Suppressors

From the 1970s well into the 1980s, the oncogene-driven paradigm of cancer dominated, which suggested that it is the over-expression or abberant activation of certain genes which mainly drives transformation. However, the "seed" for the tumor suppressor era was already in place with the work of Henry Harris and his contemporaries demonstrating that the fusion of a cancer cell with a normal cell could suppress tumorigenicity. This seminal work suggested that normal cells contributed "*something*" to the hybrid cells which suppresses the highly malignant character of the tumor cells" [3]. The next big challenge, and the likely explanation for the long latency for the evolution of a tumor suppressor-driven model of cancer, was the identification of Harris' "*something*."

Although not immediately recognized, a breakthrough came very shortly after in the form of an epidemiological study of patients with retinoblastoma, where Alfred Knudson developed a statistical model that suggested that two distinct genes must acquire mutations for retinoblastoma tumors to develop [4]. This was soon followed by the discovery that the 13q14 locus was frequently deleted in these tumors, and the eventual discovery of the *retinoblastoma* (*Rb*) gene [5–7]. Importantly, it was the discovery that both alleles of *Rb* were lost, which formally validated Knudson's two-hit hypothesis. This hypothesis proposed that mutation or loss of one allele of *Rb* is not sufficient to initiate tumor growth; however, deletion or disabling of the other *Rb* allele results in tumorigenesis [4].

The tumor suppressor era took great strides forward from the mid-1980s until well into the 1990s due to the identification of tumor-suppressor genes that were important in the pathogenesis of a variety of tumors. This was heralded by the ability to designate the localization of allelic loss to specific chromosomal regions. Candidate tumor suppressors identified in this manner included the previously described *Rb* gene on chromosome 13q14 [5–7], the p53 gene on chromosome 17p13 [8], the *Wilms' tumor gene* on chromosome 11pl3 [9, 10], the *DCC* gene on 18q21 [11], and the *BRCA1* gene [12, 13]. In a similar fashion, cytogenetic and molecular studies in the 1980s revealed the frequent presence of partial or complete loss of chromosome 10q23 in brain and prostate cancers [14–18]; this suggested the existence of yet another important tumor-suppressor gene on chromosome 10.

It was not until 1997, almost 20 years ago now, that the identity of a frequently lost tumor suppressor on human chromosome 10q23 was revealed. That year, two independent groups, one led by Ramon Parsons and the other by Peter Steck, cloned and characterized *phosphatase and tensin homologue* (*PTEN*) or *mutated in multiple advanced cancers* (*MMAC*), respectively [19, 20]. Although Steck et al. did hint at a role for PTEN in hereditary disease, it was a collaboration between Charis Eng and Ramon Parson's group that formally recognized *PTEN* as being targeted by germline mutations in patients with the cancer predisposition syndrome known as Cowden disease [21]. In that same year, in their search for novel protein tyrosine phosphatases (PTK), Li and Sun used degenerate oligonucleotide PCR or conserved sequence motif searches designed based on known phosphatase catalytic domains to identify *TEPI* (*TGFβ-regulated and epithelial cell-enriched phosphatase*), which also turned out to be *PTEN* [22]. With those discoveries, the race was on to identify the functions of PTEN in cancer.

3 The Race to Assign a Function to PTEN

By the end of 1997, at least 25 papers on PTEN were published. Already, these provided strong evidence that PTEN was a *bona fide* tumor suppressor, including a flurry of papers demonstrating that *PTEN* was mutated in several different cancers including glioblastoma, prostate cancer, melanoma, endometrial carcinoma, high-grade gliomas, thyroid tumors, and infrequently in breast cancer [23–29]. Concurrently, various groups validated the presence of germline mutations of PTEN in three related, inheritable, neoplastic disorders such as Cowden disease, Lhermitte-Duclos disease, and Bannayan-Zonana syndrome [21, 30–33].

Although it was reported that PTEN shared homology with protein tyrosine phosphatases (PTPs) and the cytoskeletal protein tensin in the original discovery papers [19, 20], a study from Nicholas Tonks' group was the first to demonstrate experimentally that PTEN was functional as a dual-specificity phosphatase, and that naturally occurring PTEN mutations observed in heritable and somatic cancer specimens could disrupt that catalytic activity on synthetic substrates [34]. Additional evidence that PTEN functions as a tumor suppressor was obtained by Furnari et al. who showed that PTEN had a growth suppressor activity in glioma cells [35]. At this point, PTEN represented the first PTP to be implicated as a *bona fide* tumor suppressor. This was a major breakthrough, because at the time several protein tyrosine kinases (PTKs) were known to act as oncogenes by phosphorylating, and thereby activating, numerous substrates and pathways that led to proliferation, survival, and cell growth. Thus the notion that a PTP could oppose such signals was generally predicted; the discovery of PTEN finally fulfilled that foresight.

In 1998, PTEN was the focus of at least 106 publications and this trend of increasing publications continues through to today (*see* Fig. 1). Great leaps forward in the understanding of PTEN function were made in this year. The most important discovery, second only to the discovery of PTEN, was the nature of its substrate. The study by Tonks already clearly showed that PTEN could dephosphorylate phosphoserine, phosphothreonine, and phosphotyrosine residues on highly acidic peptide substrates, suggesting that PTEN was a PTP [34]. However, a potential problem was that such highly acidic phosphorylated domains were not common in nature.

Fundamental insights into the biological function of PTEN would not be revealed until its physiological substrates were identified. Indeed, accepting a substrate contingency, Myers and Tonks debated that PTEN may be a phosphatidylinositol phosphatase in a review paper published in 1997 [36]. In a landmark PTEN paper in 1998, Tomohiko Maehama and Jack Dixon determined that purified PTEN protein was generally poor at catalyzing the dephosphorylation of phosphoproteins [37]. Instead they explored the possibility that PTEN could catalyze an acidic nonprotein substrate. They went on to demonstrate that phosphatidylinositol 3,4,5-trisphosphate (PI(3,4,5)P3) was a much better substrate for PTEN catalysis [37]. Although biologists searching for the elusive PTP tumor-suppressor gene may have been disheartened, this unexpected function for the PTEN tumor suppressor was actually very exciting and immediately shaped a working model where PTEN was in a critical cancer-associated pathway where it opposed the function of the proto-oncogene Akt, which was activated by PI(3,4,5)P3 downstream of PI-3 kinase [38]. That the lipid phosphatase activity of PTEN is critical for its tumor suppressor function was also published that year by the Tonks lab [39].

Later that year, Tamura et al. published two papers describing that PTEN was able to mitigate cell migration through an ability to act as a protein phosphatase by reducing tyrosine phosphorylation on focal adhesion kinase (FAK) [40, 41]. These findings once

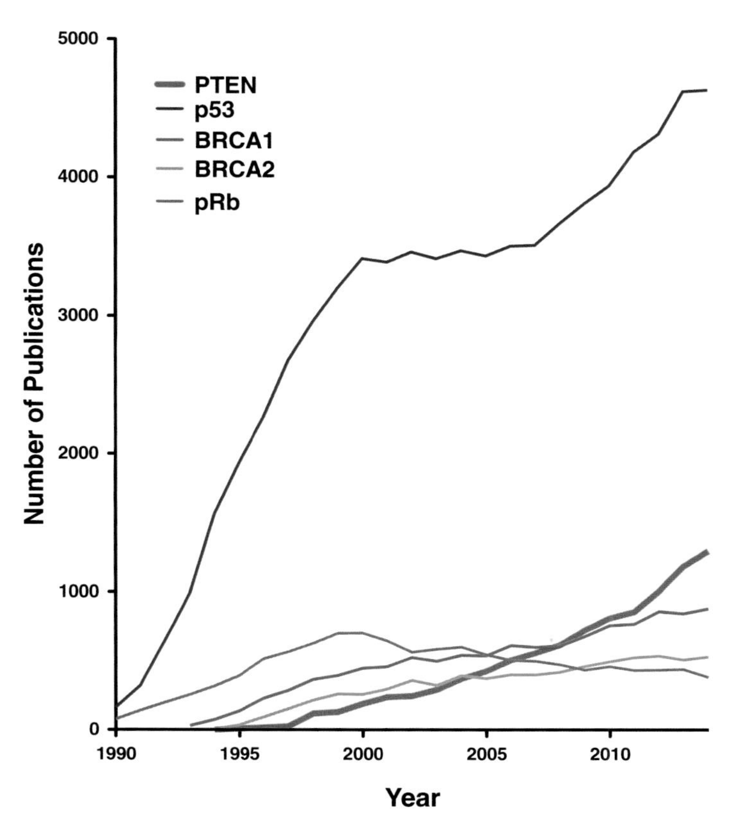

Fig. 1 Annual frequency of publication plot for five most published tumor-suppressor genes

again opened up the possibility that PTEN may function as both a lipid and a protein phosphatase. Despite a hand full of putative protein substrates, it is still, to this day, unclear whether this is truly the case. In fact, there is compelling data demonstrating that PTEN may actually be its own substrate [42]. Needless to say, it is still debated as to whether it is also a protein phosphatase, mainly because of the large majority of data demonstrating that PTEN dephosphorylates inositol phospholipids in vivo, as demonstrated by mouse models later that year.

The first of these mouse models of *Pten* was reported by Pier Paolo Pandolfi, where it was determined that *Pten* is an essential gene that demonstrated tumor suppressive qualities by controlling cellular differentiation and anchorage-independent growth [43]. Shortly thereafter, Tak Mak's group utilized their own iteration of Pten knockout mice to demonstrate experimentally that PTEN exerts its role as a tumor suppressor by negatively regulating the PI3K/PKB/Akt signaling pathway [44], and confirm that Pten was indeed a *bona fide* tumor suppressor as mice heterozygous for the Pten gene had an increased susceptibility to develop various types of tumors [45]. A third mouse developed in Ramon Parsons lab confirmed the importance of Pten tumor suppression in multiple tissue types [46]. Curiously, none of the mouse models recapitulated any of the features of Cowden disease, Lhermitte-Duclos disease, or Bannayan-Zonana syndrome. Nor did the mice develop brain tumors, despite the frequent loss of PTEN in glioblastoma. These data point to possible species-specific differences in mice and humans.

By the next year the group of Nikola Pavletich solved the crystal structure of PTEN, leading to important functional insights into how PTEN was regulated [47]. This confirmed the previous work demonstrating that PTEN was indeed a phosphoinositide phosphatase, and added functional and structural characterization of the tumor-derived mutations in this gene. With this newfound understanding of the catalytic site the question of a putative phosphoprotein substrate was again suggested as the catalytic pocket of PTEN is deep and wide enough to accommodate phosphotyrosine or a phosphoserine/threonine residue. Moreover, biochemical inroads exploring downstream pathways of PTEN function were continuously being reported including an ability to activate apoptosis or anoikis, regulate cell cycle and mitigate migration, and regulate cell growth and metabolism (reviewed in ref. 48). Importantly, in addition to all the functional discoveries, a furious stream of studies were published describing more human cancers and hereditary tumor syndromes that were found to have defective or mutated PTEN gene, firmly establishing PTEN as one of the most commonly mutated genes in human cancer.

4 PTEN and the Future

PTEN continues to be an important and oftentimes unique paradigm for tumor suppression. For instance, although PTEN was discovered during the height of Knudson's two-hit hypothesis, it has been an instrumental example in leading the discussion about the more recent concept of haploinsufficiency and tumor suppression [49]. The concept that PTEN dosage has a profound effect on tumor initiation and progression was demonstrated in a series of hypomorphic Pten mice, in which different levels of PTEN inactivation dictated prostate and breast cancer progression, incidence, latency, and biology in a dose-dependent fashion [50, 51].

Given that the focus was mostly on mutation and function in early days, we have only recently begun to appreciate other critical features of PTEN. For instance the exquisite transcriptional, posttranscriptional, and posttranslational regulation of PTEN is now understood to occur through numerous different mechanisms [52]. PTEN relocalization and its specific function in the nucleus and in extracellular space have identified novel paradigms for

PTEN function [53, 54] [*see* Chapter 2 by Gorbenko and Stambolic]. Moreover, the importance of PTEN in integral cellular processes including cellular senescence, stem cell maintenance, DNA damage repair, and cellular metabolism have also only recently come to light and add to its more established roles in cell cycle, cell survival, and cellular polarity [55]. Important roles of PTEN in non-tumor-associated diseases such as diabetes and autism are also emerging [48].

5 Closing Remarks

To conclude, the histories of PTEN and other tumor suppressors have origins near the beginning of the previous century from the work of Boveri; however a specific understanding of genes with tumor suppressor function accelerated in the last quarter of the century and into the new millennium with the innovative techniques developed by pioneers of the tumor-suppressor gene field including Henry Harris and Alfred Knudson. The research to date indicates that PTEN is a protein that has a function which is exquisitely controlled at numerous levels of cellular regulation. This is not surprising for a protein that plays such a critical cellular role in normal function and in the suppression of tumorigenesis. We await many more major discoveries in the understanding of PTEN biology and expect the trajectory of the PTEN field to continue on its upward trend for years to come.

Acknowledgements

The research in the author's lab is supported by funds from Canada Research Chairs, Human Frontier Science Program Organization, Leukemia and Lymphoma Society of Canada, Canadian Cancer Society, and the Natural Sciences and Engineering Research Council of Canada. Critical reading by J. Kotsopoulos and J.F. Woolley is greatly appreciated.

References

1. Boveri T (1914) In Zur Frage der Entstehung Maligner Tumoren. Gustav Fischer, Jena, pp 1–64
2. Boveri T (2008) Concerning the origin of malignant tumours by Theodor Boveri. translated and annotated by Henry Harris. J Cell Sci 121(Suppl 1):1–84. doi:10.1242/jcs.025742
3. Harris H, Miller OJ, Klein G et al (1969) Suppression of malignancy by cell fusion. Nature 223:363–368. doi:10.1038/223363a0
4. Knudson AG (1971) Mutation and cancer: statistical study of retinoblastoma. Proc Natl Acad Sci U S A 68:820–823. doi:10.1073/pnas.68.4.820
5. Yunis JJ, Ramsay N (1978) Retinoblastoma and subband deletion of chromosome 13. Am J Dis Child 132:161–163. doi:10.1001/archpedi.1978.02120270059012
6. Cavenee WK, Dryja TP, Phillips RA et al (1983) Expression of recessive alleles by

chromosomal mechanisms in retinoblastoma. Nature 305:779–784
7. Lee WH, Bookstein R, Hong F et al (1987) Human retinoblastoma susceptibility gene: cloning, identification, and sequence. Science 235:1394–1399. doi:10.1126/science.3823889
8. Baker SJ, Markowitz S, Fearon ER et al (1990) Suppression of human colorectal carcinoma cell growth by wild-type p53. Science 249:912–915
9. Haber DA, Buckler AJ, Glaser T et al (1990) An internal deletion within an 11p13 zinc finger gene contributes to the development of Wilms' tumor. Cell 61:1257–1269
10. Call KM, Glaser T, Ito CY et al (1990) Isolation and characterization of a zinc finger polypeptide gene at the human chromosome 11 Wilms' tumor locus. Cell 60:509–520
11. Fearon ER, Cho KR, Nigro JM et al (1990) Identification of a chromosome 18q gene that is altered in colorectal cancers. Science 247:49–56
12. Hall JM, Lee MK, Newman B et al (1990) Linkage of early-onset familial breast cancer to chromosome 17q21. Science 250:1684–1689
13. Narod SA, Feunteun J, Lynch HT et al (1991) Familial breast-ovarian cancer locus on chromosome 17q12-q23. Lancet 338:82–83
14. Carter BS, Ewing CM, Ward WS et al (1990) Allelic loss of chromosomes 16q and 10q in human prostate cancer. Proc Natl Acad Sci U S A 87:8751–8755
15. Sonoda Y, Murakami Y, Tominaga T et al (1996) Deletion mapping of chromosome 10 in human glioma. Jpn J Cancer Res 87:363–367
16. Murakami YS, Albertsen H, Brothman AR et al (1996) Suppression of the malignant phenotype of human prostate cancer cell line PPC-1 by introduction of normal fragments of human chromosome 10. Cancer Res 56:2157–2160
17. Fujimoto M, Fults DW, Thomas GA et al (1989) Loss of heterozygosity on chromosome 10 in human glioblastoma multiforme. Genomics 4:210–214. doi:10.1016/0888-7543(89)90302-9
18. Bigner SH, Mark J, Mahaley MS, Bigner DD (1984) Patterns of the early, gross chromosomal changes in malignant human gliomas. Hereditas 101:103–113
19. Li J, Yen C, Liaw D et al (1997) PTEN, a putative protein tyrosine phosphatase gene mutated in human brain, breast, and prostate cancer. Science 275:1943–1947
20. Steck PA, Pershouse MA, Jasser SA et al (1997) Identification of a candidate tumour suppressor gene, MMAC1, at chromosome 10q23.3 that is mutated in multiple advanced cancers. Nat Genet 15:356–362. doi:10.1038/ng0497-356
21. Liaw D, Marsh DJ, Li J et al (1997) Germline mutations of the PTEN gene in Cowden disease, an inherited breast and thyroid cancer syndrome. Nat Genet 16:64–67. doi:10.1038/ng0597-64
22. Li DM, Sun H (1997) TEP1, encoded by a candidate tumor suppressor locus, is a novel protein tyrosine phosphatase regulated by transforming growth factor beta. Cancer Res 57:2124–2129
23. Wang SI, Puc J, Li J et al (1997) Somatic mutations of PTEN in glioblastoma multiforme. Cancer Res 57:4183–4186
24. Cairns P, Okami K, Halachmi S et al (1997) Frequent inactivation of PTEN/MMAC1 in primary prostate cancer. Cancer Res 57:4997–5000
25. Guldberg P, Thor Straten P, Birck A et al (1997) Disruption of the MMAC1/PTEN gene by deletion or mutation is a frequent event in malignant melanoma. Cancer Res 57:3660–3663
26. Tashiro H, Blazes MS, Wu R et al (1997) Mutations in PTEN are frequent in endometrial carcinoma but rare in other common gynecological malignancies. Cancer Res 57:3935–3940
27. Risinger JI, Hayes AK, Berchuck A, Barrett JC (1997) PTEN/MMAC1 mutations in endometrial cancers. Cancer Res 57:4736–4738
28. Rasheed BK, Stenzel TT, McLendon RE et al (1997) PTEN gene mutations are seen in high-grade but not in low-grade gliomas. Cancer Res 57:4187–4190
29. Dahia PL, Marsh DJ, Zheng Z et al (1997) Somatic deletions and mutations in the Cowden disease gene, PTEN, in sporadic thyroid tumors. Cancer Res 57:4710–4713
30. Marsh DJ, Dahia PL, Zheng Z et al (1997) Germline mutations in PTEN are present in Bannayan-Zonana syndrome. Nat Genet 16:333–334. doi:10.1038/ng0897-333
31. Nelen MR, van Staveren WC, Peeters EA et al (1997) Germline mutations in the PTEN/MMAC1 gene in patients with Cowden disease. Hum Mol Genet 6:1383–1387
32. Arch EM, Goodman BK, Van Wesep RA et al (1997) Deletion of PTEN in a patient with Bannayan-Riley-Ruvalcaba syndrome suggests allelism with Cowden disease. Am J Med Genet 71:489–493
33. Lynch ED, Ostermeyer EA, Lee MK et al (1997) Inherited mutations in PTEN that are

associated with breast cancer, cowden disease, and juvenile polyposis. Am J Hum Genet 61:1254–1260. doi:10.1086/301639
34. Myers MP, Stolarov JP, Eng C et al (1997) P-TEN, the tumor suppressor from human chromosome 10q23, is a dual-specificity phosphatase. Proc Natl Acad Sci U S A 94:9052–9057
35. Furnari FB, Lin H, Huang HS, Cavenee WK (1997) Growth suppression of glioma cells by PTEN requires a functional phosphatase catalytic domain. Proc Natl Acad Sci U S A 94:12479–12484
36. Myers MP, Tonks NK (1997) PTEN: sometimes taking it off can be better than putting it on. Am J Hum Genet 61:1234–1238. doi:10.1086/301659
37. Maehama T, Dixon JE (1998) The tumor suppressor, PTEN/MMAC1, dephosphorylates the lipid second messenger, phosphatidylinositol 3,4,5-trisphosphate. J Biol Chem 273:13375–13378
38. Carpenter CL, Cantley LC (1996) Phosphoinositide kinases. Curr Opin Cell Biol 8:153–158. doi:10.1016/S0955-0674(96)80060-3
39. Myers MP, Pass I, Batty IH et al (1998) The lipid phosphatase activity of PTEN is critical for its tumor suppressor function. Proc Natl Acad Sci U S A 95:13513–13518
40. Tamura M, Gu J, Matsumoto K et al (1998) Inhibition of cell migration, spreading, and focal adhesions by tumor suppressor PTEN. Science 280:1614–1617
41. Gu J, Tamura M, Yamada KM (1998) Tumor suppressor PTEN inhibits integrin- and growth factor-mediated mitogen-activated protein (MAP) kinase signaling pathways. J Cell Biol 143:1375–1383
42. Zhang XC, Piccini A, Myers MP et al (2012) Functional analysis of the protein phosphatase activity of PTEN. Biochem J 444:457–464. doi:10.1042/BJ20120098
43. Di Cristofano A, Pesce B, Cordon-Cardo C, Pandolfi PP (1998) Pten is essential for embryonic development and tumour suppression. Nat Genet 19:348–355. doi:10.1038/1235
44. Stambolic V, Suzuki A, de la Pompa JL et al (1998) Negative regulation of PKB/Akt-dependent cell survival by the tumor suppressor PTEN. Cell 95:29–39. doi:10.1016/S0092-8674(00)81780-8
45. Suzuki A, de la Pompa JL, Stambolic V et al (1998) High cancer susceptibility and embryonic lethality associated with mutation of the PTEN tumor suppressor gene in mice. Curr Biol 8:1169–1178. doi:10.1016/S0960-9822(07)00488-5
46. Podsypanina K, Ellenson LH, Nemes A et al (1999) Mutation of Pten/Mmac1 in mice causes neoplasia in multiple organ systems. Proc Natl Acad Sci U S A 96:1563–1568
47. Lee JO, Yang H, Georgescu MM et al (1999) Crystal structure of the PTEN tumor suppressor: implications for its phosphoinositide phosphatase activity and membrane association. Cell 99:323–334
48. Worby CA, Dixon JE (2014) PTEN. Annu Rev Biochem 83:641–669. doi:10.1146/annurev-biochem-082411-113907
49. Berger AH, Knudson AG, Pandolfi PP (2011) A continuum model for tumour suppression. Nature 476:163–169. doi:10.1038/nature10275
50. Trotman LC, Niki M, Dotan ZA et al (2003) Pten dose dictates cancer progression in the prostate. PLoS Biol 1:E59. doi:10.1371/journal.pbio.0000059
51. Alimonti A, Carracedo A, Clohessy JG et al (2010) Subtle variations in Pten dose determine cancer susceptibility. Nat Genet 42:454–458. doi:10.1038/ng.556
52. Salmena L, Carracedo A, Pandolfi PP (2008) Tenets of PTEN tumor suppression. Cell 133:403–414. doi:10.1016/j.cell.2008.04.013
53. Bassi C, Stambolic V (2013) PTEN, here, there, everywhere. Cell Death Differ 20:1595–1596. doi:10.1038/cdd.2013.127
54. Hopkins BD, Hodakoski C, Barrows D et al (2014) PTEN function: the long and the short of it. Trends Biochem Sci 39:183–190. doi:10.1016/j.tibs.2014.02.006
55. Song MS, Salmena L, Pandolfi PP (2012) The functions and regulation of the PTEN tumour suppressor. Nat Rev Mol Cell Biol 13:283–296. doi:10.1038/nrm3330

Chapter 2

PTEN at 18: Still Growing

Olena Gorbenko and Vuk Stambolic

Abstract

Discovered in 1997, PTEN remains one of the most studied tumor suppressors. In this issue of Methods in Molecular Biology, we assembled a series of papers describing various clinical and experimental approaches to studying PTEN function. Due to its broad expression, regulated subcellular localization, and intriguing phosphatase activity, methodologies aimed at PTEN study have often been developed in the context of mutations affecting various aspects of its regulation, found in patients burdened with PTEN loss-driven tumors. PTEN's extensive posttranslational modifications and dynamic localization pose unique challenges for studying PTEN features in isolation and necessitate considerable development of experimental systems to enable controlled characterization. Nevertheless, ongoing efforts towards the development of PTEN knockout and knock-in animals and cell lines, antibodies, and enzymatic assays have facilitated a huge body of work, which continues to unravel the fascinating biology of PTEN.

Key words PTEN, PI3K signaling, Tumor suppressor, Phosphatase, DNA damage repair, Genome integrity

1 Introduction

1.1 PTEN and Disease

PTEN was originally identified as a candidate tumor-suppressor gene frequently deleted at chromosome 10q23 in a number of advanced tumors such as glioblastoma, prostate, kidney, and breast carcinoma [1, 2]. A high rate of PTEN mutations can also be found in high-grade but not low-grade glioblastomas [3–5] and thyroid [6] tumors, as well as in breast [1, 2, 7] and melanoma [8] cell lines. PTEN also appears to be a frequent mutation target in human tumors associated with genomic instability [9]. In addition to genetic alterations in sporadic tumors, germline mutations of PTEN cause the PTEN hamartoma tumor syndromes (PHTS) characterized by high susceptibility for benign hamartomatous tumors throughout the body and increased incidence of cancers of the breast, thyroid, and brain [10–12].

Changes in PTEN function have also been implicated in non-neoplastic disease, including Alzheimer's and Parkinson's diseases,

Leonardo Salmena and Vuk Stambolic (eds.), *PTEN: Methods and Protocols*, Methods in Molecular Biology, vol. 1388, DOI 10.1007/978-1-4939-3299-3_2, © Springer Science+Business Media New York 2016

and other neuronal disorders [13–15]. A few PTEN mutations were found to associate with autism spectrum disorder (ASD) in a group of patients who also display macrocephaly, with and without additional developmental phenotypes characteristic of PHTS [16, 17]. Taken together, these observations emphasize the diverse roles of PTEN in vivo and highlight the importance of understanding its function and the mechanisms by which it exerts its effects.

1.2 PTEN, a Phosphatidylinositol Phosphatase

The human PTEN gene encodes a 403-amino acid polypeptide that belongs to a family of dual-specificity protein phosphatases and more specifically to the Class I Cys-based protein tyrosine phosphatase (PTP), the VH1-like subfamily [1, 2, 18]. Similarly to all PTPs, PTEN contains the catalytic signature motif HCXXGXXRS/T [19]. The cysteine residue within this motif is absolutely required for catalysis, as it acts as a nucleophile to attack the phosphorous atom in the phosphate moiety of its substrate, forming a thiol-phosphate intermediate [19, 20]. Mutation of this cysteine to serine or alanine results in the complete loss of PTEN phosphatase activity [19, 20]. Three functionally selective mutants of PTEN have been discovered and represent an essential toolset for dissection of its lipid and phosphatase activity: catalytically phosphatase-dead PTEN-C124S [20], lipid phosphatase-dead-only PTEN-G129E [21], and protein phosphatase-dead-only PTEN-Y138L [22].

Work from many laboratories has demonstrated that PTEN functions as a key negative regulator of phosphatidylinositol 3′ kinase (PI3K) signaling by counteracting the activity of PI3′K, acting as a 3′ phosphatase for the second messenger products of PI3′K activity, the phosphatidylinositol (3,4,5) trisphosphate (PI(3,4,5)P_3) [20, 21, 23]. By virtue of binding proteins containing pleckstrin homology domains (PHDs), these plasma membrane-embedded second messengers act as hubs promoting functional interactions between various PHD-containing molecules [24]. These include phosphatidylinositol-dependent kinase 1 (PDK1) and protein kinase B/Akt (PKB/Akt), the main cellular mediators of PI3K signals [24]. Acting via a variety of cellular substrates, these kinases have been implicated in control of cell metabolism, growth, proliferation, survival and migration, processes invariably found aberrant in many human tumors [24–26].

1.3 Protein Phosphatase Activity of PTEN

Early experiments with synthetic substrates revealed PTEN protein phosphatase activity in vitro towards highly acidic substrates, such as phospho-tyrosines in polyGlu4Tyr1 and phospho-serine/threonines in DSD and ETE [19, 27]. More recent interest in this aspect of PTEN function has revealed roles for PTEN protein phosphatase activity in control of cell motility, invasion, migration, and maintenance of genome integrity, acting in the nucleus. Some of the findings in this area appear contradictory, if not controversial

[22, 28–31]. The selective use of PTEN lipid-phosphatase-deficient and protein-phosphatase-deficient mutants suggests that PTEN may possess both lipid and protein substrates in control of cell invasion [28, 32, 33].

PTEN also displays considerable auto-dephosphorylation activity [34, 35]. It is thought that PTEN's auto-dephosphorylation of its extensively phosphorylated C-terminal tail counters its folding over the PTEN catalytic domain, thus relieving this auto-inhibitory mechanism [36]. PTEN auto-activation may be particularly relevant in control of spine density in neurons and regulation of neuronal function in normal and disease states [34].

1.4 Means of PTEN Regulation

PTEN appears to be active in most cells and tissues providing tonic downregulation of its targets. PTEN is a generally stable protein with half-life in various cell lines ranging from 4 to 6 h [37]. Its expression appears constitutive [38]; however several stimuli can lead to an acute increase of its transcription [39, 40]. PTEN expression is also subject to negative regulation by promoter methylation [41–43] and silencing via the action of a series of microRNAs [44].

PTEN is a phosphoprotein harboring target sites for a number of serine/threonine protein kinases. Specifically, phosphorylation of a series of sites within the C-terminal 40 amino acids of PTEN by the constitutively active kinases CK2 and GSK-3 is requisite for its stability and tumor suppressor function [45–48]. In addition to these kinases, Rho-dependent kinase ROCK can phosphorylate PTEN on multiple sites within the C2 domain and possibly affect its plasma membrane interactions [49, 50]. PTEN is also acetylated [51], oxidated [52], ubiquitylated [53, 54], and poly(ADP)-ribosylated [55], with these modifications contributing to aspects of PTEN biology beyond the scope of this chapter.

Although initially described as a cytoplasmic protein, PTEN displays complex and dynamic subcellular localization. Through its C2 domain, which resides within the C-terminal half of the protein, PTEN interacts with the hydrophobic moieties within the plasma membrane [56]. Moreover, SUMOylation of PTEN within the C2 domain (lysine 266 in human PTEN) may further contribute to membrane association [57]. PTEN is also associated with the endosomal membranes, via a direct interaction with the endosomal resident lipid, phosphatidylinositol 3 phosphate [58].

PTEN is also readily found in the nuclei of many cultured cells and tissues, including normal breast epithelium [59], proliferating endometrium [60], normal pancreatic islet cells [61], vascular smooth muscle cells [62], follicular thyroid cells [63], squamous cell carcinoma [64], and primary cutaneous melanoma [65]. Significantly, PTEN nuclear exclusion has been associated with more aggressive cancers [66]. Nuclear phosphatidylinositols have been reported, but they exist as part of distinct, partially detergent-resistant proteolipid complexes that are not dynamically regulated and not likely PTEN substrates [67]. SUMOylation of PTEN on lysine 254 is a key to its nuclear retention [68]. Nuclear PTEN plays a prominent role in the cellular DNA damage response, utilizing its protein phosphatase activity and acting in concert with another tumor suppressor, ATM [68]. Moreover, PTEN status is a strong predictor of cellular sensitivity to PARP inhibitors, a class of cytotoxic drugs that target cells with impaired HR repair [69–71].

Recent work reveals that PTEN also exists in the extracellular space. PTEN-L, a PTEN protein originating from an alternate translation start site and containing an additional 173 amino acids at the N-terminus, is secreted, can be found in plasma, and is taken up by the recipient cells, where it downregulates PI3K signaling [72]. PTEN-L may also have functions in mitochondrial respiration [73]. Similarly, PTEN has been found in the exosomes, secreted microvesicles that have been found to contain proteins, lipids, DNA, and microRNAs and transport them between cells. PTEN mono-ubiquitylation appears to be required for PTEN secretion via the exosomes [74]. To reflect the existence of the longer PTEN form, new nomenclature for PTEN has been proposed [75].

2 Summary/Emerging Questions

Approaching 20 years of investigation in PTEN biology, a number of outstanding questions remain unanswered and should ensure an exciting future in PTEN research. Targets of PTEN protein phosphatase activity remain of major interest, especially in the context of control of genome stability and cell invasion and motility. Potential phosphatase-independent functions of PTEN also require further attention. Finally, the differences in the therapeutic response between the human tumors harboring PTEN loss and the activating mutations of the PIK3CA gene (PTEN-loss-associated tumors have generally shown poor response) remain a major challenge in the clinic. The likely dependence of PTEN-deficient tumors on ligand stimulation, as opposed to ligand-independent upregulation of PI3K signaling in mutant PIK3CA-driven cancers, should drive targeted therapeutic development in this area.

Acknowledgements

The work in the authors' laboratory is supported by the funds from the Canadian Institutes of Health Research and the Canadian Cancer Society. OG is supported by the Excellence in Radiation Research for the 21st century (EIRR21st) postdoctoral fellowship.

References

1. Li J et al (1997) PTEN, a putative protein tyrosine phosphatase gene mutated in human brain, breast, and prostate cancer. Science 275(5308):1943–1947
2. Steck PA et al (1997) Identification of a candidate tumour suppressor gene, MMAC1, at chromosome 10q23.3 that is mutated in multiple advanced cancers. Nat Genet 15(4):356–362
3. Bostrom J et al (1998) Mutation of the PTEN (MMAC1) tumor suppressor gene in a subset of glioblastomas but not in meningiomas with loss of chromosome arm 10q. Cancer Res 58(1):29–33
4. Liu W et al (1997) PTEN/MMAC1 mutations and EGFR amplification in glioblastomas. Cancer Res 57(23):5254–5257
5. Rasheed BK et al (1997) PTEN gene mutations are seen in high-grade but not in low-grade gliomas. Cancer Res 57(19):4187–4190
6. Dahia PL et al (1997) Somatic deletions and mutations in the Cowden disease gene, PTEN, in sporadic thyroid tumors. Cancer Res 57(21):4710–4713
7. Teng DH et al (1997) MMAC1/PTEN mutations in primary tumor specimens and tumor cell lines. Cancer Res 57(23):5221–5225
8. Guldberg P et al (1997) Disruption of the MMAC1/PTEN gene by deletion or mutation is a frequent event in malignant melanoma. Cancer Res 57(17):3660–3663
9. Saal LH et al (2008) Recurrent gross mutations of the PTEN tumor suppressor gene in breast cancers with deficient DSB repair. Nat Genet 40(1):102–107
10. Liaw D et al (1997) Germline mutations of the PTEN gene in Cowden disease, an inherited breast and thyroid cancer syndrome. Nat Genet 16(1):64–67
11. Marsh DJ et al (1998) Mutation spectrum and genotype-phenotype analyses in Cowden disease and Bannayan-Zonana syndrome, two hamartoma syndromes with germline PTEN mutation. Hum Mol Genet 7(3):507–515
12. Eng C (1998) Genetics of Cowden syndrome: through the looking glass of oncology. Int J Oncol 12(3):701–710
13. Gupta A, Dey CS (2012) PTEN, a widely known negative regulator of insulin/PI3K signaling, positively regulates neuronal insulin resistance. Mol Biol Cell 23(19):3882–3898
14. Haas-Kogan D, Stokoe D (2008) PTEN in brain tumors. Expert Rev Neurother 8(4):599–610
15. Domanskyi A et al (2011) Pten ablation in adult dopaminergic neurons is neuroprotective in Parkinson's disease models. FASEB J 25(9):2898–2910
16. Spinelli L et al (2015) Functionally distinct groups of inherited PTEN mutations in autism and tumour syndromes. J Med Genet 52(2):128–134
17. Redfern RE et al (2010) A mutant form of PTEN linked to autism. Protein Sci 19(10):1948–1956
18. Alonso A et al (2004) Protein tyrosine phosphatases in the human genome. Cell 117(6):699–711
19. Myers MP et al (1997) P-TEN, the tumor suppressor from human chromosome 10q23, is a dual-specificity phosphatase. Proc Natl Acad Sci U S A 94(17):9052–9057
20. Maehama T, Dixon JE (1998) The tumor suppressor, PTEN/MMAC1, dephosphorylates the lipid second messenger, phosphatidylinositol 3,4,5-trisphosphate. J Biol Chem 273(22):13375–13378
21. Myers MP et al (1998) The lipid phosphatase activity of PTEN is critical for its tumor supressor function. Proc Natl Acad Sci U S A 95(23):13513–13518
22. Davidson L et al (2010) Suppression of cellular proliferation and invasion by the concerted lipid and protein phosphatase activities of PTEN. Oncogene 29(5):687–697
23. Stambolic V et al (1998) Negative regulation of PKB/Akt-dependent cell survival by the tumor suppressor PTEN. Cell 95(1):29–39
24. Manning BD, Cantley LC (2007) AKT/PKB signaling: navigating downstream. Cell 129(7):1261–1274
25. Vivanco I, Sawyers CL (2002) The phosphatidylinositol 3-Kinase AKT pathway in human cancer. Nat Rev Cancer 2(7):489–501
26. Engelman JA, Luo J, Cantley LC (2006) The evolution of phosphatidylinositol 3-kinases as regulators of growth and metabolism. Nat Rev Genet 7(8):606–619
27. Li DM, Sun H (1997) TEP1, encoded by a candidate tumor suppressor locus, is a novel protein tyrosine phosphatase regulated by transforming growth factor beta. Cancer Res 57(11):2124–2129
28. Poon JS, Eves R, Mak AS (2010) Both lipid- and protein-phosphatase activities of PTEN contribute to the p53-PTEN anti-invasion pathway. Cell Cycle 9(22):4450–4454
29. Trotman LC et al (2007) Ubiquitination regulates PTEN nuclear import and tumor suppression. Cell 128(1):141–156
30. Fouladkou F et al (2008) The ubiquitin ligase Nedd4-1 is dispensable for the regulation of

PTEN stability and localization. Proc Natl Acad Sci U S A 105(25):8585–8590
31. Wang X et al (2007) NEDD4-1 is a proto-oncogenic ubiquitin ligase for PTEN. Cell 128(1):129–139
32. Tibarewal P et al (2012) PTEN protein phosphatase activity correlates with control of gene expression and invasion, a tumor-suppressing phenotype, but not with AKT activity. Sci Signal 5(213):ra18
33. Liliental J et al (2000) Genetic deletion of the Pten tumor suppressor gene promotes cell motility by activation of Rac1 and Cdc42 GTPases. Curr Biol 10(7):401–404
34. Zhang XC et al (2012) Functional analysis of the protein phosphatase activity of PTEN. Biochem J 444(3):457–464
35. Williams R, Masson G, Vadas O, Burke J, Perisic O (2015) Structural mechanisms of PI3K and PTEN regulation. FASEB J 29(1):1–6
36. Bolduc D et al (2013) Phosphorylation-mediated PTEN conformational closure and deactivation revealed with protein semisynthesis. Elife 2:e00691
37. Tang Y, Eng C (2006) p53 down-regulates phosphatase and tensin homologue deleted on chromosome 10 protein stability partially through caspase-mediated degradation in cells with proteasome dysfunction. Cancer Res 66(12):6139–6148
38. Han B et al (2003) Regulation of constitutive expression of mouse PTEN by the 5′-untranslated region. Oncogene 22(34):5325–5337
39. Stambolic V et al (2001) Regulation of PTEN transcription by p53. Mol Cell 8(2):317–325
40. Tamguney T, Stokoe D (2007) New insights into PTEN. J Cell Sci 120(Pt 23):4071–4079
41. Garcia JM et al (2004) Promoter methylation of the PTEN gene is a common molecular change in breast cancer. Genes Chromosomes Cancer 41(2):117–124
42. Goel A et al (2004) Frequent inactivation of PTEN by promoter hypermethylation in microsatellite instability-high sporadic colorectal cancers. Cancer Res 64(9):3014–3021
43. Kang YH, Lee HS, Kim WH (2002) Promoter methylation and silencing of PTEN in gastric carcinoma. Lab Invest 82(3):285–291
44. Poliseno L, Pandolfi PP (2015) PTEN ceRNA networks in human cancer. Methods 77–78C: 41–50
45. Vazquez F et al (2000) Phosphorylation of the PTEN tail regulates protein stability and function. Mol Cell Biol 20(14):5010–5018
46. Al-Khouri AM et al (2005) Cooperative phosphorylation of the tumor suppressor phosphatase and tensin homologue (PTEN) by casein kinases and glycogen synthase kinase 3beta. J Biol Chem 280(42):35195–35202
47. Cordier F et al (2012) Ordered phosphorylation events in two independent cascades of the PTEN C-tail revealed by NMR. J Am Chem Soc 134(50):20533–20543
48. Cordier F, Chaffotte A, Wolff N (2015) Quantitative and dynamic analysis of PTEN phosphorylation by NMR. Methods 77–78: 82–91
49. Fragoso R, Barata JT (2015) Kinases, tails and more: regulation of PTEN function by phosphorylation. Methods 77–78:75–81
50. Vemula S et al (2010) ROCK1 functions as a suppressor of inflammatory cell migration by regulating PTEN phosphorylation and stability. Blood 115(9):1785–1796
51. Okumura K et al (2006) PCAF modulates PTEN activity. J Biol Chem 281(36):26562–26568
52. Kwon J et al (2004) Reversible oxidation and inactivation of the tumor suppressor PTEN in cells stimulated with peptide growth factors. Proc Natl Acad Sci U S A 101(47): 16419–16424
53. Wang X et al (2008) Crucial role of the C-terminus of PTEN in antagonizing NEDD4-1-mediated PTEN ubiquitination and degradation. Biochem J 414(2):221–229
54. Bar N, Dikstein R (2010) miR-22 forms a regulatory loop in PTEN/AKT pathway and modulates signaling kinetics. PLoS One 5(5):e10859
55. Li N et al (2015) Poly-ADP ribosylation of PTEN by tankyrases promotes PTEN degradation and tumor growth. Genes Dev 29(2):157–170
56. Lee JO et al (1999) Crystal structure of the PTEN tumor suppressor: implications for its phosphoinositide phosphatase activity and membrane association. Cell 99(3):323–334
57. Huang J et al (2012) SUMO1 modification of PTEN regulates tumorigenesis by controlling its association with the plasma membrane. Nat Commun 3:911
58. Naguib A et al (2015) PTEN functions by recruitment to cytoplasmic vesicles. Mol Cell 58(2):255–268
59. Perren A et al (1999) Immunohistochemical evidence of loss of PTEN expression in primary ductal adenocarcinomas of the breast. Am J Pathol 155(4):1253–1260
60. Mutter GL et al (2000) Changes in endometrial PTEN expression throughout the human menstrual cycle. J Clin Endocrinol Metab 85(6):2334–2338
61. Perren A et al (2000) Mutation and expression analyses reveal differential subcellular compart-

mentalization of PTEN in endocrine pancreatic tumors compared to normal islet cells. Am J Pathol 157(4):1097–1103

62. Deleris P et al (2003) SHIP-2 and PTEN are expressed and active in vascular smooth muscle cell nuclei, but only SHIP-2 is associated with nuclear speckles. J Biol Chem 278(40): 38884–38891
63. Gimm O et al (2000) Differential nuclear and cytoplasmic expression of PTEN in normal thyroid tissue, and benign and malignant epithelial thyroid tumors. Am J Pathol 156(5):1693–1700
64. Tachibana M et al (2002) Expression and prognostic significance of PTEN product protein in patients with esophageal squamous cell carcinoma. Cancer 94(7):1955–1960
65. Whiteman DC et al (2002) Nuclear PTEN expression and clinicopathologic features in a population-based series of primary cutaneous melanoma. Int J Cancer 99(1):63–67
66. Fridberg M et al (2007) Protein expression and cellular localization in two prognostic subgroups of diffuse large B-cell lymphoma: higher expression of ZAP70 and PKC-beta II in the non-germinal center group and poor survival in patients deficient in nuclear PTEN. Leuk Lymphoma 48(11):2221–2232
67. Lindsay Y et al (2006) Localization of agonist-sensitive PtdIns(3,4,5)P3 reveals a nuclear pool that is insensitive to PTEN expression. J Cell Sci 119(Pt 24):5160–5168
68. Bassi C et al (2013) Nuclear PTEN controls DNA repair and sensitivity to genotoxic stress. Science 341(6144):395–399
69. Dedes KJ et al (2010) PTEN deficiency in endometrioid endometrial adenocarcinomas predicts sensitivity to PARP inhibitors. Sci Transl Med 2(53):53ra75
70. McEllin B et al (2010) PTEN loss compromises homologous recombination repair in astrocytes: implications for glioblastoma therapy with temozolomide or poly(ADP-ribose) polymerase inhibitors. Cancer Res 70(13): 5457–5464
71. Mendes-Pereira AM et al (2009) Synthetic lethal targeting of PTEN mutant cells with PARP inhibitors. EMBO Mol Med 1(6–7): 315–322
72. Hopkins BD et al (2013) A secreted PTEN phosphatase that enters cells to alter signaling and survival. Science 341(6144):399–402
73. Liang H et al (2014) PTENalpha, a PTEN isoform translated through alternative initiation, regulates mitochondrial function and energy metabolism. Cell Metab 19(5): 836–848
74. Li Y et al (2014) Rab5 and Ndfip1 are involved in Pten ubiquitination and nuclear trafficking. Traffic 15(7):749–761
75. Pulido R et al (2014) A unified nomenclature and amino acid numbering for human PTEN. Sci Signal 7(332):pe15

Part II

Methods to Detect *PTEN* in Disease

Chapter 3

Immunopathologic Assessment of PTEN Expression

Mireia Castillo-Martin, Tin Htwe Thin, Ana Collazo Lorduy, and Carlos Cordon-Cardo

Abstract

Immunohistochemistry (IHC) is an excellent technique used routinely to define the phenotype in pathology laboratories through the analysis of molecular expression in cells and tissues. The PTEN protein is ubiquitously expressed in the majority of human tissues, and allelic or complete loss of PTEN is frequently observed in different types of malignancies leading to an activation of the AKT/mTOR pathways. IHC-based analyses are best to determine the level of PTEN expression in histological samples, but not to assess partial or heterozygous deletions, for which FISH analyses are more appropriate. Interpretation of the IHC results is the most critical point in the assessment of PTEN expression, since it is used both as a prognostic factor and as a tool to guide therapeutic intervention and response to therapy. Importantly, analyses of well-known downstream markers, such as AKT or mTOR, may be used to further analyze PTEN functional status.

Key words Immunohistochemistry, Immunofluorescence, Formalin-fixed tissue samples, Cultured cells, Fresh-frozen tissue samples, PTEN protocol

1 Introduction

1.1 PTEN and the PI3K Pathway

The phosphatase and tensin homolog on chromosome ten (PTEN) gene is a tumor suppressor mapped to 10q23 and initially characterized in 1997 [1]. It encodes a phosphatase that acts as an antagonist of the PI3K/AKT/mTOR signaling pathways, which are involved in several cellular programs such as proliferation, apoptosis, and metabolism. Recently, the study of The Cancer Genome Atlas (TCGA) reported that PTEN is one of the most commonly mutated tumor suppressors in human cancer [2]. Although somatic alterations of PTEN including point mutations, insertions, and deletions occur throughout the whole gene, there are hot spot mutations at specific amino acids, which are not specific for a particular type of cancer. In certain solid tumors, such as prostate or breast cancers, allelic or complete loss of PTEN is frequently observed [3]. A less abundant longer PTEN protein (described as

Leonardo Salmena and Vuk Stambolic (eds.), *PTEN: Methods and Protocols*, Methods in Molecular Biology, vol. 1388, DOI 10.1007/978-1-4939-3299-3_3, © Springer Science+Business Media New York 2016

PTEN-Long) was recently reported, and it contains 173 additional amino acids in the amino-terminal end, as a result of the usage of an alternative CUG translation initiation site upstream to the canonical AUG sequence used to produce the shorter 403-amino-acid form [4, 5].

The PTEN protein is ubiquitously expressed in the majority of human tissues, and it is readily measurable by immunohistochemistry (IHC). Thus normal tissues can serve as good internal positive control during immunopathological analyses. Through its C2 domain, PTEN is bound to the cell membrane and the phosphatase domain is active in the cytoplasm, dephosphorylating phosphatidylinositol (3,4,5)-trisphosphate (PIP_3), resulting in the inhibition of the AKT signaling pathway. The standard expression pattern of PTEN in normal tissues is localized to the cell cytoplasm, and it corresponds to a homogeneous and solid expression, sometimes with a prominent cell membrane enhancement. PTEN mutation or deletion usually produces impaired protein synthesis, with the subsequent loss of PTEN expression. Due to its inhibitory function, PTEN loss can also be assessed by the evaluation of expression of downstream proteins, such as AKT, mTOR, or pS6-kinase, and their activated, phosphorylated forms. Thus, overexpression of these biomarkers can be used as an indirect indicator of PTEN alteration [6] (see below).

1.2 Immunohistochemistry Procedures

IHC was introduced during the early 1940s, but it was not until the 1980s when development of efficient antigen retrieval methods and the systems of secondary antibody detection allowed an enormous increase in the applicability of the technique in the routine setting of diagnostic anatomic pathology. Hence, IHC is currently an excellent technique used routinely to define the phenotype in pathology laboratories through the analysis of molecular expression in cells and tissues of all their components, including protein and lipid profiles, among others. Compared to other protein analysis techniques, such as Western Blot, IHC has the tremendous advantage of being able to reveal exact protein location within the examined tissue, and even more importantly its subcellular location (e.g., nucleus, cytoplasm, cell membrane, or even intracellular compartments such as Golgi apparatus, nucleolus). In addition, it allows for assessment of functionality, since active (e.g., phosphorylated protein isoforms) or inactive (e.g., deglycosylated proteins) molecules can be defined at the microanatomical detail. Today it represents an important tool for scientific research as well as a fundamental complementary technique in the elucidation of differential diagnosis in the clinical settings which are not determinable by conventional hematoxylin and eosin analyses.

The immunostaining procedure is based in the use of antibodies that recognize specific cell and tissue structures, such as protein epitopes (antigens), in histological specimens. Following a set of steps (see protocols below), the final visualization of the reaction between the antibody and the antigen can be accomplished in different ways. In IHC staining, a high-affinity molecule is conjugated to an enzyme, such as peroxidase or alkaline phosphatase, which in turn catalyzes a color-producing reaction that can be then visualized in the cells or tissue using a bright-field microscope. As an alternative, in immunofluorescence (IF) staining, the antibody (primary or secondary) or the high-affinity molecule can be labeled with a fluorophore, which can be then visualized using a fluorescent microscope with the specific corresponding filters.

Advancement in the past years in antibody generation, in IF protocols, and in microscope systems has allowed the development of a "multiplex immunofluorescence" platform, which permits a quantitative analyses of several antibodies labeled with different fluorophores in the light spectrum (from 350 nm to 700 nm) stained in the same slide [7].

2 Materials

A list of the needed materials to perform these procedures is summarized below:

1. Square glass cover slips (24 × 24).
2. 6-well plates.
3. Phosphate-buffered saline (PBS).
4. Acetone (100 %).
5. Methanol (100 %).
6. Ethanol (95 %, 100 %).
7. Permeabilization buffer: 1 % Triton X100 in PBS.
8. Blocking serum (normal serum).
9. Antibody diluent: 2 % BSA in PBS.
10. Primary antibody.
11. Secondary antibody.
12. Wash buffer: 0.5 % Triton X100 in PBS.
13. Avidin-Biotin complexes (Vectastatin, Vector Laboratories).
14. VECTASTAIN ABC Kit (Standard) (Vector Laboratories, Burlingame, CA).
15. Diaminobenzidine (DAB).
16. Hematoxylin (Harris-modified solution).
17. Xylene.
18. Heating block.
19. Hydrogen peroxide.

20. 10 % Formalin solution.
21. Agarose.
22. Steamer.
23. Pretreatment buffer (pH 6: Citric buffer, pH 8–9: EDTA buffer).
24. Blank sections for positive control tissues with known expression of the target protein.
25. Cytoseal XYL (Richard-Allan Scientific, Kalamazoo, MI).
26. Wet chamber, or a plastic box with lid equipped with a moistened paper towel (*see* **Note 1**).
27. Fixatives for fresh tissue and cells (*see* **Note 2**).
28. 4',6-Diamidino-2-phenylindole, dihydrochloride (DAPI) in mounting medium.
29. Hydrophobic pen.

3 Methods

Characterization of antibodies for immunopathology is mandatory to obtain optimal results. Figure 1 illustrates the method for the optimization of an antibody used in our laboratory. Briefly, protein expression of PTEN is determined in cell lines with well-known PTEN status both by Western blot and immunofluorescence (IF). Once the specificity and sensitivity of the antibody have been assessed, we may try different fixation procedures in fresh-frozen tissue sections and go on with the titration of the antibody (testing different primary antibody dilutions) and antigen retrieval determination in agarose cell blocks. Once the antibody is working in these settings, we are ready to move to FFPE tissue sections for clinical or research assessment of PTEN expression, always using in parallel the specified positive and negative controls.

There is a great variety of procedures for immunohistochemistry staining; however, below we summarize the protocols that we have been using in our laboratory, which have led to successful and consistent results in the different settings of immunopathology included in several publications from our group.

3.1 Immunofluorescence Protocol for Cell Lines Fixed on Glass Cover Slips

First, cells growing on cover slips in 6-well plates with a 60–70 % confluence need to be fixed following these steps:

1. Aspirate the medium and wash with PBS.
2. Fix cells in cold 100 % methanol for 15 min at −20 °C (*see* **Note 3**).
3. Aspirate.
4. Fix cells in cold 100 % acetone for 2 min at −20 °C.

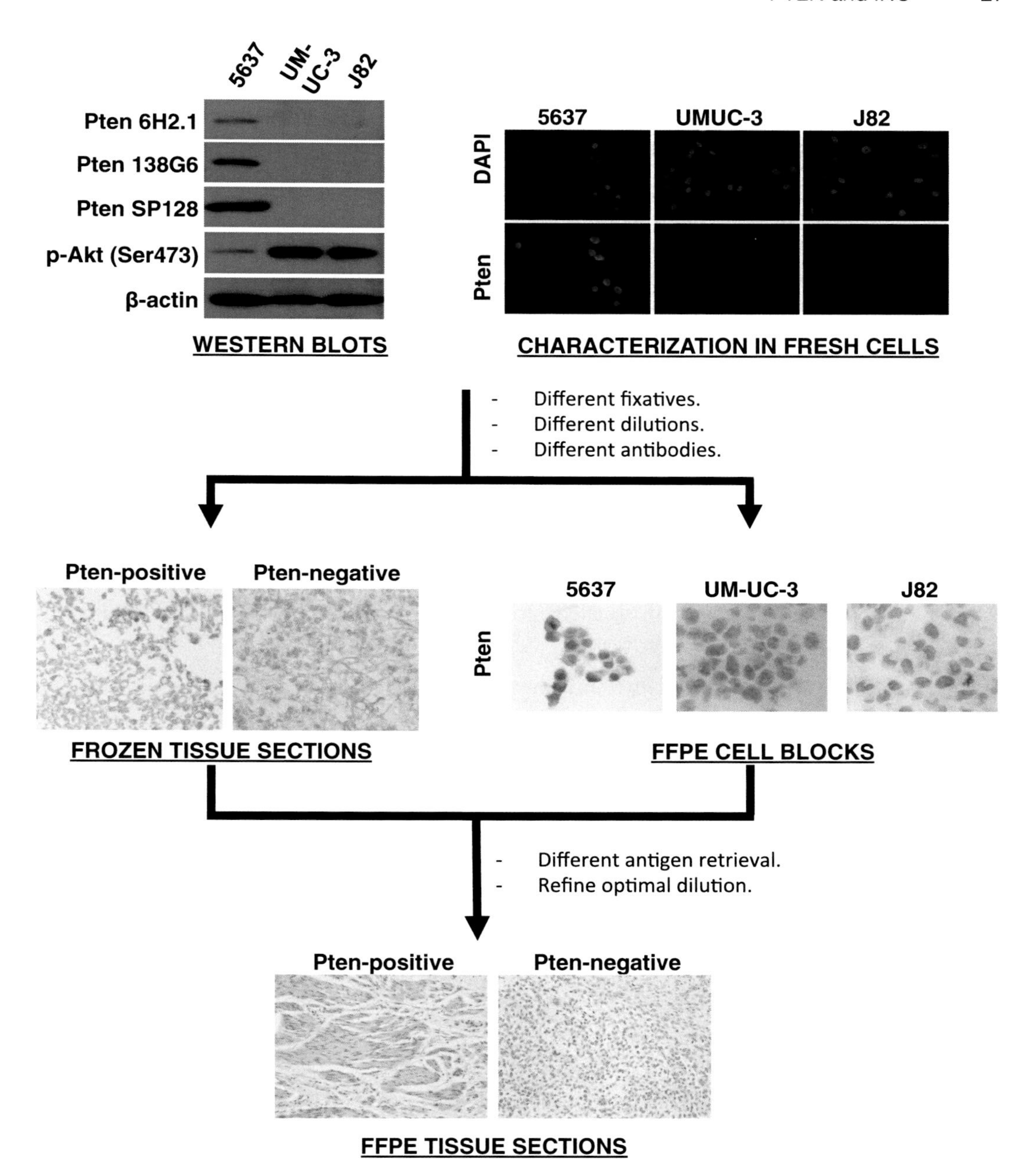

Fig. 1 Antibody characterization and optimization protocol for immunohistochemistry in formalin-fixed paraffin-embedded tissue samples

5. Wash in PBS for 5 min, three times at room temperature (RT).
6. Store in 3 ml PBS. Wrap plates tightly and store at 4 °C until ready to use.

   Then, immunofluorescence staining is performed on the glass cover slips containing the cells as follows:

7. Permeabilize with 1 % Triton X100-PBS at RT for 5 min.
8. Wash in PBS for 5 min, two times at RT.
9. Apply 10 % donkey serum in antibody diluent at RT for 30 min (*see* **Notes 4, 5 and 6**).
10. Aspirate.
11. Apply primary antibody 1.30 h, at RT or overnight at 4 °C.
12. Aspirate.
13. Wash in PBS for 5 min, three times at RT.
14. Apply secondary antibody at 1:200 dilution at RT for 45 min (*see* **Note 7**).
15. Aspirate.
16. Wash coverslips in 0.5%Triton X100-PBS for 5 min, two times at RT.
17. Wash in PBS for 5 min at RT.
18. Counterstain with DAPI (see **Note 8**).

3.2 IHC/IF in Tissue Samples

We propose the following procedures to perform IHC staining in frozen sections.

3.2.1 IHC on Frozen Tissue Samples with Peroxidase

1. Let the slides warm to RT and fix (*see* **Note 2**) for 10 min at RT.
2. Circle the tissue on the slide with a hydrophobic pen.
3. Apply 10 % blocking serum in antibody diluent for 45 min at RT (see **Note 9**).
4. Aspirate.
5. Apply primary antibody and incubate overnight at 4 °C.
6. Aspirate.
7. Rinse well in running tap water for 5 min at RT.
8. Wash in PBS for 5 min, three times at RT.
9. Apply secondary antibody (*see* **Note 10**) and incubate for 45 min at RT.
10. Wash in PBS for 5 min, three times at RT.
11. Apply avidin-biotin complexes (Vectastatin, Vector Laboratories) at a 1:25 dilution and incubate for 30 min at RT.
12. Wash in PBS for 5 min, three times at RT.
13. Leave in 0.5 % Triton X100-PBS for 1 min at RT.
14. Prepare and activate diaminobenzidine (DAB) in 0.5 % Triton X100-PBS for developing (*see* **Notes 11 and 12**).
15. Counterstain with hematoxylin.

16. Dehydrate: Four times in 95 % ethanol for 3 min, four times in 100 % ethanol for 3 min, and four times in xylene for 3 min.
17. Coverslip and let dry under the hood.

3.2.2 IHC on Formalin-Fixed Paraffin-Embedded Tissues with Peroxidase

The procedure is similar to the one used for frozen sections, but paraffin removal and antigen retrieval need to be added at the beginning of the protocol as follows.

1. Put slides on top of a heating block (at 100 °C) to melt paraffin for 15 s.
2. Deparaffinize: Four times in xylene for 3 min, four times in 100 % ethanol for 3 min, and four times in 95 % ethanol for 3 min.
3. Wash in water for 3 min at RT.
4. Wash in PBS for 1 min at RT.
5. Incubate in 1 % hydrogen peroxide in PBS for 15 min at RT.
6. Wash in PBS for 1 min at RT.
7. Wash in water for 1 min at RT.
8. Perform antigen retrieval (*see* **Notes 13 and 14**).
9. Wash in water for 5 min at RT.
10. Wash in PBS for 1 min at RT.
11. Follow the steps as described in antibody IHC on frozen samples (Subheading 3.2.1).

3.2.3 IF in Fresh/Frozen and FFPE Tissues

To perform immunofluorescence in fresh/frozen or FFPE tissue sections we will initially use the same procedures described above in Subheadings 3.2.1 and 3.2.2, but secondary antibody will be fluorescently labeled and protocol from then will be as follows:

1. Apply fluorescent-labeled secondary antibody at 1:200 dilution for 45 min at RT.
2. Aspirate antibody and wash in 1 % Triton X100-PBS for 3 min, two times at RT.
3. Wash with PBS for 1 min at RT.
4. Wash in water for 1 min at RT.
5. Apply mounting medium with DAPI (Vectashield, Vector Laboratories) and coverslip.
6. Keep them in a covered slide holder or box at –20 °C until imaging.

3.3 Preparation of a Cell Pellet in Agarose

To generate a cell pellet is a good technique to test antibodies in well-known positive and negative cells before proceeding to the staining of tissue sections.

1. Grow cells (approximately $1–2 \times 10^7$ cells).
2. Trypsinize cells and collect.

3. Rinse the cells with PBS once.
4. Spin down to collect the cells in a 1.5 ml Eppendorf tube.
5. Fix with 10 % formalin for 24–48 h by resuspending them and letting them pellet again.
6. Rinse cells in PBS.
7. Spin down to get a pellet.
8. Make 1 % agarose in PBS and cool down to near RT.
9. Put the agarose over the cells and wait until it is solid.
10. Place the cell pellet block into a tissue cassette and store in PBS at 4 °C until processed in the tissue processor. Process as a tissue sample and cut 5 μm sections for later use.

3.4 PTEN Immunohistochemistry

3.4.1 Commercially Available Antibodies

A wide variety of commercially available antibodies against PTEN are currently being used as biomarkers to assess loss of function of PTEN regulatory network. Table 1 summarizes the characteristics of the most frequently published PTEN antibodies, some of which have been validated in the Molecular Pathology Laboratory at the Icahn School of Medicine at Mount Sinai. Sensitivity, tissue specificity, accuracy, and reliability are critical characteristics of an antibody for it being used as a potential prognostic as well as diagnostic indicator [8]. Although diverse PTEN antibodies are routinely used for IHC in surgical pathology laboratories with excellent results, currently all these antibodies are for "research use only" (RUO) as none have been approved by the FDA yet. Recently, a rabbit monoclonal anti-PTEN (clone SP218) from Spring Bioscience has been presented as an accurate antibody for determining PTEN status, compared to other RUO antibodies, based on the evaluation in 100 cases of primary prostate and colon cancer (Media release, Ventana Translational Diagnostics CAP/CLIA Laboratory). Furthermore, Sangale et al. reported a robust PTEN antibody Myriad's PREZEONTM for IHC detection which has been implemented in a CLIA-certified laboratory [9].

3.4.2 Use of Positive and Negative Controls, and IHC Result Validation

To assess the specificity and sensitivity of an antibody to detect PTEN, it is important to use good positive and negative controls. A positive control corresponds to a cell or tissue which is well known to express PTEN, whereas there are two types of negative controls in IHC: an "experimental" negative control, which corresponds to a cell or tissue that lack expression of PTEN, and a "technical" negative control which is obtained by processing the slides without adding the primary antibody, thus not rendering any significant staining.

Several cell lines contain endogenous PTEN expression (e.g., prostate cancer DU145, breast cancer MCF7 and HT29, bladder cancer 5637), whereas others are characterized by PTEN loss (e.g., prostate cancer PC3, breast cancer U87MG, bladder cancer J82).

Table 1
Commercially available PTEN antibodies

Vendor	Source[a]	Clone	Catalog number
Abcam	Rb Mab	Y184	ab32199
BioGenex	Ms Mab	28H6	MU435-UC
Cascade Biosciences[b]	Ms Mab	11G8.11	ABM-2055
Cascade Biosciences[b]	Ms Mab	6H2.1	ABM-2052
Cell Signaling Technologies	Rb Mab	D4.3	9188
Cell Signaling Technologies[b]	Rb Mab	138G6	9559
Cell Signaling Technologies[b]	Ms Mab	26H9	9556
Dako	Ms Mab	6H2.1	M3627
Life Technologies	Rb Pab	–	51-2400
NeoMarkers	Ms Mab	Ab4	MS1601
Novus Biologicals	Rb Mab	Y184	NB110-57441
R&D Systems	Rb Pab	–	AF847
R&D Systems	Ms Mab	217702	MAB847
Spring Bioscience[b]	Rb Mab	SP218	M5180

[a]*Ms* mouse; *Rb* rabbit; *Mab* monoclonal antibody; *Pab* polyclonal antibody
[b]Validated PTEN antibodies

Cell blocks made from these cell lines can be then used as positive and negative controls for the clinical assessment of PTEN loss since they are processed similarly to formalin-fixed paraffin-embedded (FFPE) specimens [10–12]. Since PTEN deletions have been reported in a high percentage of solid tumors such as glioma, melanoma, endometrial cancer, kidney cancer, lung cancer, and upper respiratory tract cancers, these tissues can also be used as negative controls. The use of positive and negative controls along the staining experiment demonstrates the consistency of the used staining methodology, and it is obligatory to include them every time that an IHC experiment is performed. Of note, immunostaining analyses are more relevant for studies aiming to determine the level of PTEN expression, not for distinguishing heterozygous from homozygous PTEN gene loss. For that purpose fluorescent in situ hybridization (FISH) analysis with a probe to the PTEN locus would be preferable.

3.4.3 Automatized Systems and Multiple Antibody Assessment

In the past decade, a variety of automatized systems have evolved to perform IHC and IF in large cohort of cases at the same time with consistent results. Single PTEN IHC or dual IHC with different partners may be run on various automatized systems such as Bond-max autoimmunostainer (Leica Biosystem, Melbourne, VIC, Australia), with Bond Polymer refine detection system DS9, or Discovery XT automated slide staining processor system (Ventana Medical Systems Inc, Roche, Tucson, AZ, USA) using HRP and AP detection systems. Current advances are available using combination of biomarkers (with panels up to four antibodies) in an automated quantitative multiplex immunofluorescence assay to measure multi-protein expression level at the same tissue section, to determine protein signaling networks. This molecular platform produces a better optimization of diagnostic specificity and sensitivity than single IHC procedures [13]. In 2008, an assay to study localization of several biomarkers in tissue slides using mass spectrometry was developed with high expectancy in the research community [14], but this technology has not been transferred to the clinical setting yet mainly due to the extraordinary cost and high technical difficulty [15].

3.5 Interpretation and Reporting of Results

Interpretation of the immunohistochemical results is the most critical point in the assessment of phenotype, in this specific case mostly through PTEN protein expression analyses. An expert pathologist is the best person to perform this evaluation, since knowledge in anatomy, histology, and cell identification is mandatory to produce a correct report. As previously mentioned, one of the most important features of immunohistochemistry resides in the fact that it allows the study of protein expression at the single-cell level and subcellular compartment, providing also information about the functionality of the protein of interest and its regulatory networks.

3.5.1 Quantification of IHC Expression: Semiquantitative versus Computer-Assisted Analysis

Due to the widespread use of immunopathology analyses both at the research and clinical levels, quantification of the results has become a paramount. Today, a "negative–positive" is no longer accepted as consistent result of IHC evaluation [16–18]. Most of the times a semiquantitative analysis is performed by an experienced pathologist, who assesses both the intensity of expression and the percentage of cells of interest with a positive phenotype, generating a score system which is the result of multiplying the intensity by the percentage of expression [19, 20]. One of the caveats of IHC standardization and quantification when compared to IF is that development time can affect the intensity of the signal (the longer it is being developed, the higher the intensity). This problem can be overcome by the use of automatized systems, or by performing a tight timing in manual procedures.

However, computer-based methods using different systems (MetaMorph® Microscopy Automation and Image Analysis Software, Molecular Devices, Sunnyvale, CA, USA; Aperio® ePathology Systems, Leica Microsystems Inc., Buffalo Grove, IL, USA; Nuance™; CRi, Woburn, MA, USA) have generated an enormous interest in the pathology community. These systems have been standardized for specific tests such as ER, PR, and Her2Neu, but they have also been used to perform quantitative analyses of both IHC and IF with a variety of markers. Unfortunately, PTEN is not one of the markers commonly analyzed with these methods.

3.5.2 PTEN Subcellular Expression and Analysis

PTEN is mainly localized into the cell cytoplasm, but it is also present in other subcellular locations where it performs distinct functions through the activation or inhibition of specific effectors [21]. PTEN has been identified in the nucleus, where it acts as a potent tumor-suppressor gene, although it does not seem to be a stationary resident of this compartment [22]. Interestingly, it appears that PTEN residing or migrating into the nucleus is phosphorylated, mainly on a cluster of serine and threonine residues in the carboxy-terminal tail. Some controversies in the functional activity of PTEN depending on its cellular localization have arisen in the past years: whereas some reports show that PTEN localized in the cell membrane promotes higher suppressor activity [23, 24], other authors suggest that nuclear PTEN is the main functional protein [25].

3.5.3 Partial Loss of PTEN Expression

One of the major caveats of immunopathology of PTEN comes from the fact that in many cases there is only a loss of part of an exon in the gene. Since the different antibodies recognize different epitopes distributed along the protein (most of the commercially available antibodies recognize the C-terminus, but some are intragenic), protein product can be detected if that part of the protein is present even though the protein may be truncated afterwards. As described below, IHC and IF analyses are best to determine the level of expression, but not to assess partial or heterozygous deletions, for which FISH analyses are indicated.

3.5.4 Use of Downstream Biomarkers for Assessment of PTEN Function: Immunofluorescence Multiplexing

Since PTEN is a negative regulator of PI3 kinase pathway, well-known downstream markers, such as AKT or mTOR, may be used to further analyze its expression and functional status. For example, it has previously been shown that loss of PTEN is associated with high expression of the activated, phosphorylated form of AKT, the so-called p-AKT [6]. In this study, Cho et al. demonstrated that HGPIN areas with PTEN loss were characterized by high levels of p-AKT expression. In our experience, p-AKT is the best surrogate marker to determine PTEN status when the IHC results of PTEN are not conclusive (*see* Fig. 2).

3.6 Implication of PTEN Immunopathology in the Clinical Management of Patients

Most of the immunopathology reports regarding PTEN in the clinical setting are related to neoplastic pathology, due to its implication in tumor initiation and progression. An optimal PTEN reporting is not only important at the diagnostic level [26, 27], but it also goes beyond it, since it can be used as a prognostic factor in determined types of solid tumors such as bladder, pancreatic, and gastric cancers [28–30].

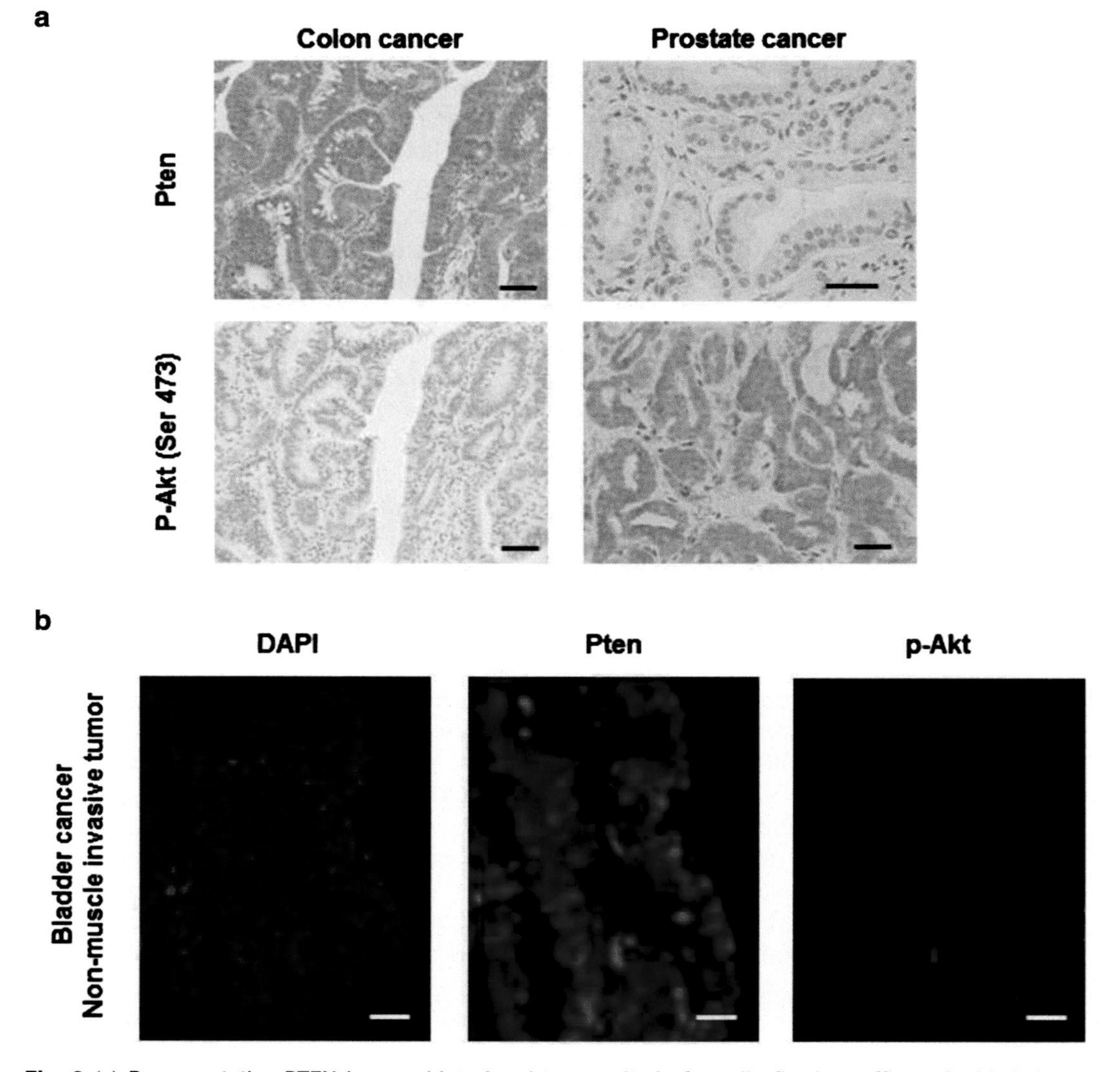

Fig. 2 (**a**) Representative PTEN immunohistochemistry results in formalin-fixed paraffin-embedded tissues specimens of certain human solid tumors such as colon and prostate cancer. (**b**) Dual immunofluorescence of PTEN, and p-AKT (Ser 473) in a representative sample of a non-muscle invasive bladder cancer. Scale bars correspond to 100 μm

Moreover, defining the PTEN phenotype is also of critical relevance in guiding therapeutic intervention. PTEN itself is not a "druggable" target; however, loss of PTEN leads to a robust activation of the AKT/mTOR pathways, for which there are several approved inhibitors which are individualized therapeutic interventions for patients with PTEN-deficient tumors. Thus, PTEN is a critical biomarker in decision-making approaches in clinical services, used as a prognostic factor as well as a tool to guide therapeutic intervention and response to therapy [30, 31]. There are currently several Phase I and II clinical trials for advanced solid tumors with PTEN deficiency, which can be accessed in the following web page: https://clinicaltrials.gov.

4 Notes

1. For incubation of slides or coverslips during these procedures, use always a humid chamber to avoid drying of the solutions.
2. Frozen tissue sections and cells can be fixed with different fixatives, such as 100 % ethanol, 10 % formaldehyde, 4 % paraformaldehyde, 50 % cold methanol-acetone, and 100 % acetone. Some antibodies will react with specific fixatives and not others, so this needs to be characterized for each antibody.
3. Before fixing the cells in Subheading 3.1 above, methanol and acetone need to be cooled down in the −20 °C freezer for at least 1 h.
4. For all these procedures, all blocking and antibody solutions are prepared with antibody diluents (2 % BSA-PBS).
5. In Subheading 3.1 above, donkey serum is used as blocking solution because all the secondary fluorescent antibodies that we are using are made in donkey. Blocking can be performed with antibody diluent alone or with another 10 % animal serum if secondary is made in this animal.
6. In Subheading 3.1 above, use 230–250 μl solution per cover slip to make sure that the whole surface of the glass is covered.
7. For IF protocols (Subheadings 3.1 and 3.2.3 above) the most commonly used secondary fluorescent antibodies are from Invitrogen, Alexafluor 488 (visualized with FITC or GFP filters), and Alexafluor 594 (visualized with Texas Red or Cy3 filters).
8. Once the mounting medium with DAPI has been applied into the cover slip/slide, we putit is best to keep them covered from direct light (in a closed folder or dark box) at −20 °C for 20 min, to enhance the IF signal. For visualization, slides need to be brought to RT before putting them under the microscope.
9. For IHC protocols, the blocking serum depends on the secondary antibody that will be used. For example, for a rabbit primary antibody, the blocking is with goat serum; for a mouse primary antibody, the blocking is with horse serum; for a goat primary antibody blocking is with rabbit; for a rat primary antibody blocking is with rabbit. All the blocking serums are used at a 1:10 dilution.
10. For IHC protocols, Vector Laboratories' secondary antibodies are used at the following dilutions: anti-rabbit = 1:1000;

anti-mouse, anti-goat, anti-hamster = 1:500; anti-rat, anti-guinea pig = 1:200 and cells.

11. Before starting the developing of slides with diaminobenzidine (DAB), the solution needs to be prepared. Add DAB powder in 250 μl of 1 % Triton X100 PBS and filter. Once filtered, put in a non-transparent container and add 30 μl of H2O2 to activate the reaction. Wait for about 30 min before use.
12. For development of DAB-based IHC, if it is the first time we try an antibody, we recommend to leave it in DAB only for 30 s, stop the reaction with distilled water, and check the intensity of the reaction under the light microscope. If it is still weak, leave it for 30 more seconds. Time needed in the DAB for the IHC signal to appear depends on the different antibodies. Check every 1–2 min. Wash in running tap water thoroughly to stop the reaction before the counterstaining.
13. There are several commercially available antigen retrieval solutions for IHC, but the most commonly used are citric buffer (at pH: 6.0) and EDTA (at pH: 9.0).
14. Heat antigen retrieval may be performed in different equipment, such as microwave, pressure-cooker, or steamer. All our protocols are based on a 20-min antigen retrieval in a steamer. It is mandatory to preheat the solution inside the steamer for at least 30 min before putting the slides in the solution for antigen retrieval.

References

1. Li J, Yen C, Liaw D et al (1997) PTEN, a putative protein tyrosine phosphatase gene mutated in human brain, breast, and prostate cancer. Science 275(5308):1943–1947
2. Kandoth C, McLellan MD, Vandin F et al (2013) Mutational landscape and significance across 12 major cancer types. Nature 502(7471):333–339
3. Song MS, Salmena L, Pandolfi PP (2012) The functions and regulation of the PTEN tumour suppressor. Nat Rev Mol Cell Biol 13(5):283–296
4. Hopkins BD, Fine B, Steinbach N et al (2013) A secreted PTEN phosphatase that enters cells to alter signaling and survival. Science 341(6144):399–402
5. Pulido R, Baker SJ, Barata JT et al (2014) A unified nomenclature and amino acid numbering for human PTEN. Sci Signal 7(332), e15
6. Cho H, Herzka T, Zheng W et al (2014) RapidCaP, a novel GEM model for metastatic prostate cancer analysis and therapy, reveals myc as a driver of Pten-mutant metastasis. Cancer Discov 4(3):318–333
7. Cordon-Cardo C, Kotsianti A, Verbel DA et al (2007) Improved prediction of prostate cancer recurrence through systems pathology. J Clin Invest 117(7):1876–1883
8. Bordeaux J, Welsh A, Agarwal S et al (2010) Antibody validation. Biotechniques 48(3):197–209
9. Sangale Z, Prass C, Carlson A et al (2011) A robust immunohistochemical assay for detecting PTEN expression in human tumors. Appl Immunohistochem Mol Morphol 19(2):173–183
10. Djordjevic B, Hennessy BT, Li J et al (2012) Clinical assessment of PTEN loss in endometrial carcinoma: immunohistochemistry outperforms gene sequencing. Mod Pathol 25(5):699–708
11. Forbes SA, Bhamra G, Bamford S et al (2008) The Catalogue of Somatic Mutations in Cancer

(COSMIC). Curr Protoc Hum Genet Chapter 10:Unit 10 1

12. Weigelt B, Warne PH, Downward J (2011) PIK3CA mutation, but not PTEN loss of function, determines the sensitivity of breast cancer cells to mTOR inhibitory drugs. Oncogene 30(29):3222–3233
13. Shipitsin M, Small C, Giladi E et al (2014) Automated quantitative multiplex immunofluorescence in situ imaging identifies phospho-S6 and phospho-PRAS40 as predictive protein biomarkers for prostate cancer lethality. Proteome Sci 12:40
14. Andersson M, Groseclose MR, Deutch AY et al (2008) Imaging mass spectrometry of proteins and peptides: 3D volume reconstruction. Nat Methods 5(1):101–108
15. Alexandrov T (2012) MALDI imaging mass spectrometry: statistical data analysis and current computational challenges. BMC Bioinformatics 13(Suppl 16):S11
16. Rizzardi AE, Johnson AT, Vogel RI et al (2012) Quantitative comparison of immunohistochemical staining measured by digital image analysis versus pathologist visual scoring. Diagn Pathol 7:42
17. Taylor CR, Levenson RM (2006) Quantification of immunohistochemistry--issues concerning methods, utility and semiquantitative assessment II. Histopathology 49(4):411–424
18. Walker RA (2006) Quantification of immunohistochemistry--issues concerning methods, utility and semiquantitative assessment I. Histopathology 49(4):406–410
19. Aytes A, Mitrofanova A, Lefebvre C et al (2014) Cross-species regulatory network analysis identifies a synergistic interaction between FOXM1 and CENPF that drives prostate cancer malignancy. Cancer Cell 25(5):638–651
20. Irshad S, Bansal M, Castillo-Martin M et al (2013) A molecular signature predictive of indolent prostate cancer. Sci Transl Med 5(202):202ra122
21. Bononi A, Pinton P (2015) Study of PTEN subcellular localization. Methods 77-78:92–103
22. Lian Z, Di Cristofano A (2005) Class reunion: PTEN joins the nuclear crew. Oncogene 24(50):7394–7400
23. Nguyen HN, Yang JM Jr, Rahdar M et al (2014) A new class of cancer-associated PTEN mutations defined by membrane translocation defects. Oncogene 34:3737–3743
24. Vazquez F, Grossman SR, Takahashi Y et al (2001) Phosphorylation of the PTEN tail acts as an inhibitory switch by preventing its recruitment into a protein complex. J Biol Chem 276(52):48627–48630
25. Trotman LC, Wang X, Alimonti A et al (2007) Ubiquitination regulates PTEN nuclear import and tumor suppression. Cell 128(1):141–156
26. Jiang K, Lawson D, Cohen C et al (2014) Galectin-3 and PTEN expression in pancreatic ductal adenocarcinoma, pancreatic neuroendocrine neoplasms and gastrointestinal tumors on fine-needle aspiration cytology. Acta Cytol 58(3):281–287
27. Mutter GL, Lin MC, Fitzgerald JT et al (2000) Altered PTEN expression as a diagnostic marker for the earliest endometrial precancers. J Natl Cancer Inst 92(11):924–930
28. Foo WC, Rashid A, Wang H et al (2013) Loss of phosphatase and tensin homolog expression is associated with recurrence and poor prognosis in patients with pancreatic ductal adenocarcinoma. Hum Pathol 44(6):1024–1030
29. Puzio-Kuter AM, Castillo-Martin M, Kinkade CW et al (2009) Inactivation of p53 and Pten promotes invasive bladder cancer. Genes Dev 23(6):675–680
30. Tapia O, Riquelme I, Leal P et al (2014) The PI3K/AKT/mTOR pathway is activated in gastric cancer with potential prognostic and predictive significance. Virchows Arch 465(1):25–33
31. Garg K, Broaddus RR, Soslow RA et al (2012) Pathologic scoring of PTEN immunohistochemistry in endometrial carcinoma is highly reproducible. Int J Gynecol Pathol 31(1):48–56

Chapter 4

Measurement of PTEN by Flow Cytometry

John F. Woolley and Leonardo Salmena

Abstract

Recent advancements have driven the development of smaller footprint, less expensive, and user-friendly flow cytometers introducing the technology to more users.

Flow cytometry is an established tool for multiparametric analysis of various important cellular characteristics. Fluorescent dyes or fluorophore-conjugated antibodies allow for measurement of protein expression, identification of cell populations, or DNA content analysis. This is combined with analysis of light-scattering detection to determine cell size and complexity to allow for the study of complex cell samples, such as whole blood. Through antibody staining for a variety of surface markers as well as intracellular proteins we can also elucidate intracellular signaling, and phosphor-signaling, on a single-cell basis.

Here we describe the application of flow cytometry analysis to the tumor suppressor PTEN in various cancer cell lines and a mouse model.

Key words PTEN, Flow cytometry, Tumor suppressor, CD45, Mouse model

1 Introduction

Flow cytometry allows for high-throughput single-cell analysis, in a quantitative as well as qualitative manner [1]. Recent advances in flow cytometry have allowed for a greater access to this technique than ever before in academic, industrial, and clinical settings. Indeed the newest flow cytometers are smaller, more robust, less expensive, have higher throughput, and combine with much more user-friendly software. Flow cytometry can be utilized for the characterization of cells in any fluid (e.g., blood, plasma, cerebrospinal fluid, urine) as well as cell suspensions made from solid tumours, bone marrow, or laboratory-cultured cells. Characteristics that can be measured include cell size, cytoplasmic complexity, DNA or RNA content, and a wide range of membrane-bound and intracellular proteins [2].

A flow cytometry system consists of four main operating units: a light source, flow cell, optical filter units for specific wavelength detection over a broad spectral range, and photodiodes or photomultiplier tubes for sensitive detection of the signals of interest.

Leonardo Salmena and Vuk Stambolic (eds.), *PTEN: Methods and Protocols*, Methods in Molecular Biology, vol. 1388, DOI 10.1007/978-1-4939-3299-3_4, © Springer Science+Business Media New York 2016

A cell suspension is injected into the flow cell where the cells pass one after another across a laser beam (or mercury lamp light) that is orthogonal to the flow. Sample delivery has been accomplished using either pumps or an increase of pressure differential between the sample and the system. The use of pumps adds complexity, but offers control over delivery rates, which enables direct counting of particles per unit volume without the addition of a particle standard of a known concentration. Typically, ensuring one particle passes the light source at a time is achieved by hydrodynamic focusing of the sample stream, wherein the sample stream is injected into the sheath stream inside the flow cell. The velocities of both fluids are in the range of laminar flow. Cells are isolated in the focused stream since the sheath flow is greater than the sample flow. This allows cells to be excited by the laser individually. Using this technology, it is possible to detect up to 30,000 events per second [3]. Upon contact with the particle or cell, the excitation light is scattered in both forward and sideways directions. The forward-scattered light (FSC) provides information on the size of the cells and can be detected without further manipulation. The sideways-scattered light (SSC) is affected by several parameters, including granularity, cell size, and cell morphology. Intrinsic fluorescence or cellular components stained with specific fluorescence dyes allow certain cell components to be measured. These signals can be combined in various ways that allow all subpopulations to be observed. Depending on the dyes that are used, many of these measurements can be combined with each other. In modern cytometers, the information from scattered light is always available and can be detected along with as many as 18 fluorescent dyes [3].

Most applications of flow cytometry are based on fluorescence monitoring. The measurable cellular parameters can be characterized as intrinsic or extrinsic, depending on their need for reagents. While usually no pretreatment is necessary for the assay of intrinsic fluorescence, studies of specific cell components with fluorescent dyes (extrinsic fluorescence) requires that a staining procedure be performed before the cells are analyzed. These procedures include fixation, staining, and several washing steps.

Fluorescent labelling of cells is typically achieved by means of fluorescent dyes or antibody-fluorophore conjugates, both of which allow for quantitation of cells or cellular components. Here we discuss the application of primary-conjugated antibodies to measure PTEN levels in cells, both cultured cells and isolated cells from murine spleens.

Through a combination of surface marker staining and intracellular protein staining, characterization of protein expression in specific cell types from complex samples (e.g., blood) can be achieved. However, care must be taken when staining both surface markers and intracellular proteins on the same sample. The issue arises with the effect on surface markers of fixation and permeabilization of the cell, which is necessary for intracellular delivery of antibodies. Some protocols that provide excellent staining of phospho-epitopes decrease staining levels of particular surface antigens, while preservation of surface epitopes leads to weak phospho-staining [4]. This balance between surface and intracellular epitopes must be kept in mind while attempting to stain particular antigens. For example, the mouse B-cell marker CD19 is exquisitely sensitive to methanol permeabilization [5]. Thus we recommend that surface markers are measured before and after fixation/permeabilization and protocols are optimized for each combination of surface markers and intracellular proteins.

The ability to probe phosphorylated products in a multiparametric assay, alongside surface marker staining, in complex samples is particularly useful. Such assays allow researchers to test the activity of multiple signaling pathways at the same time in cells, and thus the interactions between these pathways on a cell-to-cell basis, in conjunction with protein expression and surface marker data. In the context of PTEN expression level, this would allow the characterization of phosphorylation status of downstream signaling effectors such as Akt or RPS6.

Measurement of protein expression via flow cytometry offers many advantages in both clinical and research settings. Because it allows for multiparametric measurement, reduced sampling can be achieved. This is especially important in clinical monitoring or diagnosis, whereby sampling is often not trivial (e.g., blood or bone marrow draws). It also allows for separating complex/mixed samples to be separated using specific markers, and subsequent measurement of characteristics of interest. This is of particular interest for a protein such as PTEN, which is the focus of significant research interest and is associated with multiple cancers.

2 Materials

2.1 Cell Lines

The following human leukemia cell lines were used in these studies: HL-60, U937, OCI/AML-3, and MV4-11.

1. HL-60 cells were established from the peripheral blood of a 35-year-old woman with acute myeloid leukemia (AML FAB M2 [6]). HL-60 cells resemble promyelocytes but can differentiate terminally in vitro in the presence of dimethylsulfoxide (DMSO) or retinoic acid to granulocytes [7]. Other compounds like 1,25-dihydroxyvitamin D3, 12-*O*-tetradecanoylphorbol-13-acetate (TPA), and GM-CSF can induce HL-60 to differentiate to monocytic, macrophage-like, and eosinophil phenotypes, respectively.
2. U937 are myeloid cells established from the pleural effusion of a 37-year-old male patient with histiocytic lymphoma [8].

U937 cells mature in the presence of numerous soluble stimuli, adopting the characteristics and morphology of macrophages [9].

3. OCI/AML3 cells were established from the peripheral blood of a 57-year-old man with acute myeloid leukemia (AML FAB M4; [10]). OCI/AML3 are known to carry a mutation in the *NPM1* gene (type A) and in *DNMT3A* (R882C; [11]).
4. MV4-11 cells were established from a 10-year-old boy with acute monocytic leukemia (AML FAB M5; [12]). These cells are known to carry a chromosomal translocation at t(4;11) and are homozygous for the internal tandem duplication of the *FLT3* gene (FLT3-ITD; [13]).

2.2 Cell Line Maintenance

HL-60 and OCI/AML-3 cell lines were maintained at a culture density of 1×10^5–1×10^6 cell/mL in 10 mL of alpha-MEM medium supplemented with 10 % fetal calf serum (FCS, *v/v*), 100 units of penicillin per mL, and 100 μg of streptomycin per mL at 37 °C and 5 % CO_2.

U937 and MV4-11 cell lines were maintained at a culture density of 1×10^5–1×10^6 cell/mL in 10 mL of RPMI 1640 medium supplemented with 10 % fetal calf serum (FCS, *v/v*), 100 units of penicillin per mL, and 100 μg of streptomycin per mL at 37 °C and 5 % CO_2.

2.3 Mouse Models

All mouse experiments were performed using male C57/BL6 young adult mice (10 weeks). Animals were obtained from the Jackson Laboratory (Bar Harbor, MA, USA). The animals were housed in standard 12-h light:dark cycle and at a temperature of 23 °C with free access to food and water in groups of five mice. All experimental protocols were approved by the research ethics board of this university and were carried out in compliance with the Canadian Council on Animal Care recommendations.

2.4 Equipment

In this study we analyzed cell samples by means of a Cytoflex flow cytometer (*see* **Note 1**; Beckman Coulter, Mississauga, ON, Canada). The CytoFLEX is a three-laser system including lasers at wavelengths of 488 nm, 638 nm, and 405 nm. This allows for a variety of configurations of 4–13 colors with changeable bandpass filters. There are three light scatter parameters, the typical blue laser scatter yielding forward and side scatter, and an additional side scatter parameter off the 405 nm laser. Pulse height and area are collected for all parameters, and a width signal can be selected for any one of the parameters. Emitted light is collected by fiber-optic bundles and carried to fiber array photo detector modules, where it exits to pass through wavelength division multiplexers, which replace conventional dichroics, to partially split the light into distinct ranges, which is then further refined by band-pass filters. The light then is detected by an avalanche photodiode, rather than a conventional photomultiplier tube.

The fluidics system is controlled via peristaltic pumps for both the sheath and sample lines, and the sample volume flow rates can range up to 240 μL/min. Here we will utilize the 488 nm laser with the one forward scatter and one side scatter parameter.

The software used to analyze flow cytometry data was CytExpert (Beckman Coulter, Canada) and FlowJo Vx (FlowJo LLC, Ashland, OR, USA).

2.5 General Equipment

Centrifuge, microcentrifuge, Gilson pipetman (P10, P20, P200, P1000), dissection kit, 70 μM cell strainer, 26-gauge needles and syringes, 15 mL centrifuge tubes, 1.5 mL microcentrifuge tubes.

2.6 Antibodies

1. PTEN-PE (624048; BD Biosciences, San Jose, CA, USA).
2. CD45.2-FITC (*see* **Note 2**; 11-0454-81; eBiosciences, San Diego, CA, USA).

2.7 Reagents

1. Red blood cell (RBC) lysis buffer: 155 mM Ammonium chloride (NH_4Cl), 12 mM sodium bicarbonate ($NaHCO_3$), and 0.1 mM EDTA were prepared in double-distilled H_2O.
2. Phosphate-buffered saline (PBS): 8 g NaCl, 0.2 g KCl, 1.44 g Na_2HPO_4, 0.24 g KH_2PO_4 in 1 L of double-distilled H_2O. pH 7.4.
3. Fixation buffer: 8 % Formaldehyde (*v/v*) in PBS. Should be made fresh from paraformaldehyde.
4. Permeabilization buffer: 0.2 % Triton-x100 (*v/v*) in PBS.
5. Staining buffer: 5 % FCS in PBS, made fresh. If not made fresh, an antimicrobial agent such as sodium azide (at a concentration of 0.1 % *v/v*) should be added.

3 Methods

3.1 Cell Line Collection for Flow Cytometry

1. Cell lines were cultured to a density of 5×10^5 cells/mL and then collected in 15 mL centrifuge tubes.
2. Cells were centrifuged at 400 × *g* for 5 min.
3. Supernatant medium was removed.
4. Cells were washed with 1 mL of PBS.
5. Cells were resuspended in 1 mL of PBS.

3.2 Mouse Splenocyte Collection for Flow Cytometry

1. C57/BL6 mice were sacrificed at 10 weeks.
2. Spleens were isolated *postmortem* and placed in PBS on ice.
3. Splenocytes were isolated utilizing the thumb piece of a plunger removed from a 1 mL syringe, mashing the spleen through a 70 μM cell strainer into ice-cold PBS.

4. Splenocyte preparation was passed through 70 μM cell strainer to remove debris.
5. Cells were then transferred to a 15 mL centrifuge tube.
6. This was then centrifuged at 400 ×*g* for 5 min.
7. Splenocytes were resuspended in RBC lysis buffer and vortexed briefly (*see* **Note 3**). The cells were allowed to incubate for 10 min.
8. Cells were then centrifuged at 400 ×*g* for 5 min and supernatant was removed.
9. Cells were resuspended in PBS and kept on ice.

3.3 Cell Fixation and Permeabilization (See Note 4)

1. 1 mL of freshly made 8 % paraformaldehyde fixation buffer was added dropwise to 1 mL of cell suspension (from Subheading 3.1 or 3.2) in a 15 mL centrifuge tube (*see* **Note 5**). This 1:1 ratio gave a final concentration of 4 % paraformaldehyde to fix cells.
2. This cell fixation mixture was briefly vortexed and then placed in a 37 °C water bath for 10 min.
3. Cells were removed from water bath and to this 2 mL of fix cells, 2 mL of 0.2 % Triton-X100 permeabilization buffer was added dropwise while gently vortexing to prevent clumping of cells. This 1:1 ratio gives a final concentration of 0.1 % Triton-X100 to permeabilize cells.
4. Cells were allowed to permeabilize for 30 min on ice.
5. Cells were centrifuged at 400 ×*g* for 15 min.
6. Cells were resuspended in flow cytometry staining buffer, made fresh on the day of staining.
7. Cells were stored on ice for 30 min prior to staining.

3.4 Cell Staining for Flow Cytometry

1. 50 μL of fixed cell suspension in staining buffer (from Subheading 3.3) was transferred to microcentrifuge tubes.
2. Antibodies were added in the appropriate concentrations (*see* **Note 6**). PTEN-PE was stained at 10 ng in 100 μL PBS; CD45.2-FITC was stained at 10 ng in 100 μL PBS.
3. Cells were stained for 30 min in the dark at room temperature.
4. Cells were centrifuged 400 ×*g* for 5 min.
5. Cells were washed once with staining buffer.
6. Cells were resuspended in 200 μL of staining buffer.

3.5 Flow Cytometric Analysis

1. Cells (from Subheading 3.4) were analyzed for fluorescence on a Cytoflex flow cytometer (Beckman Coulter).
2. Cells were gated on FSC vs. SSC to identify the correct, viable cell population (*see* Fig. 1a).

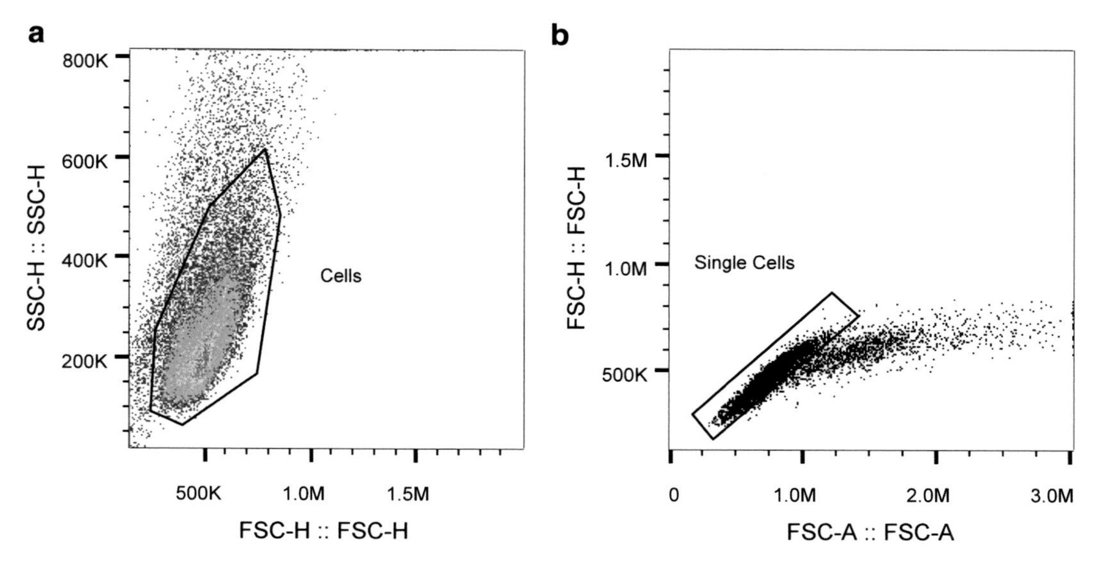

Fig. 1 Cell line gating strategy. (**a**) 10,000 cells were gating on FSC vs. SSC parameters. Threshold was set to remove debris. (**b**) Cells were then gated on FSC-area vs. -height to identify singlet cells and remove any doublet cells from further analysis

3. Gated cells were further gated on FSC-area vs. -height to discriminate singlet cells from doublet cells (*see* Fig. 1b).
4. Single cells from cell lines were analyzed for fluorescence signal of PTEN conjugated to PE (*see* Fig. 2), using the U937 cell line as a negative control for PTEN expression (*see* **Note 7**).
5. Mouse spleen cells stained with CD45.2-FITC and PTEN-PE were gated for mononuclear cells according to FSC vs. SSC intensity (*see* Fig. 3).
6. Mononuclear splenocytes were then displayed as FL1 vs. SSC to determine CD45.2 staining intensity (*see* Fig. 4a). Cells were gated as positive/bright vs. negative/dim for CD45.
7. Cells were plotted as CD45-positive and CD45-negative on a histogram overlay to display PTEN-PE staining in these two populations (*see* Fig. 4b). From this we identified CD45-positive cells as high expressers of PTEN.
8. Splenocytes stained solely with PTEN-PE were gated on FSC vs. SSC showing three populations (*see* Fig. 5a). These cells correspond to lymphocytes (P1), monocytes (P2), and granulocytes (P3). Plotting these populations on a histogram overlay for PTEN-PE shows that PTEN is expressed highly in the granulocyte population (*see* Fig. 5b).

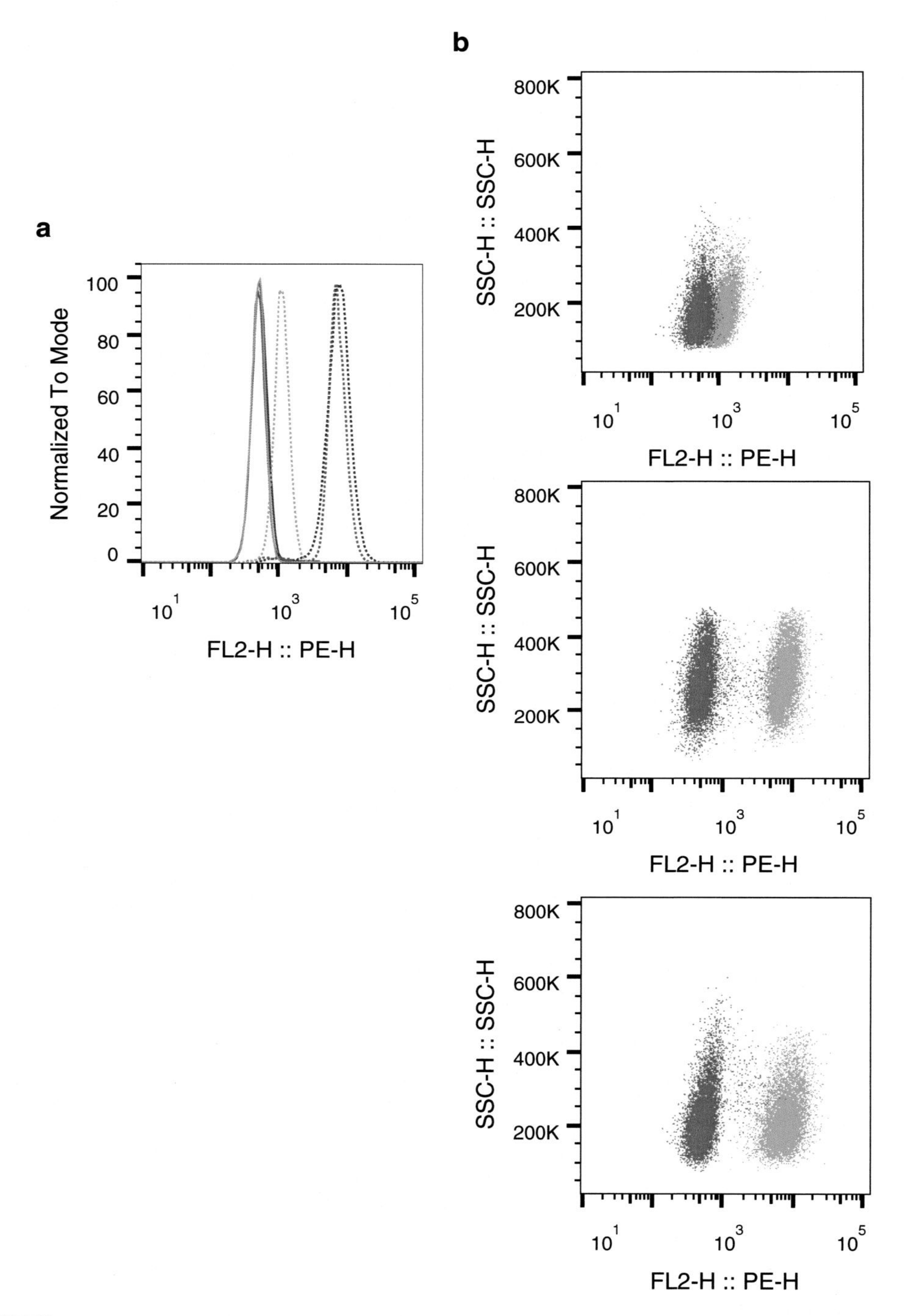

Fig. 2 PTEN expression in cell lines. (**a**) U937 (*green*), AML2 (*red*), and HL-60 singlet cells (*blue*) were stained and were examined for PTEN expression. Unstained cells (*solid lines*) were compared against cells stained with a PTEN-PE antibody (*dotted lines*) for FL2 fluorescence on a histogram overlay. (**b**) U937 (*top*), AML2 (*middle*), and HL-60 (*bottom*) cells were examined for PTEN expression. Unstained cells (*red*) were plotted with PTEN-PE-stained cells (*blue*) on dot-plots of FL2 vs. SSC

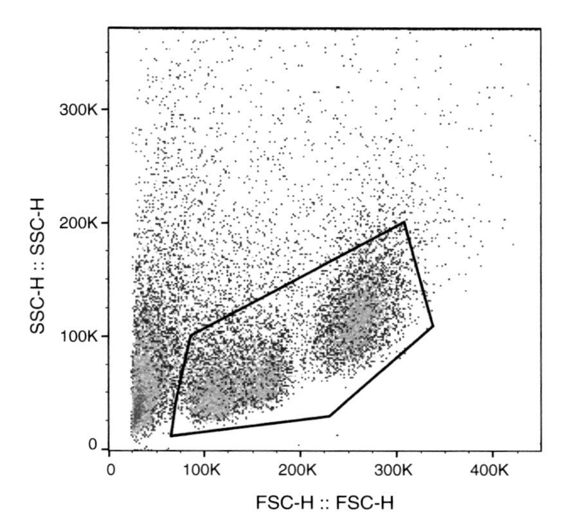

Fig. 3 Mouse splenocyte gating. Isolated spleen cells from C57/BL6 cells are displayed on a FSC vs. SSC plot. Mononuclear cells were gated, excluding debris from further analysis

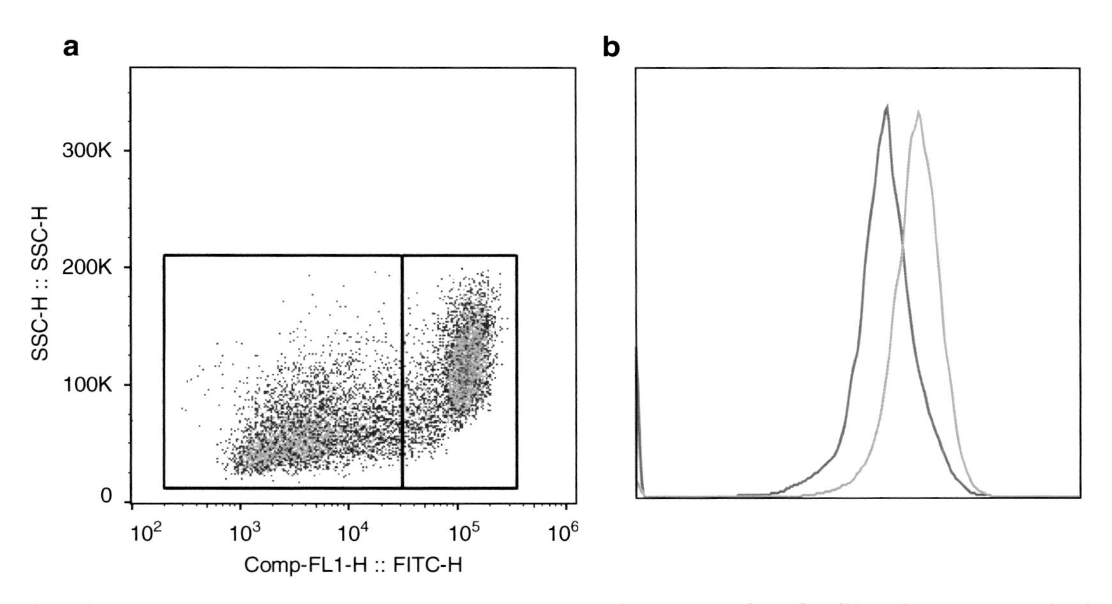

Fig. 4 CD45.2 and PTEN co-expression in murine splenocytes. Cells isolated from C57/BL6 mice were co-stained with CD45.2-FITC and PTEN-PE. (**a**) Cells were gated as CD45.2-positive and CD45.2-negative. (**b**) CD45.2-negative cells (*red*) and CD45.2-positive cells (*blue*) were plotted on an overlay histogram for PE fluorescence, demonstrating that CD45.2-positive cells display a higher PTEN expression than CD45.2-negative cells

4 Notes

1. In this work we make use of the Cytoflex flow cytometer (Beckman Coulter); however the methodology described here is general and can be undertaken with almost any flow cytometer capable of quantifying fluorescein isothiocyanate (FITC) and phycoerythrin (PE) fluorescence.

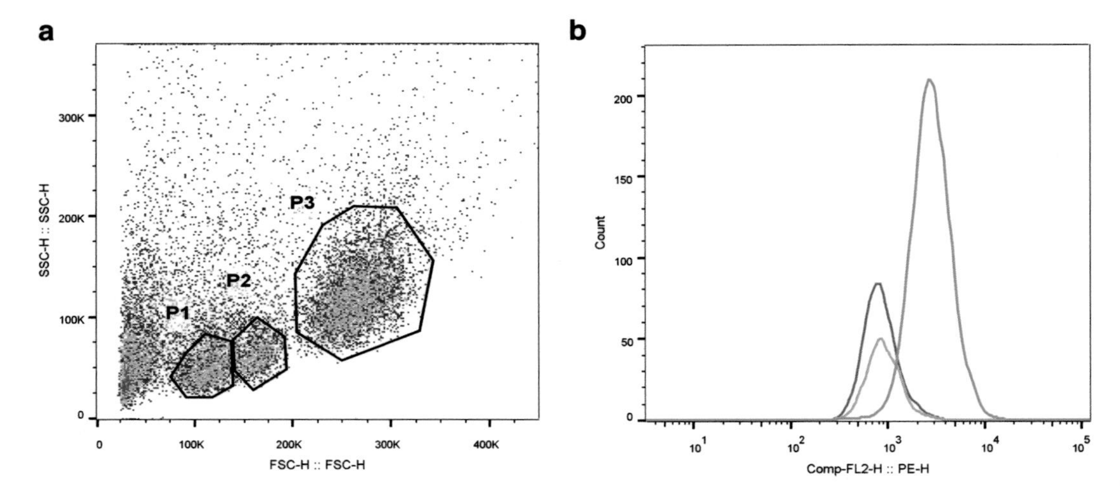

Fig. 5 PTEN expression in murine spleen. Cells isolated from C57/BL6 mice were stained for PTEN-PE only. (**a**) Cells were gated as lymphocytes (P1), monocytes (P2), and granulocytes (P3). (**b**) These populations were then plotted on a histogram overlay showing P1 (*red*), P2 (*blue*), and P3 (*green*). From this we can see that granulocytes express a noticeably higher level of PTEN than either monocytes or lymphocytes which both express similar levels

2. The C57/BL6 mouse spleen is a source of both myeloid and lymphoid cells. The wild-type C57/BL6 mice express the CD45.2 allele as opposed to the CD45.1 allele expressed on, for example, SJL mice [14]. CD45 vs. SSC plots allow for better identification of blood cell populations than simply using FSC vs. SSC, with separation of blasts and plasma cell, precursors, and erythroid cells [15].
3. It is desirable to remove erythrocytes from spleen mononuclear cell preparations prior to flow cytometry experiments as large numbers of RBCs in the sample can occlude populations of interest [16]. A small number of RBCs remaining in the sample will not prove difficult to gate out however, so partial lysis of RBCs is sufficient and should be optimized depending on the individual experiment being performed.
4. In order to detect intracellular antigens, cells first must be permeabilized especially after fixation with cross-linking agents such as formaldehyde and glutaraldehyde [17, 18]. Permeabilization is required to allow complex macromolecules such as antibodies to cross the cell membrane. Two general types of reagents are commonly used: organic solvents or detergents. The organic solvents dissolve lipids from cell membranes making them permeable to antibodies. Because the organic solvents also coagulate proteins, they can be used to fix and permeabilize cells at the same time. However, these solvents may also extract lipidic antigens or lipid-associated antigens from cells. The other large class of permeabilization

agents is detergents. Saponin, a plant glycoside, interacts with membrane cholesterol, selectively removing cholesterol and making holes in the membrane. In addition to forming membrane holes, antibodies may be incorporated into saponin/cholesterol micelles facilitating their entry into the cells. Saponin permeabilization is not effective on mitochondrial membranes or the nuclear envelope. Nonionic detergents such as Triton X-100 and Tween 20 are also widely used to permeabilize cells and tissues [19]. The disadvantage of these detergents is that they are nonselective in nature and may extract proteins along with the lipids, resulting in a false negative during immunostaining. Depending on the antigen, a combination of permeabilizing agents may be preferable. It is absolutely essential to optimize fixation/permeabilization protocols for each antibody-antigen pair being tested to ensure efficient staining and to minimize false-negative staining.

5. Here we fix cells in a 4 % formaldehyde solution. There are numerous methods to prepare this fixative from stock solutions [4]. The most common options are to begin with paraformaldehyde, a polymerized formaldehyde, or formalin. Paraformaldehyde is usually sourced as a powder, and can be prepared as a 4 % solution (*w/v*) directly with heating to break down polymers. Formalin is generally sourced as 37–40 % formaldehyde solution, as the formaldehyde gas is dissolved in water at a saturating concentration somewhere between these concentrations. This formalin can be diluted 1:10 to achieve a 4 % fixative solution. It should also be noted that typically formalin will contain methanol, to prevent polymerization, so the final fixative will contain a small amount of methanol.

6. The importance of antibody titrations in flow cytometry should not be underestimated. Efficient antibody titration identifies the concentration of the antibody which, under specific staining conditions, will deliver optimal separation between positive and negative populations. Supersaturating concentrations will increase background and nonspecific binding and is not cost effective. Non-saturating concentration may cause sample-to-sample variation and decrease resolution and sensitivity. At saturating staining concentrations the amount of antibody present is not limiting and is sufficient to stain all relevant antigens without significantly lowering the concentration of free antibody. Therefore the antibody concentration but not the number of cells is critical for optimal staining. The optimal antibody concentration must be determined for each application and set of experimental conditions (including staining time and temperature) and is determined by using a series of dilutions. Prior to running these experiments, we determined the optimal concentration for PTEN-PE in these leukemia

cells to be 10 ng in 100 μL, stained for 30 min, by titration (data not shown).

7. In this work we utilized the U937 cell line as a negative control for PTEN antibody staining. U937 has previously been shown to display low or undetectable PTEN expression, and indeed has been characterized with hemizygous deletion of PTEN [20].

Acknowledgements

The research in the Salmena lab is supported by funds from Canada Research Chairs, Human Frontier Science Program Organization, Leukemia and Lymphoma Society of Canada, Canadian Cancer Society, and the Natural Sciences and Engineering Research Council of Canada.

References

1. Brown M, Wittwer C (2000) Flow cytometry: principles and clinical applications in hematology. Clin Chem 46:1221–1229
2. Shapiro H (2003) Practical flow cytometry, 4th edn. Wiley-Liss, New York
3. Picot J, Guerin CL, Le Van Kim C, Boulanger CM (2012) Flow cytometry: retrospective, fundamentals and recent instrumentation. Cytotechnology 64:109–130. doi:10.1007/s10616-011-9415-0
4. Krutzik PO, Irish JM, Nolan GP, Perez OD (2004) Analysis of protein phosphorylation and cellular signaling events by flow cytometry: techniques and clinical applications. Clin Immunol 110:206–221. doi:10.1016/j.clim.2003.11.009
5. Krutzik PO, Clutter MR, Nolan GP (2005) Coordinate analysis of murine immune cell surface markers and intracellular phosphoproteins by flow cytometry. J Immunol 175:2357–2365. doi:10.4049/jimmunol.175.4.2357
6. Collins S, Gallo R, Gallagher R (1977) Continuous growth and differentiation of human myeloid leukaemic cells in suspension culture. Nature 270:347–349. doi:10.1038/270347a0
7. Breitman TR, Collins SJ, Keene BR (1980) Replacement of serum by insulin and transferrin supports growth and differentiation of the human promyelocytic cell line, HL-60. Exp Cell Res 126:494–498. doi:10.1016/0014-4827(80)90296-7
8. Sundström C, Nilsson K (1976) Establishment and characterization of a human histiocytic lymphoma cell line (U-937). Int J Cancer 17:565–577
9. Phillips TA, Ni J, Pan G et al (1999) TRAIL (Apo-2L) and TRAIL receptors in human placentas: implications for immune privilege. J Immunol 162:6053–6059
10. Wang C, Curtis J, Minden M, McCulloch E (1989) Expression of a retinoic acid receptor gene in myeloid leukemia cells. Leukemia 3:264–269
11. Hollink IHIM, Feng Q, Danen-van Oorschot AA et al (2012) Low frequency of DNMT3A mutations in pediatric AML, and the identification of the OCI-AML3 cell line as an in vitro model. Leukemia 26:371–373. doi:10.1038/leu.2011.210
12. Lange B, Valtieri M, Santoli D et al (1987) Growth factor requirements of childhood acute leukemia: establishment of GM-CSF-dependent cell lines. Blood 70:192–199
13. Quentmeier H, Reinhardt J, Zaborski M, Drexler HG (2003) FLT3 mutations in acute myeloid leukemia cell lines. Leukemia 17:120–124. doi:10.1038/sj.leu.2402740
14. Shen FW, Saga Y, Litman G et al (1985) Cloning of Ly-5 cDNA. Proc Natl Acad Sci U S A 82:7360–7363. doi:10.1073/pnas.82.21.7360
15. Haycocks NG, Lawrence L, Cain JW, Zhao XF (2011) Optimizing antibody panels for efficient and cost-effective flow cytometric diagnosis of acute leukemia. Cytometry B Clin Cytom 80B:221–229. doi:10.1002/cyto.b.20586

16. Bossuyt X, Marti GE, Fleisher TA (1997) Comparative analysis of whole blood lysis methods for flow cytometry. Cytometry 30:124–133
17. Fox C, Johnson F, Whiting J, Roller P (1985) Formaldehyde fixation. J Histochem Cytochem 33:845–853. doi:10.1056/NEJMra1313875
18. Pollice AA, McCoy JP, Shackney SE et al (1992) Sequential paraformaldehyde and methanol fixation for simultaneous flow cytometric analysis of DNA, cell surface proteins, and intracellular proteins. Cytometry 13:432–444. doi:10.1002/cyto.990130414
19. Le Maire M, Champeil P, Møller JV (2000) Interaction of membrane proteins and lipids with solubilizing detergents. Biochim Biophys Acta 1508:86–111. doi:10.1016/S0304-4157(00)00010-1
20. Dahia PLM, Aguiar RCT, Alberta J et al (1999) PTEN is inversely correlated with the cell survival factor Akt/PKB and is inactivated via multiple mechanisms in haematological malignancies. Hum Mol Genet 8:185–193. doi:10.1093/hmg/8.2.185

Chapter 5

Detecting PTEN and PI3K Signaling in Brain

Guo Zhu and Suzanne J. Baker

Abstract

The central nervous system is comprised of multiple cell types including neurons, glia, and other supporting cells that may differ dramatically in levels of signaling pathway activation. Immunohistochemistry in conjunction with drug interference are powerful tools that allow evaluation of signaling pathways in different cell types of the mouse central nervous system in vivo. Here we provide detailed protocols for immunohistochemistry to evaluate three essential components in the PI3K pathway in mouse brain: Pten, p-Akt, and p-4ebp1, and for rapamycin treatment to modulate mTOR signaling in vivo.

Key words PI3K, PTEN, p-AKT, p-4EBP1, Rapamycin, Immunohistochemistry

1 Introduction

The phosphoinositide 3-kinase (PI3K) pathway is an essential signaling cascade that regulates multiple processes in the central nervous system including proliferation, survival, metabolism, and cell migration [1]. When extracellular growth signals are relayed to the PI3K pathway through receptor tyrosine kinases, AKT is fully activated by recruitment to the cell membrane by binding of its pleckstrin homology domain (PH domain) to increased levels of PIP3 and subsequent phosphorylation at amino acid residue threonine 308 by PDK1 and serine 473 by mTORC2 complex [2, 3]. Among a variety of downstream targets, mTORC1 is a master regulator of cell growth and proliferation as well as many other cellular processes [4]. mTORC1 promotes cell growth and proliferation through phosphorylation of two of its downstream effectors, the S6 kinases (S6Ks) and 4E-BPs which enhance ribosome assembly and mRNA translation. Upon activation by growth factors, mTORC1 limits the extent of upstream growth factor signaling by a negative feedback loop through IRS-1 phosphorylation by S6K1 [5, 6]. PTEN is the major negative regulator of the PI3K pathway. Inactivation of PTEN results in unopposed and constitutive activation of the PI3K-AKT-mTORC1 pathway in the presence of

Leonardo Salmena and Vuk Stambolic (eds.), *PTEN: Methods and Protocols*, Methods in Molecular Biology, vol. 1388, DOI 10.1007/978-1-4939-3299-3_5, © Springer Science+Business Media New York 2016

upstream pathway activation. Increases in phosphorylation of AKT serine 473 and threonine 308, phospho-S6 serine 235/236 (substrate of S6K), and phospho-4EBP1 threonine 37/46 are often used as the readouts of pathway activation. Pten loss in the mouse brain could result in diverse phenotypes such as neuronal hypertrophy and premature differentiation, astrocyte hypertrophy, or increased neural stem cell self-renewal and proliferation given different cellular and developmental contexts [7–14].

Rapamycin is a small-molecule inhibitor of mTOR complexes. It inhibits mTORC1 activity in acute treatment while it also inhibits mTORC2 activity in long-term or high-dose treatment by interfering with its assembly [15, 16]. In many settings, inhibition of mTORC1 by rapamycin releases the negative feedback loop and results in enhanced PI3K signaling to AKT [17], although this is not always observed.

Here we provide detailed immunohistochemistry protocols for Pten, p-Akt, and p-4EBP1 in mouse brain and a protocol for rapamycin treatment of mice to manipulate PI3K signaling in vivo.

2 Materials

2.1 Immunohisto chemistry for Pten, p-Akt, and p-4EBP1 in FFPE Tissue

1. Anti-Pten antibody, rabbit monoclonal, Cell Signaling #9559.
2. Anti-p-Akt (Ser473) antibody, rabbit polyclonal, Cell Signaling #9271.
3. Anti-p-4EBP1 (T37/46) antibody, rabbit monoclonal, Cell Signaling #2855.
4. Biotinylated anti-rabbit-IgG, included in Elite ABC reagent kit, Vector Lab PK-6101.
5. Elite ABC reagent, Vector Lab PK-6101.
6. NovaRED kit, Vector lab SK-4800 (DAB kit SK-4100).
7. Hematoxylin QS, Vector Lab H-3404.
8. Permount, Fisher SP15-100.
9. Tyramide Signal Amplification (TSA) Kit, Perkin Elmer NEL746A.
10. Antigen-retrieval solution: 18 ml of 0.1 M citric acid, 82 ml of 0.1 M sodium citrate, 850 ml of H_2O. Adjust to pH 6.0 by addition of 1 N NaOH, and then add H_2O for final volume of 1 l.
11. 10× TBS buffer (10× Tris-buffered saline): 60.6 g Tris, 87.6 g NaCl in 800 ml ultrapure water. Adjust pH to 7.5 with 1 M HCl, and add ultrapure water to final volume of 1 l.
12. TBS-T (1× TBS + 0.01 % Tween-20).
13. 0.6 % H_2O_2 in TBS (freshly made): 1 ml of 30 % H_2O_2 in 49 ml of TBS.

14. Blocking solution: 10 % goat serum in TBS-T, with 0.01 % thimerosal, store at 4 °C.
15. Blocking solution: 2 % goat serum in TBS-T, with 0.01 % thimerosal, store at 4 °C.
16. Plastic Coplin jars, at least four for antigen retrieval.
17. Microwave oven with rotating tray.
18. Wax pen.
19. Belly dancer rocking platform.
20. Cover slips.

2.2 Rapamycin Treatment

1. Rapamycin (LC Laboratories).
2. Sterile dimethyl sulfoxide (DMSO).
3. Tween80.
4. 1 ml Syringe (BD).

3 Methods

3.1 Immunohisto chemistry for Pten, p-Akt, and p-4EBP1 in FFPE Tissue

For cryosections, mice were anesthetized and perfused transcardially with PBS followed by 4 % paraformaldehyde (PFA) in PBS. Following dissection, tissues were post-fixed overnight in 4 % PFA in PBS at 4 °C, and then equilibrated in 25 % sucrose in PBS for an additional 24 h at 4 °C. Tissues were embedded in embedding media OCT (Triangle Biomedical Sciences) on dry ice and cut into 12 μm thick cryosections. Tissue slides were equilibrated at room temperature for 20 min and then washed three times in PBS prior to staining.

For paraffin sections, tissue was processed the same way as above except after dissection, tissue was post-fixed for 24 h in 4 % PFA in PBS at 4 °C, then processed and embedded in paraffin, and cut into 5 μm sections.

For optimal IHC, tissue should not be left in fixing solution for more than 4 days before being processed for paraffin-embedding or equilibrated in 25 % sucrose in PBS for cryoprotection followed by embedding in OCT for cryosections.

3.1.1 Deparaffinization (Only for Paraffin Sections)

Transfer slides through the following series of solutions (*see* **Notes 1** and **2**):

1. Xylenes 5–10 min.
2. Xylenes 3 min × 2.
3. 100 % EtOH, 2 min × 2.
4. 95 % EtOH, 2 min.
5. 70 % EtOH, 1 min.

6. 50 % EtOH, 1 min.
7. 20 % EtOH, 1 min.
8. H_2O, 2 min × 2.

3.1.2 Antigen Retrieval

Saturation: Place sections in plastic Coplin jar(s) filled with the antigen-retrieval solution completely to the top as below:

1. For deparaffinized sections: antigen-retrieval solution 10 min.
2. Microwave. Prepare a total of four plastic Coplin jars either with or without slides (jars without slides can be filled with H_2O).
3. Seal the jar(s) as tightly as possible and position them at the center of a microwave oven with rotating tray.
4. After each microwave interval, refill Coplin jars if any solution has leaked out.
5. 2.5 min at power 100 %.
6. 2.5 min at power 50 %.
7. 2.5 min at power 50 %.
8. 2.5 min at power 50 %.
9. Open the lid and cool the solution to room temperature (at least 30 min; *see* **Note 3**).

3.1.3 Blocking Endogenous Peroxidase

1. Wash sections with TBS in Coplin jars (up to 13 slides/jar) on belly dancer, 10 min.
2. Treat sections with freshly made 1 % H_2O_2 (in TBS) for 30 min at room temperature.
3. Wash with TBS, 5 min × 2, and with TBS-T, 5 min × 1, in Coplin jars on belly dancer.
4. After the second wash draw a boundary with a wax pen around the area of the slide to be stained (*see* **Note 4**).

3.1.4 Blocking and Primary Antibody

1. Block nonspecific binding sites with 10 % goat serum (in TBS-T) for 30 min to 1 h, usually 300–500 μl for each slide in a humidified chamber at room temperature.
2. Dilute primary antibody in 2 % goat serum (in TBS-T) (~300 μl/full slide) in a humidified chamber at room temperature. A. anti-Pten: 1:100 dilution. B. anti-p-Akt antibody 1:50 dilution. C. anti-p-4EBP1 1:500 dilution.
3. Remove the blocking solution and incubate the sections with the primary antibody in a humidified chamber overnight at 4 °C.

3.1.5 Secondary Antibody and Signal Amplification

1. Wash the sections with TBS-T in Coplin jars on belly dancer, 5 min × 3 (*see* **Note 5**). Incubate the sections with biotinylated anti-rabbit IgG antibody diluted to 1:200 in 2 % goat serum (in TBS-T) for 1 h in a humidified chamber at room temperature.

2. Prepare Elite ABC solution at least 30 min before use: To 2.5 ml of TBS-T, add two drops of solution A. Mix by vortexing and add two drops of solution B.
3. Mix by vortexing and leave the solution at room temperature.
4. Wash the sections with TBS-T, 5 min × 2, and 10 min × 1, in a humidified chamber at room temperature.
5. Optional tyramide amplification of p-Akt signal: (1) Wash the sections with TBS-T, 5 min × 3. (2) Add the biotinylated-tyramide (1:175 dilution in TBS-T) and incubate for 10 min at room temperature (300–500 μl/slide). (3) Wash the sections with TBS-T, 5 min × 3 (*see* **Note 6**).
6. Add the ABC solution and incubate for 1 h in a humidified chamber at room temperature (300–500 μl/slide), or for 30 min if tyramide amplification was used (*see* **Notes** 7 and **8**).

3.1.6 Color Development

IHC signals are visualized by incubating with NovaRED or diaminobenzidine (DAB), HRP substrates that produce a red or dark brown reaction product, respectively. NovaRED is generally more sensitive than DAB. The optimal substrate depends on the level of signal associated with the phenotype.

1. Wash with TBS-T, 5 min × 2, and 10 min × 1.
2. During the last wash prepare NovaRED solution (or DAB solution). To 5 ml of water, add three drops of reagent 1 and mix well by vortexing. Add two drops of reagent 2 and mix by vortexing. Add two drops of reagent 3 and mix by vortexing. Add two drops of H_2O_2 reagent and mix by vortexing.
3. Apply the solution onto washed sections. Incubate until signals become distinguishable between sample and control. Pten usually takes approximately 7–10 min. p-Akt takes approximately 10–15 min without tyramide amplification and 4 min with tyramide amplification. p-4EBP1 takes approximately 10 min.
4. Optimal incubation time will vary with tissue preparation. Also, if detecting pathological activation of the pathway, the optimal detection time will depend on the level of pathway activation. Observe under the microscope to select the optimal time to stop the development reaction. Representative immunohistochemistry images are shown in Fig. 1.
5. Discard the solution into a beaker and stop the development by washing sections with water 5 min × 2 (*see* **Note 9**).

3.1.7 Counter Staining with Hematoxylin

1. Incubate the sections in hematoxylin QS solution for 15 s.
2. Rinse the sections in a Coplin jar with running tap water until the rinse is colorless (5–10 min).
3. Dip the slides in water ten times.

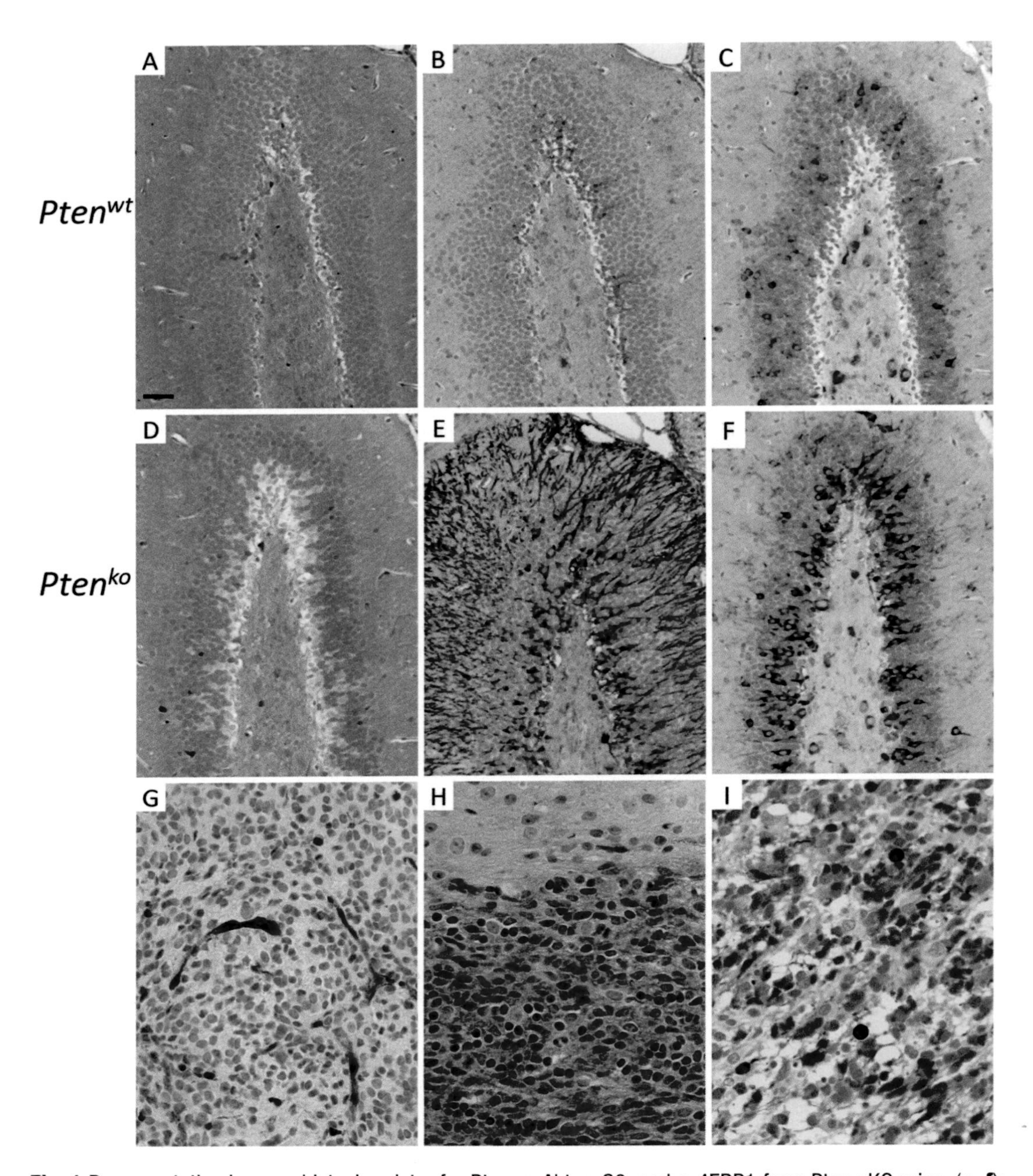

Fig. 1 Representative immunohistochemistry for Pten, p-Akt, p-S6, and p-4EBP1 from Pten cKO mice. (**a–f**): *Nestin-creER*; *Ptenflox/flox* with cre activity induced by tamoxifen administration at postnatal days 0 and 1, with tissue collected at day 30. (**a–c**): *Ptenwt* mouse; (**d–f**): *Ptenko*. (**g–h**): *Nestin-creER*; *Ptenflox/flox*; *Tp53flox/flox* mouse medulloblastoma. (**i**): *GFAP-creER*; *Ptenflox/flox*; *Tp53flox/flox* mouse astrocytoma. (**a** and **d**): Pten IHC: mouse dentate gyrus, Pten wild-type cells are red while Pten-null cells in the inner layer of the dentate gyrus stain only with hematoxylin counter stain. (**b** and **e**): p-Akt IHC,: Pten-null dentate gyrus neurons are strongly positive (*red*) for p-Akt. (**c** and **f**): p-S6 IHC: Pten-null dentate gyrus cells are more strongly positive (*red*) for p-S6 than the baseline p-S6 level in wild-type mice. (**g**) Pten IHC, mouse medulloblastoma: Pten-null tumor cells stain only with hematoxylin counter stain while wild-type blood vessel endothelial cells are brown. (**h**) p-Akt IHC, mouse medulloblastoma, and Pten-null tumor cells are strongly positive (*red*) while the adjacent normal brain is negative for p-Akt. (**i**) p-4ebp1 IHC, GFAP-creER; Pten; Tp53 double-knockout mouse astrocytoma are strongly positive for p-4ebp1. Scale bar = 50 μm. Substrate: Nova red, counter stain, hematoxylin

3.1.8 Dehydration and Mounting

1. Dehydration: 20 % EtOH 1 min, 50 % 1 min, 70 % 1 min, 95 % 2 min, 100 % 2 min × 2, xylenes 5 min × 3.
2. Coverslip with Permount in the fume hood.
3. Leave mounted slides in the hood at least overnight to dry.

3.2 Rapamycin Treatment

The inhibitory effects of rapamycin on mTOR complexes are dose and time dependent [15, 16]. Low-dose and short-term rapamycin treatment selectively inhibits mTorc1 activity, while high-dose and long-term treatment inhibits both mTorc1 and mTorc2 activity. We use loss of p-S6 as a downstream indication of rapamycin effects on mTorc1 and use p-Akt pSer473 as an indication of rapamycin effects on mTorc2. Increased phosphorylation of pS6 is more strongly associated with hypertrophic effects of mTorc1, while hyperproliferation is more strongly associated with p-4ebp1. In our Pten conditional knockout models of hypertrophy versus brain tumor proliferation, rapamycin appears to have much greater efficacy in blocking downstream S6 phosphorylation, but not p-4ebp1. Given the fact that rapamycin inhibition of mTorc1 signaling in brain increases with time, we typically dose a minimum of 5 days to see maximal pathway inhibition [18]. Tissue is collected 2 h after the last dose is administered.

3.2.1 Reagent Preparation

1. Dissolve rapamycin powder in sterile DMSO at 20 mg/ml, stored at −20 °C as stock solution.
2. Prepare 5.2 % Tween80 in MilliQ (make sure Tween80 is completely dissolved).
3. Rapamycin working solution is prepared by diluting stock solution in 5.2 % Tween80 to 10–20 μg/ml immediately before use.
4. Vehicle control is prepared by dissolving the same volume of DMSO in 5.2 % Tween80 immediately before use (*see* **Note 10**).

3.2.2 Injection of Rapamycin

1. To selectively inhibit mTorc1 in the subventricular zone and rostral migratory stream in mice from postnatal day 8 (P8) to P31, low-dose rapamycin (1.5 μg rapamycin/g body weight) is injected i.p. daily in Pten cKO or wild-type control mice from postnatal day 8 to 31.
2. At higher doses of rapamycin (7.5 μg rapamycin/g body weight) from P8-P31, both mTorc1 and mTorc2 are inhibited. Representative IHC to show pathway inhibition following rapamycin treatment is shown in Fig. 2. Rapamycin toxicity varies with age. Young mice cannot tolerate the higher doses used in adults, even when adjusted for body weight. 7.5 μg/g is the highest dose that FVB mice from P8–P31 could tolerate.
3. The dose of rapamycin required to block mTorc1 signaling varies between different brain regions. In adult mice, 25 μg

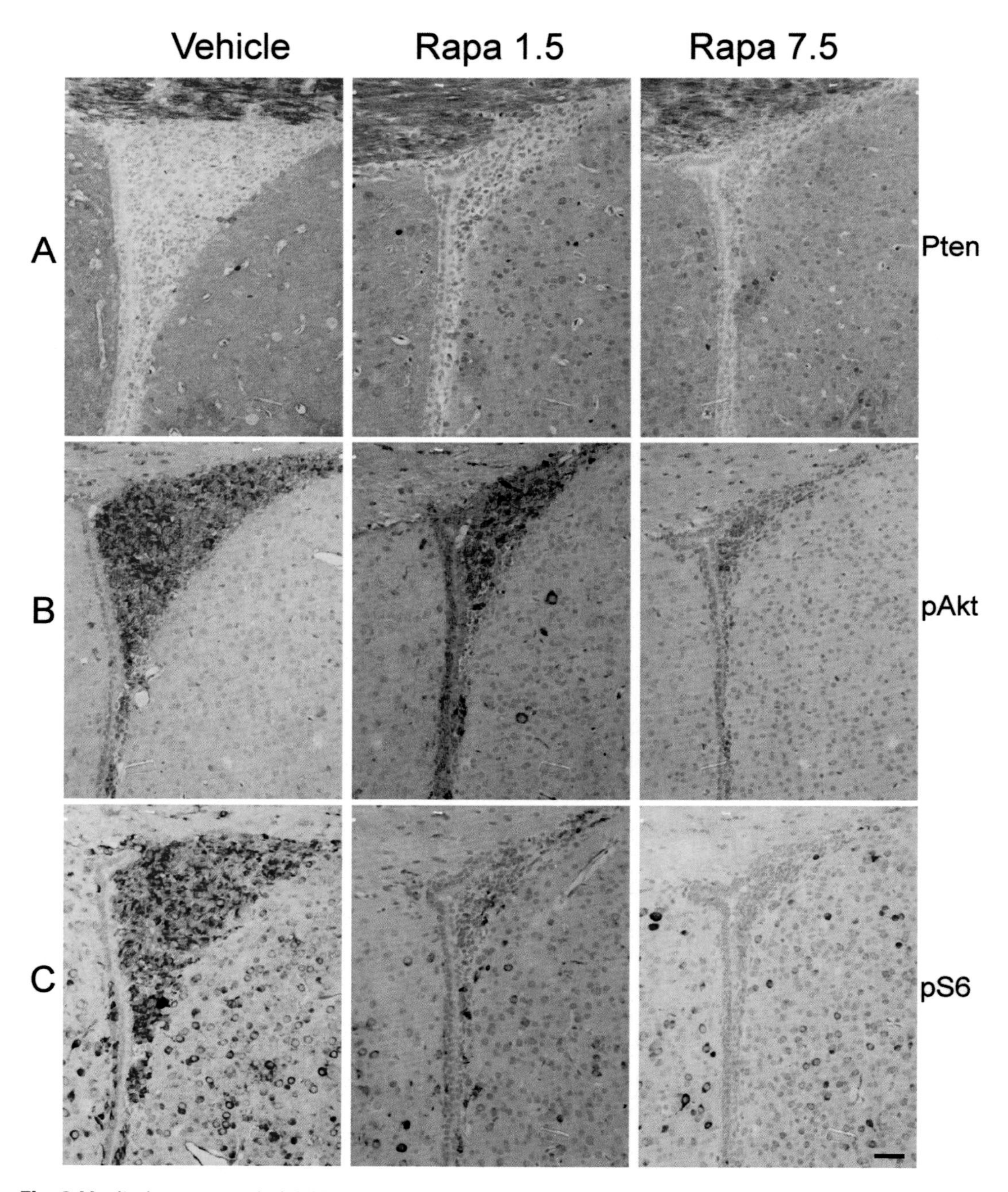

Fig. 2 Monitoring rapamycin inhibition of mTorc1 and mTorc2 activity by IHC. Representative Pten (**a**), p-Akt (**b**), and p-S6 (**c**) IHC images in the subventricular zone (SVZ) of vehicle- or rapamycin-treated *Nestin-creER*; *Ptenflox/flox* mice induced to delete Pten by tamoxifen administration at P0, P1 [7]. Pten conditional loss in the SVZ causes an expanded SVZ with elevated levels of pAkt and pS6. High-dose rapamycin (7.5 μg/g) inhibits both p-Akt (S473) and p-S6 while low-dose (1.5 μg/g) inhibits p-S6 only. Both doses rescue the expanded size of the SVZ in Pten cKO mice.FF

rapamycin/g body weight was sufficient to block mTorc1 signaling in the dentate gyrus of Pten conditional knockout mice. Increasing the dose to 50 μg rapamycin/g body weight was necessary to block the downstream signaling in cerebellum [18].

4 Notes

1. Prepare all solutions using ultrapure water (MilliQ, prepared by purifying deionized water to attain a sensitivity of 18 MΩ at 25 °C) and analytical grade reagents.
2. Work with xylene in a fume hood.
3. Labeling of microwave power output is not accurate, and does not necessarily translate from one microwave to the next, even if they are labeled as equivalent power output. The correct power output should be empirically determined by placing filled Coplin jars and monitoring time to boiling. The key point in this antigen retrieval is to boil the sections for at least the last 7.5 min with minimal air exposure. Therefore new conditions (power levels) may need to be set up for each microwave oven.
4. The area inside the wax boundary can be adjusted based on the size of sections as long as the entire section is completely contained inside the boundary. The amount of primary and secondary antibody incubation solutions can be reduced accordingly based on the area encircled inside the boundary.
5. To minimize cross-contamination among antibodies, keep slides with different primary antibodies in separate jars.
6. The p-Akt signal can be significantly enhanced with tyramide amplification without increasing background. This step is optional depending on the intensity of the p-Akt stain associated with the phenotype under study.
7. Elite ABC is a kit developed by Vector Laboratory, which utilizes the biotin-avidin complex to substantially amplify immunohistochemistry signals and increase detection sensitivity. Solution A contains avidin which complexes with biotinylated horseradish peroxidase (HRP) in Solution B. This complex binds with extremely high affinity to the biotinylated secondary antibody. HRP will react with substrate for colorimetric detection.
8. Secondary Ab incubation and Elite ABC incubation time can be extended to total 2–3 h. In this case, wash 5 more minutes for each 30-min incubation extension.
9. NovaRed and DAB (a known mutagen) are hazardous materials that are harmful if inhaled, ingested, or through direct con-

tact with skin. Handle according to the materials safety data sheet, and collect and dispose as hazardous waste.

10. Adjust rapamycin working solution concentration to make final injection volume around 0.5–1 ml/mouse.

Acknowledgments

We thank Sherri Rankin for the p-4ebp1 IHC image. SJB was supported in part by grants from the NIH (CA096832 and CA188516) and by the American Lebanese and Syrian Associated Charities of St Jude.

References

1. Chalhoub N, Baker SJ (2009) PTEN and the PI3-kinase pathway in cancer. Annu Rev Pathol 4:127–150
2. Alessi DR et al (1997) 3-Phosphoinositide-dependent protein kinase-1 (PDK1): structural and functional homology with the Drosophila DSTPK61 kinase. Curr Biol 7(10):776–789
3. Sarbassov DD et al (2005) Phosphorylation and regulation of Akt/PKB by the rictor-mTOR complex. Science 307(5712):1098–1101
4. Sengupta S, Peterson TR, Sabatini DM (2010) Regulation of the mTOR complex 1 pathway by nutrients, growth factors, and stress. Mol Cell 40(2):310–322
5. Shah OJ, Hunter T (2006) Turnover of the active fraction of IRS1 involves raptor-mTOR- and S6K1-dependent serine phosphorylation in cell culture models of tuberous sclerosis. Mol Cell Biol 26(17):6425–6434
6. Harrington LS et al (2004) The TSC1-2 tumor suppressor controls insulin-PI3K signaling via regulation of IRS proteins. J Cell Biol 166(2):213–223
7. Zhu G et al (2012) Pten deletion causes mTorc1-dependent ectopic neuroblast differentiation without causing uniform migration defects. Development 139(18):3422–3431
8. Chalhoub N et al (2009) Cell type specificity of PI3K signaling in Pdk1- and Pten-deficient brains. Genes Dev 23(14):1619–1624
9. Endersby R, Baker SJ (2008) PTEN signaling in brain: neuropathology and tumorigenesis. Oncogene 27(41):5416–5430
10. Fraser MM et al (2004) Pten loss causes hypertrophy and increased proliferation of astrocytes in vivo. Cancer Res 64(21):7773–7779
11. Kwon CH et al (2001) Pten regulates neuronal soma size: a mouse model of Lhermitte-Duclos disease. Nat Genet 29(4):404–411
12. Groszer M et al (2001) Negative regulation of neural stem/progenitor cell proliferation by the Pten tumor suppressor gene in vivo. Science 294(5549):2186–2189
13. Backman SA et al (2001) Deletion of Pten in mouse brain causes seizures, ataxia and defects in soma size resembling Lhermitte-Duclos disease. Nat Genet 29(4):396–403
14. Sperow M et al (2012) Phosphatase and tensin homologue (PTEN) regulates synaptic plasticity independently of its effect on neuronal morphology and migration. J Physiol 590 (Pt 4):777–792
15. Huang S, Bjornsti MA, Houghton PJ (2003) Rapamycins: mechanism of action and cellular resistance. Cancer Biol Ther 2(3):222–232
16. Sarbassov DD et al (2006) Prolonged rapamycin treatment inhibits mTORC2 assembly and Akt/PKB. Mol Cell 22(2):159–168
17. O'Reilly KE et al (2006) mTOR inhibition induces upstream receptor tyrosine kinase signaling and activates Akt. Cancer Res 66(3):1500–1508
18. Kwon CH et al (2003) mTor is required for hypertrophy of Pten-deficient neuronal soma in vivo. Proc Natl Acad Sci U S A 100(22): 12923–12928

Chapter 6

Germline *PTEN* Mutation Analysis for *PTEN* Hamartoma Tumor Syndrome

Joanne Ngeow and Charis Eng

Abstract

Clinically, deregulation of PTEN function resulting in reduced PTEN expression and/or activity is implicated in human disease. Cowden syndrome (CS) is an autosomal dominant disorder characterized by benign and malignant tumors. CS-related individual features occur commonly in the general population. Approximately 25 % of patients diagnosed with CS have pathogenic germline *PTEN* mutations, which increase lifetime risks of breast, thyroid, uterine, renal, and other cancers. *PTEN* testing and intensive cancer surveillance allow for early detection and treatment of these cancers for mutation-positive patients and their relatives. In this methods chapter, we highlight our protocol for identifying patients at risk of harboring a germline *PTEN* mutation.

Key words *PTEN* hamartoma tumor syndrome, Cancer

1 Introduction

Cowden syndrome (CS; MIM 158350) is an autosomal dominant disorder and part of the *PTEN* hamartoma tumor syndrome (PHTS), which also includes subsets of Bannayan-Riley-Ruvalcaba syndrome, Proteus syndrome, and Proteus-like syndrome [1, 2]. PHTS is molecularly defined as having a germline *PTEN* mutation irrespective of clinical syndrome. PHTS is a highly penetrant genetic disorder. More than 90 % of individuals with *PTEN* mutations are believed to manifest some feature of the syndrome (although rarely cancer) by age 20, and by age 30 nearly 100 % of carriers are believed to have developed at least some of the mucocutaneous signs. Affected individuals also have an increased risk for several malignancies, including female breast cancer (85 % lifetime risk), epithelial thyroid cancer with an overrepresentation of follicular thyroid carcinoma, endometrial cancer, renal cell carcinoma and colorectal carcinoma, as well as melanoma [3–5].

Leonardo Salmena and Vuk Stambolic (eds.), *PTEN: Methods and Protocols*, Methods in Molecular Biology, vol. 1388, DOI 10.1007/978-1-4939-3299-3_6, © Springer Science+Business Media New York 2016

1.1 Identifying Patients for Genetics Risk Assessment

Consensus diagnostic criteria for CS were first developed in 1996 by the International Cowden Consortium (ICC) and form the basis for National Comprehensive Cancer Network Guidelines (Table 1) [6]. In part because of phenotypic heterogeneity, and in part because of the high frequency of de novo *PTEN* germline mutations [7], a family history of associated cancers may not be apparent. Additionally, many of the benign features of CS are common in the general population making the diagnosis of CS a challenge for most clinicians. To aid clinical diagnosis, a clinical predictor (Cleveland Clinic *PTEN* Risk Calculator) was developed

Table 1
Revised *PTEN* hamartoma tumor syndrome clinical diagnostic criteria (J Natl Cancer Inst;2013;105:1607–1616)

Major criteria	
Breast cancer Endometrial cancer (epithelial) Thyroid cancer (follicular) Gastrointestinal hamartomas (including ganglioneuromas, but excluding hyperplastic polyps; ≥3) Lhermitte-Duclos disease (adult) Macrocephaly (≥97 percentile: 58 cm for females, 60 cm for males) Macular pigmentation of the glans penis Multiple mucocutaneous lesions (any of the following): Multiple trichilemmomas (≥3, at least one biopsy proven) Acral keratoses (≥3 palmoplantar keratotic pits and/or acral hyperkeratotic papules) Mucocutaneous neuromas (≥3) Oral papillomas (particularly on tongue and gingiva), multiple (≥3) OR biopsy proven OR dermatologist diagnosed	
Minor criteria	
Autism spectrum disorder Colon cancer Esophageal glycogenic acanthosis (≥3) Lipomas (≥3) Mental retardation (i.e., IQ≤75) Renal cell carcinoma Testicular lipomatosis Thyroid cancer (papillary or follicular variant of papillary) Thyroid structural lesions (e.g., adenoma, multinodular goiter) Vascular anomalies (including multiple intracranial developmental venous anomalies)	
Operational diagnosis in an individual (either of the following):	**Operational diagnosis in a family where one individual meets revised PTEN hamartoma tumor syndrome clinical diagnostic criteria or has a PTEN mutation:**
1. Three or more major criteria, but one must include macrocephaly, Lhermitte-Duclos disease, or gastrointestinal hamartomas; or 2. Two major and three minor criteria	1. Any two major criteria with or without minor criteria; or 2. One major and two minor criteria; or 3. Three minor criteria

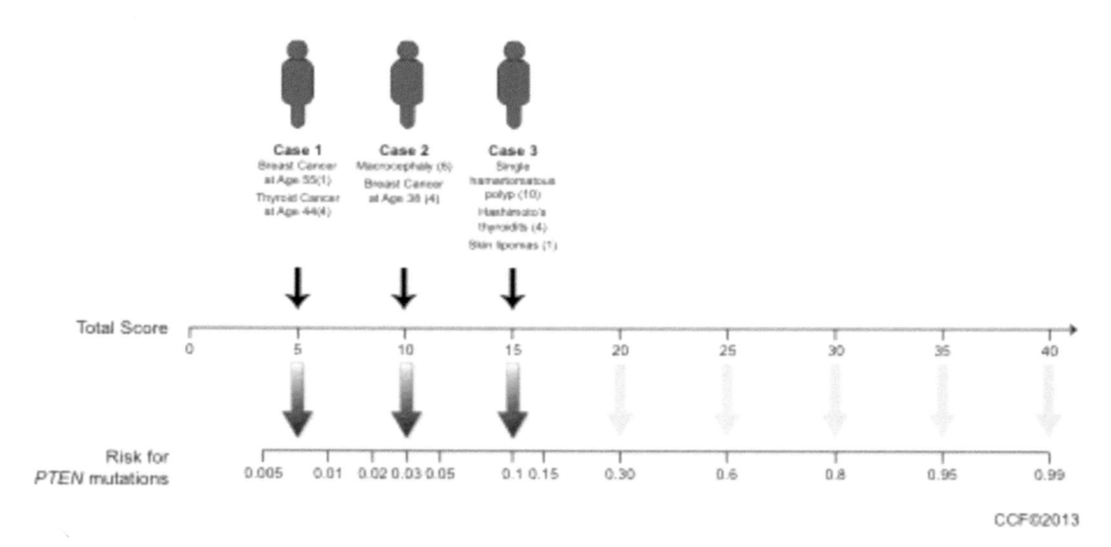

Fig. 1 Cleveland Clinic (CC) PTEN Risk Calculator: three illustrative cases. The CC score (total score) is first derived by a sum of the weights of positive clinical features. To illustrate, three hypothetical cases are presented, each corresponding to a CC score of 5 points, 10 points, and 15 points, respectively. Case 1 may present with breast cancer at age 55 (one point), with background of thyroid cancer at age 44 (four points), with a final score of 5 and corresponding point probability <1 %. Case 2 may present with breast cancer at age 38 (four points) and concurrent macrocephaly (six points), with a final score of 10 and corresponding point probability of 3 %. Case 3 may present with a single hamartomatous gastrointestinal polyp (ten points) found on endoscopy, Hashimoto's thyroiditis (four points), and lipomas (one point), for a final score of 15 and corresponding point probability of 10 % (adapted from Tan et al., 2011 with Permission from Cell Press [3])

based on clinical features derived from a prospective study of over 3000 patients suspected of having CS [3]. The questionnaire-based clinical decision tool is available online (http://www.lerner.ccf.org/gmi/ccscore/) to assist clinicians at the point of patient care. Based on a patient's presentation, a score will be derived which corresponds to an estimated risk for germline *PTEN* mutation (Fig. 1) to guide clinicians for referral to genetics professionals for genetic testing.

1.2 PTEN Mutation Analysis

In this methods chapter we describe methods for detecting germline *PTEN* mutations in genomic DNA samples. All research participants clinically suspected of having PHTS in our cohort underwent *PTEN* (NM_000314.4) mutation analysis. Depending on specific phenotypic features and phenotypic load, germline *PTEN* mutations are found in 25–80 % of patients presenting with CS-related features [3, 8]. Germline *PTEN* mutations in PHTS are found throughout most of the *PTEN* coding region, with the exception of exon 9, which encodes the carboxy-terminal 63 amino acids [9]. The N-terminal domain contains the phosphatase domain (the enzymatic activity of PTEN) and it is, therefore, not surprising that the majority of *PTEN* mutations occur within this domain [10, 11]. Forty percent of *PTEN* pathogenic mutations occur in the C2 domain and the tail sequence, suggesting an important role for the C terminal in maintaining PTEN function and stability [2, 12].

A subset of patients with PHTS carries germline mutations in the *PTEN* promoter or in potential splice donor and acceptor sites [3, 13]. Complete mutation analysis for PHTS therefore involves mutation screening across all nine exons and the *PTEN* promoter. Scanning of genomic DNA samples for *PTEN* mutations was performed with a combination of high-resolution melting (HRM) curve analysis with direct Sanger sequencing [14] as described below.

2 Materials

2.1 Reagents

1. 10× PCR buffer: 30 mM $MgCl_2$, 2.5 mg/ml bovine serum albumin (BSA).
2. 500 mM Tris pH 8.3 (Idaho Technology Inc., cat. no. 1770).
3. TaqStart antibody (1100 ng/μl; Clontech).
4. KlenTaq1 (25 U/μl; AB Peptides).
5. Enzyme diluent: 10 mM Tris, pH 8.3, 2.5 mg/ml, BSA.
6. 10× dNTP mix: 2 mM dATP, dCTP, dGTP, dTTP.
7. 10× LCGreen Plus (Idaho Technology Inc.).
8. Genomic DNA (25 ng/μl): For best results, all human genomic DNA samples compared should be prepared by the same method and the concentration adjusted to an $A_{260} = 1.0$ (~25 ng/μl).
9. Primer mix: 10× (5 mM) Reverse direction oligonucleotide primer, 10× (1 mM) forward direction oligonucleotide primer, 10× (5 mM) 3′-blocked forward direction oligonucleotide probe.

2.2 Equipment

1. 96- or 384-well plates 96- or 384-well LightScanner (Idaho Technology Inc.) (*see* **Note 1**).

3 Methods

3.1 Sample Format for High-Resolution Melting Analysis

PCR and DNA melting can be performed either in capillaries or plates (96- or 384-well). The main difference between capillary and plate formats is the throughput. For high-throughput applications, 96- or 384-well plate format is recommended and is described here.

1. DNA preparation:

   Genomic DNA can be isolated and purified using any method ranging from conventional phenol/chloroform extraction and

Table 2
PCR master mix

Reagent	Volume per reaction (µl)
10× PCR buffer	1
10× Taq polymerase/antibody	1
10× dNTPs	1
10× LC Green plus	1
10× Forward primer	1
10× Reverse primer	1
10× Probe	1
H_2O	2
Final	9

ethanol precipitation to modern automated solid-phase-binding method. However, all DNA samples that will be compared should be dissolved in the same buffer solution because different buffers will affect the melting temperature.

2. Prepare a PCR master mix for each sample to be amplified, as follows (Table 2):

   Alternatively use the commercially available master mix (Table 3):

3. Perform PCR in a 96- or 384-well plate format in a plate thermal cycler:
   - Aliquot 9 µl PCR master mix into an appropriate number of wells of a 96- or 384-well plate, one for each sample to be tested.
   - Add 1 µl genomic DNA to each well.
   - Add 10–20 µl mineral oil to each well.
   - Centrifuge plate at 20–25 °C for 2 min at 1500 × *g*.
   - Place 96- or 384-well plate in thermal cycler.
   - Perform PCR with appropriate cycling conditions, as follows (Tables 4 and 5):
   - *Pause point*: PCR products may be stored at 20–25 °C overnight or at 2–4 °C for up to a week.
4. Melting acquisition from the 96- or 384-well plate using a LightScanner:
   - Briefly spin down plate at 1500 × *g*.
   - Insert 96- or 384-well plate into the LightScanner.
   - Melt samples from 50 to 90 °C at a rate of 0.1 °C/s.

Table 3
PCR cycles

Reagent	Volume per reaction (μl)
2.5× LightScanner master mix	4 μl
Primer Mix[a]	1
H_2O	4
Final	9

[a]*see* **Note 2**

Table 4
LightSanner PCR program

	95 °C	2 min
37 cycles* ×	95 °C	30 s
	Annealing temp (°C)	30 s
	95 °C	30 s
	25 °C	30 s

5. Melting analysis:
   - Starting with the original melting curve data, remove the background fluorescence by fitting a decreasing exponential to the slope of the curve in regions where no melting occurs.
   - Normalize all plots between 0 and 100 % fluorescence.
   - To scan the amplicon for sequence variants, analyze the shape of the amplicon melting transition (using the background-subtracted, normalized data) with and without curve overlay.
   - Cluster each genotype by standard, unbiased hierarchal clustering.
   - Display the melting curves as difference plots (Fig. 2) (*see* **Note 6**).

TIMING:

Steps 1 and 2: ~2.5 h.

Step 3: 10 min per plate.

Step 4: 2–10 min.

Table 5
***PTEN* LightScanner primers and annealing temperatures (°C)**

Name	Sequence 5′ → 3′	Annealing temp (°C)[a]
PTEN LS E1F	GCAGCTTCTGCCATCTC	66
PTEN LS E1R	GCATCCGTCTACTCCCAC	–
PTEN LS E2F	TGATTGCTGCATATTTCAGA	60
PTEN LS E2R	CTAAATGAAAACACAACATGAATATAAACA	–
PTEN LS E3F	ATGTTAGCTCATTTTTGTTAATGGTG	60
PTEN LS E3R	CAAGCAGATAACTTTCACTTAATAGTTG	–
PTEN LS E4F	TTTTTTCTTCCTAAGTGCAAAAGATAAC	60
PTEN LS E4R	CAGTAAGATACAGTCTATCGGGT	–
PTEN LS E5F	ACCTACTTGTTAATTAAAAATTCAAGAGTT	60
PTEN LS E5R	ATCCAGGAAGAGGAAAGGAAA	–
PTEN LS E6F	CCCAGTTACCATAGCAATTTAGTGA	60
PTEN LS E6R	TAGATATGGTTAAGAAAACTGTTCCAATAC	–
PTEN LS E7F[b]	CAGTTTGACAGTTAAAGGCATTTC	61
PTEN LS E7R[b]	AATATAGCTTTTAATCTGTCCTTATTTTGG	–
PTEN LS E8F	TTTGTTGACTTTTTGCAAATGTTTAACATA	61
PTEN LS E8R	ATTTCTTGATCACATAGACTTCCA	–
PTEN LS E9F	AAGATGAGTCATATTTGTGGGTT	61
PTEN LS E9R	TTTCAGTTTATTCAAGTTTATTTTCATGG	–
PTEN Promoter F[c]	GCG TGG TCA CCT GGT CCT TT	62
PTEN Promoter R[c]	GCT GCT CAC AGG CGC TGA	

[a]*see* **Note 3**
[b]*see* **Note 5**
[c]*see* **Note 4**

3.2 Sample Format for High-Resolution Melting Analysis

PCR products showing different melting curves from the reference groups were selected for PCR cleanup with Exo-SAP (*see* **Note 7**).

1. Exo-Sap treatment:
   - 10 μl PCR_1 product with 2.5 μl Exo-Sap mix (0.5 μl Exonuclease I; 2 μl shrimp alkaline phosphatase mix).
   - 1 cycle of 37 °C for 20 min, 82 °C for 15 min.
2. Sequencing mix:

   Add 0.5 μl of PCR product after Exo-SAP clean up (PCR_2) with 4.5 μl of sequencing primer (1 μM) for direct Sanger sequencing (ABI 3730xl, Life Technologies, Carlsbad, CA).

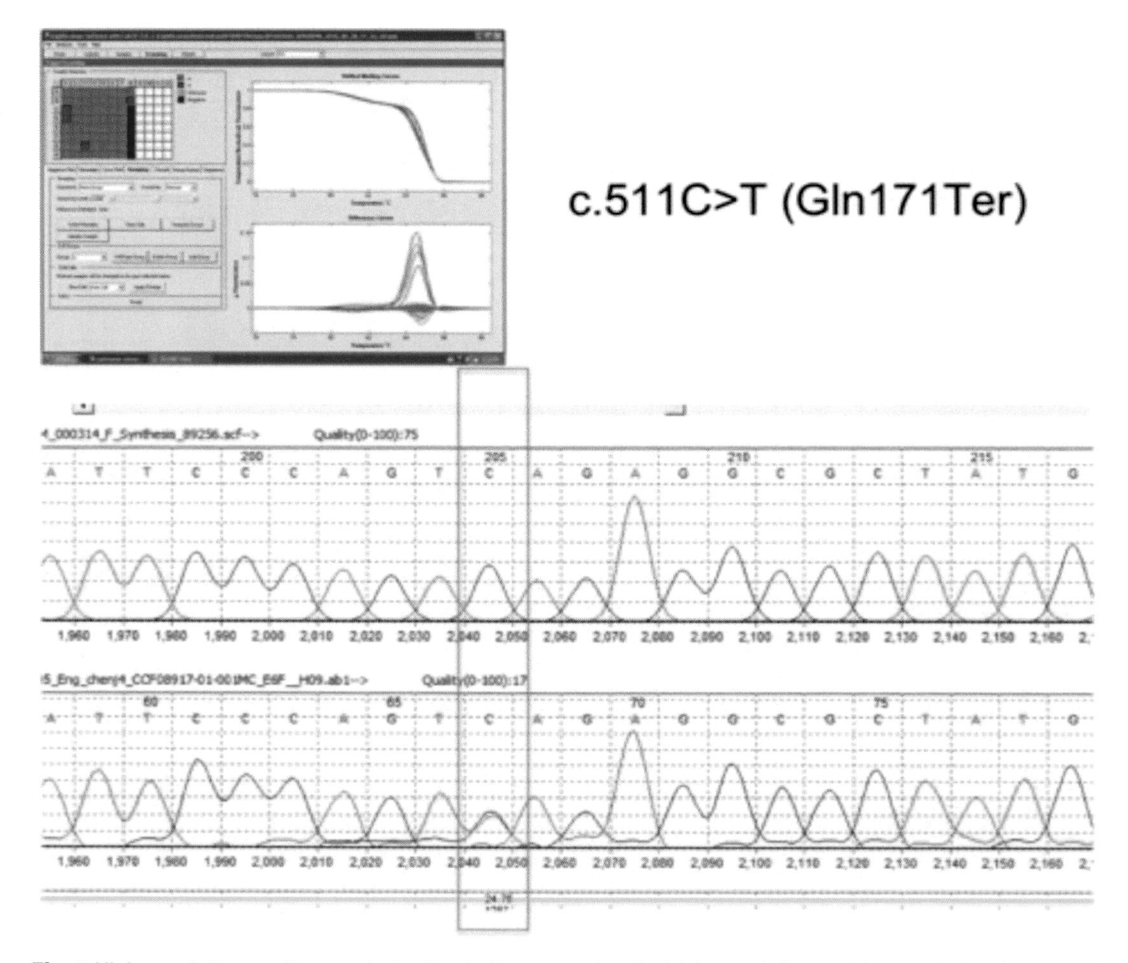

Fig. 2 High-resolution melting analysis. Illustrative example of a high-resolution melting analysis of exon 6 of *PTEN*: results for a set of samples. Difference plots (*top panel*) of the amplicon melting profiles allowed identification of a total of four different genotypes in *red* (including the positive control). Sanger sequencing (*bottom panel*) revealed one of the mutations to be c.511C>T

4 Notes

1. The accuracy of scanning and genotyping by PCR product melting analysis is dependent on the instrument resolution. Instruments specifically designed for high-resolution melting are used here and produce the best results.
2. Primer mix for exons includes 2 μM of both forward and reverse primers.
3. Annealing temperatures may change with specific lots of LightScanner mastermix. Do gradient PCR to find out optimal annealing temperatures if there is a problem with amplifications.
4. *PTEN* promoter sequencing requires the following modifications (Table 6) to reaction setup (**step 1**) and PCR program (**step 2**).

Table 6
***PTEN* promoter sequencing modifications**

Reagent	Volume per reaction (μl)	PCR program		
2.5× LightScanner master mix	12.5		95 °C	10 min
Promoter forward primer [20 μM]	0.8	35 cycles ×	95 °C	30 s
Promoter reverse primer [20 μM]	0.8		62 °C	45 s
GC enhancer	4		72 °C	1 min
H_2O	2.9		72 °C	10 min
Final	24		4 °C	10 min

5. Initial scanning for germline mutations in *PTEN* in individuals carrying the clinical diagnosis of CS noted an IVS7-15 to -53del39 alteration. Because of its proximity to the intron 7-exon 8 splicing boundary and the number of nucleotides deleted, it was initially considered a deleterious *PTEN* mutation. This heterozygous 39 bp deletion was subsequently found in almost one-third of African-American control individuals. RT-PCR experiments demonstrated no effect on the transcript. Taken together with its frequent occurrence in a normal control population, specifically of ethnic African origin, our observations suggest that the *PTEN* IVS7-15 to -53del39 is a novel polymorphism possibly unique to African-American individuals. This observation is worthy of note and clinical cancer geneticists, genetic counselors, and laboratory directors should be aware that the IVS7-15 to -53del39 should be treated as a normal polymorphic variant in the African-American population.
6. Mutation scanning using high-resolution melting curve analysis described above is an effective and sensitive method to detect sequence variations. However, the presence of a common SNP within a mutation scanning amplicon may considerably complicate the interpretation of results and increase the number of samples flagged for sequencing by interfering with the clustering of samples according to melting profiles. Systematic resequencing of all variant samples is the most common approach to this issue. However, when applied on large series, this latter approach is expensive, laborious, and time consuming. For amplicons that contain a common SNP, it has been shown that stratification of high-resolution-melt data by common SNP genotype prior to mutation scanning analysis would increase the detection sensitivity for those rare variants [15]. For example in exon 8 of *PTEN*, c.1026+32 T>G (rs555895), interferes with the protocol above. By performing an additional asymmetric PCR using an SNP-specific unlabeled probe (Table 7), we can stratify samples according to their

Table 7
SNP-specific unlabeled probes

Name	Sequence 5′→3′	Annealing temp (°C)
PTEN LS SNP E8.F	GCAAATAAAGACAAAGCCAACCGA	60
PTEN LS SNP E8.R	AGCTGTACTCCTAGAATTAAACACACATC	
PTEN LS SNP E8.2 Probe	CATACAAGTCACCAACCCCCAC-block	

probe-target melting profiles. This approach improves identification of rare known and unknown variants while dramatically reducing the sequencing effort.

7. The ExoSAP protocol is the simplest way to clean up PCR products before sequencing. The exonuclease I removes leftover primers, while the shrimp alkaline phosphatase removes the dNTPs. The advantage of this protocol is that you eliminate the need to pipette your PCR products out of their original tubes, thus minimizing the potential for PCR contamination of your lab and equipment.
8. Deletion analysis should be performed for patients to exclude any large deletions if mutation analysis is normal and PHTS is suspected. This can be done using the multiplex ligation-dependent probe amplification (MLPA) assay [16] according to the manufacturer's protocol.

Acknowledgements

We apologize to authors whose works or relevant references were not cited due to word limitations. American Cancer Society, Breast Cancer Research Foundation, Arthur Blank Foundation, US Department of Defense Breast Cancer Research Program, Doris Duke Charitable Foundation, William Randolph Hearst Foundations, Susan G. Komen for the Cure, Ambrose Monell Foundation, National Cancer Institute, and National Institutes of Health are gratefully acknowledged for funding C.E.'s patient-oriented research over the last 17 years. C.E is the Sondra J. and Stephen R. Hardis Chair of Cancer Genomic Medicine at the Cleveland Clinic and is an American Cancer Society Clinical Research Professor.

References

1. Zhou XP, Woodford-Richens K, Lehtonen R et al (2001) Germline mutations in BMPR1A/ALK3 cause a subset of cases of juvenile polyposis syndrome and of Cowden and Bannayan-Riley-Ruvalcaba syndromes. Am J Hum Genet 69(4):704–711
2. Waite KA, Eng C (2002) Protean PTEN: form and function. Am J Hum Genet 70(4): 829–844
3. Tan MH, Mester J, Peterson C et al (2011) A clinical scoring system for selection of patients for PTEN mutation testing is proposed on the basis of a prospective study of 3042 probands. Am J Hum Genet 88(1):42–56
4. Ngeow J, Mester J, Rybicki LA et al (2011) Incidence and clinical characteristics of thyroid cancer in prospective series of individuals with Cowden and Cowden-like syndrome characterized by germline PTEN, SDH, or KLLN alterations. J Clin Endocrinol Metab 96(12): E2063–E2071
5. Tan MH, Mester JL, Ngeow J et al (2012) Lifetime cancer risks in individuals with germline PTEN mutations. Clin Cancer Res 18(2): 400–407
6. Eng C (2000) Will the real Cowden syndrome please stand up: revised diagnostic criteria. J Med Genet 37(11):828–830
7. Mester J, Eng C (2012) Estimate of de novo mutation frequency in probands with PTEN hamartoma tumor syndrome. Genet Med 14(9):819–822
8. Marsh DJ, Coulon V, Lunetta KL et al (1998) Mutation spectrum and genotype-phenotype analyses in Cowden disease and Bannayan-Zonana syndrome, two hamartoma syndromes with germline PTEN mutation. Hum Mol Genet 7(3):507–515
9. Orloff MS, Eng C (2008) Genetic and phenotypic heterogeneity in the PTEN hamartoma tumour syndrome. Oncogene 27(41): 5387–5397
10. Eng C (2003) PTEN: one gene, many syndromes. Hum Mutat 22(3):183–198
11. Zbuk KM, Eng C (2007) Cancer phenomics: RET and PTEN as illustrative models. Nat Rev Cancer 7(1):35–45
12. Georgescu MM, Kirsch KH, Kaloudis P et al (2000) Stabilization and productive positioning roles of the C2 domain of PTEN tumor suppressor. Cancer Res 60(24):7033–7038
13. Pezzolesi MG, Li Y, Zhou XP et al (2006) Mutation-positive and mutation-negative patients with Cowden and Bannayan-Riley-Ruvalcaba syndromes associated with distinct 10q haplotypes. Am J Hum Genet 79(5): 923–934
14. van der Stoep N, van Paridon CD, Janssens T et al (2009) Diagnostic guidelines for high-resolution melting curve (HRM) analysis: an interlaboratory validation of BRCA1 mutation scanning using the 96-well LightScanner. Hum Mutat 30(6):899–909
15. Nguyen-Dumont T, Calvez-Kelm FL, Forey N et al (2009) Description and validation of high-throughput simultaneous genotyping and mutation scanning by high-resolution melting curve analysis. Hum Mutat 30(6):884–890
16. Schouten JP, McElgunn CJ, Waaijer R et al (2002) Relative quantification of 40 nucleic acid sequences by multiplex ligation-dependent probe amplification. Nucleic Acids Res 30(12):e57

Chapter 7

Methods for Assessing the In Vivo Role of PTEN in Glucose Homeostasis

Cynthia T. Luk, Stephanie A. Schroer, and Minna Woo

Abstract

PTEN plays an important role in diabetes pathogenesis not only as a key negative regulator of the PI3K/Akt pathway required for insulin action, but also via its role in other cell processes required to maintain metabolic homeostasis. We describe the generation of tissue-specific PTEN knockout mice and models of both type 1 and type 2 diabetes, which we have found useful for the study of diabetes pathogenesis. We also outline common methods suitable for the characterization of glucose homeostasis in rodent models, including techniques to measure beta cell function and insulin sensitivity.

Key words PTEN, Diabetes, Glucose, Insulin sensitivity

1 Introduction

Diabetes mellitus is characterized by disruption of glucose homeostasis and hyperglycemia, with type 1 diabetes associated with a loss of pancreatic beta cells, and type 2 diabetes associated with both beta cell dysfunction and insulin resistance [1,2]. Hyperglucagonemia and inappropriate hepatic glucose production are also key features of diabetes. With the growing epidemic of diabetes, there is an urgent need to better understand diabetes pathogenesis. Animal models are particularly well suited to the study of whole-body glucose homeostasis, and the methods described here focus on mouse models, which are commonly used for the study of PTEN in vivo.

PTEN was initially discovered as a tumor-suppressor gene, and is well recognized as a potent negative regulator of the phosphoinositide 3-kinase (PI3K)/Akt pathway [3, 4]. The PI3K/Akt pathway regulates a variety of cell processes, such as proliferation, differentiation, and survival. Notably, this pathway is an important effector of insulin signaling, mediating glucose uptake and glycogen synthesis in muscle and fat, and inhibiting glycogenolysis and glucose output in the liver [5–8]. While whole-body deletion of PTEN is embryonic lethal [9], specific deletion of PTEN in tissues

Leonardo Salmena and Vuk Stambolic (eds.), *PTEN: Methods and Protocols*, Methods in Molecular Biology, vol. 1388, DOI 10.1007/978-1-4939-3299-3_7, © Springer Science+Business Media New York 2016

important for glucose regulation such as beta cells, alpha cells, hepatocytes, muscle, adipose tissue, or RIP-cre neurons improves glucose homeostasis and has advanced our understanding of the role of PTEN in diabetes pathogenesis [10–16] (Table 1). PTEN knockdown by antisense oligonucleotide protects db/db mice from diabetes [17], and PTEN polymorphisms have been described in association with a type 2 diabetes cohort [18], suggesting a key role for PTEN in glucose homeostasis.

We describe the generation of mice with tissue-specific deletion of PTEN, specifically the generation of RIP-cre^{+} PTEN$^{fl/fl}$ mice from transgenic RIP-cre^{+} and PTEN$^{fl/fl}$ mice [9, 20]. These methods can be adapted by using various transgenic mouse strains with conditional expression of Cre to generate a variety of models (Table 1). RIP-cre^{+} PTEN$^{fl/fl}$ mice have PTEN-deleted in both pancreatic islets and RIP-cre neurons, and have improved insulin sensitivity. These mice are protected from both multiple low-dose streptozotocin (MLDS)-induced model of type 1 diabetes and high fat diet (HFD) feeding-induced insulin resistance or type 2 diabetes, and we describe protocols for these induced diabetes

Table 1
Diabetes pathogenesis and PTEN

Tissue	Role in glucose homeostasis	Related Cre-expressing strain(s)	Role of PTEN deletion	Related cell lines (organism)
Beta cells	Secrete insulin	Pdx-Cre [19] RIP-Cre [20]	Preserves beta cell mass and function [10, 11, 21]	INS-1 (rat) MIN6 (mouse)
Alpha cells	Secrete glucagon	Glc-Cre [22]	Attenuates hyperglucagonemia [23]	In-R1-G9 (hamster) αTC (mouse)
Liver	Major source of glucose production	Alb-Cre [20]	Fatty liver and improved insulin sensitivity [13, 14]	Hep G2 (human)
Muscle	Major source of glucose uptake	Mck-Cre [24]	Improved insulin sensitivity and glucose uptake [15]	C2C12 (mouse)
Adipose	Energy storage, and regulation of FFA and secreted factors	Adipoq-Cre [25] Ap2-Cre [26]	Improved insulin sensitivity and resistance to STZ [16]	3T3-L1 (mouse)
Hypothalamus	Senses and responds to hyperglycemia	RIP-Cre [20]	Deletion in RIP-Cre neurons reduces metabolic inflammation [12]	

models. Further discussion of diabetes models is provided in an earlier volume [27]. The study of RIP-cre^{+} PTEN$^{fl/fl}$ mice has identified the important in vivo regulatory role of PTEN in pancreatic beta cells and RIP-cre neurons in these models [10–12].

There are many important points to consider when utilizing Cre-loxP recombinant technology, which have been discussed in detail [28]. Cre is a DNA recombinase used to excise target genes surrounded by recognition loxP sites. Controlling the expression of Cre enables gene inactivation in various cell types or in response to promoter activation, for example, by estrogen receptor ligand. Pancreas-specific Cre lines have also recently been reviewed by Magnuson and Osipovich [29]. When using Cre-loxP systems, it is particularly important to confirm the distribution and efficiency of gene deletion; timing of gene inactivation, for example during development or during cell differentiation, may also play a role in interpreting results.

We further describe commonly used methods appropriate for initial evaluation of glucose homeostasis. Fed and fasting blood glucose measurements can readily be used to identify differences or changes in blood glucose. These measurements guide further testing, and timing of experiments. Simultaneous measurement of serum insulin can also help provide basic information on insulin sensitivity. We describe intraperitoneal glucose tolerance testing (GTT), which provides general information on glucose homeostasis, and insulin tolerance testing (ITT), which is a complementary test to examine insulin sensitivity. While the "gold standard" for determining insulin sensitivity is hyperinsulinemic euglycemic clamping, this technique can be more technically challenging and invasive [30, 31]. This clamp is a consideration for further testing; essentially, exogenous insulin is infused intravenously, and glucose infused to maintain euglycemia. Glucose infusion rates provide information on whole-body insulin sensitivity.

Further information specifically on pancreatic beta cell function can be obtained by measurement of in vivo glucose-stimulated insulin secretion, which we also describe. Isolated islets can be used for further study and mechanistic elucidation. This technique can require some practice and has been previously described, along with methods for the characterization of islet morphology [32, 33]. Muscle is also the major organ of glucose uptake and we describe the measurement of insulin-stimulated glucose uptake in this tissue. Further data on the direct role of PTEN in specific cell types can be obtained by studying primary cells or tissue ex vivo. We also list a number of related cell lines that may offer a less technically challenging alternative (Table 1). Together the methods described facilitate the study of PTEN in diabetes pathogenesis.

2 Materials

2.1 Generation of RIP-cre+ PTEN$^{fl/fl}$ Mice

1. Appropriate mouse husbandry environment, including caging system, food, water, and controlled facility (*see* **Note 1**).
2. RIP-cre+ mice (Jackson Laboratories, Bar Harbor, Maine).
3. PTEN$^{fl/fl}$ mice (*see* **Note 2**).
4. Laboratory notebook or record system for breeding and genotyping.

2.2 Genotyping RIP-cre+ PTEN$^{fl/fl}$ Mice (See Note 2)

1. Ear punch (Thumb Type Punch).
2. 1.5 mL Eppendorf tubes.
3. Permanent marker.
4. Ear buffer: 25 mM Tris (pH 7.5), 50 mM EDTA, 1 % SDS.
5. 10 mg/mL Proteinase K: Reconstitute with water.
6. 10× Taq Polymerase Buffer.
7. Betaine.
8. dNTP.
9. Taq Polymerase.
10. PTEN primers: Forward 5′-ctc ctc tac tcc att ctt ccc-3′, reverse 5′-act ccc acc aat gaa caa ac-3′ (*see* **Note 2**).
11. CRE primers: Forward 5′-ggc agt aaa aac tat cca gca a-3′, reverse 5′-gtt ata agc aat ccc cag aaa tg3′.
12. Ice bucket.
13. Thermocycler.
14. Nucleic acid gel electrophoresis system including power supply.
15. Agarose.
16. Ethidium bromide.
17. TBE buffer: To prepare 5× stock solution, add 54 g Tris base, 27.5 g boric acid, and 20 mL 0.5 M EDTA to 1 L H_2O. pH stock solution to 8.4.
18. Loading dye: 10 mM Tris HCl (pH 7.6), 10 mM EDTA, 0.015 % bromophenol blue, 0.015 % xylene cyanol, 15 % glycerol.
19. Molecular Imager.

2.3 Blood Glucose Measurement

1. Glucometer (*see* **Note 3**).
2. Glucometer strips (*see* **Note 3**).
3. Scalpel blade.
4. Mouse restraint.
5. Lab notebook or appropriate record system.

2.4 Glucose Tolerance Test

1. Supplies for blood glucose measurement.
2. Mouse scale.
3. 0.5 g/mL Glucose solution (*see* **Note 4**).
4. Tuberculin or insulin syringes.
5. Timer.

2.5 Insulin Tolerance Test (ITT)

1. Supplies as for glucose tolerance test (GTT).
2. 0.5 U/mL Human regular insulin (*see* **Note 5**).

2.6 Glucose-Stimulated Insulin Secretion (GSIS)

1. Microvette® CB 300 blood collection tubes: three per mouse (Sarstedt).
2. Rodent insulin ELISA kit (Ultra Sensitive Rat Insulin ELISA Kit, 90060, Crystal Chem, Downers Grove, IL) (*see* **Note 6**).
3. Spectrophotometer for reading 96-well plates.
4. Multichannel pipette.

2.7 Measurement of Glucose Uptake in Muscle

1. Shaking water bath warmed to 30 °C.
2. 95 % O_2-5 % CO_2.
3. Bell jar.
4. Kimwipes.
5. Isoflurane.
6. Dissecting board.
7. 70 % Ethanol for dissection.
8. Dissecting tools: scissors, forceps.
9. 25 mL flasks.
10. Krebs-Ringer bicarbonate (KRB) buffer supplemented with 8 mM glucose, 32 mM mannitol, and 0.1 % BSA (radioimmunoassay (RIA) grade).
11. KRB buffer supplemented with 40 mM mannitol and 0.1 % BSA.
12. Human regular insulin (*see* **Note 7**).
13. KRB buffer supplemented with 8 mM 2-[3H]DG (2.25 μCi/mL), 32 mM [14C]mannitol (0.3 μCi/mL), 2 mM sodium pyruvate, and 0.1 % BSA.
14. Hot plate or heating block.
15. Ice bath.
16. Vortex.
17. Centrifuge.
18. Scintillation counter.

2.8 Multiple Low-Dose Streptozotocin Model of Type 1 Diabetes

1. Streptozotocin (STZ).
2. 0.1 M Citrate buffer: 29.41 g Sodium citrate in 1 L PBS, pH to 4.5.

2.9 High Fat Diet Feeding Model of Obesity-Associated Insulin Resistance

1. High-fat diet (Fat Calories (60 %), Bio-serv).

3 Methods

3.1 Generation of RIP-cre^{+} PTEN$^{fl/fl}$ Mice

1. In keeping with institutional animal care guidelines, cross-breed RIP-cre^{+} mice to Pten$^{fl/fl}$ mice (*see* **Note 1**).
2. Genotype offspring (Tables 2, 3, and 4; *see* **Note 2**).
3. At 6–8 weeks of age, interbreed RIP-cre^{+} Pten$^{fl/+}$ mice (*see* **Note 8**).
4. Genotype offspring, which consist of RIP-cre^{+} Pten$^{+/+}$, RIP-cre^{+} Pten$^{fl/+}$, RIP-cre^{+} Pten$^{fl/fl}$, RIP-cre^{-} Pten$^{+/+}$, RIP-cre^{-} Pten$^{fl/+}$, RIP-cre^{-} Pten$^{fl/fl}$.
5. RIP-cre^{+} Pten$^{+/+}$ and RIP-cre^{+} Pten$^{fl/fl}$ mice are typically used as control and tissue-specific knockout mice, respectively, in experiments. RIP-cre^{+} Pten$^{fl/+}$ can also be characterized to examine gene dosage effects (*see* **Note 9**).

Table 2
PCR master mix recipes

	Cre	Pten$^{fl/fl}$
Reaction component	**Volume (μL) per reaction**	**Volume (μL) per reaction**
ddH$_2$O	18.75	12.75
10× reaction buffer	2.5	2.5
Betaine	–	6
dNTP	0.5	1
Primer 1	0.5	0.25
Primer 2	0.5	0.25
Taq polymerase	0.25	0.25
Volume of master mix per tube	23	23
Volume of DNA to add per tube	1	2

Table 3
Cre cycling

Step#	Temp °C	Time	Note
1	95	5 min	
2	95	30 s	
3	60	30 s	
4	72	45 s	Repeat steps 2–4 for 35 cycles
5	72	7 min	
6	4	HOLD	

Table 4
Pten cycling

Step#	Temp °C	Time	Note
1	94	5 min	
2	94	1 min	
3	62	1 min	
4	72	1 min	Repeat steps 2–4 for 40 cycles
5	72	10 min	
6	4	HOLD	

3.2 Genotyping RIP-cre^{+} PTEN$^{fl/fl}$ Mice (see Note 2)

1. Place 2 mm ear punch in labeled Eppendorf tube.
2. Add 47 μL ear buffer and 3 μL proteinase K and ensure that ear punch is fully immersed.
3. Digest in 55 °C water bath for 4–16 h.
4. Remove samples from water bath and add 950 μL ddH_2O to each DNA sample. Vortex well.
5. Keeping mastermix and reaction tubes on ice, prepare PCR master mix, and aliquot into PCR reaction tubes.
6. Aliquot DNA into PCR reaction tubes and attach lids.
7. Place tubes in thermocycler and run appropriate PCR cycling protocol.
8. Prepare 1 % agarose gel in gel-casting system.
9. Add 10 μL loading dye to each PCR product and carefully pipette into wells of agarose gel.

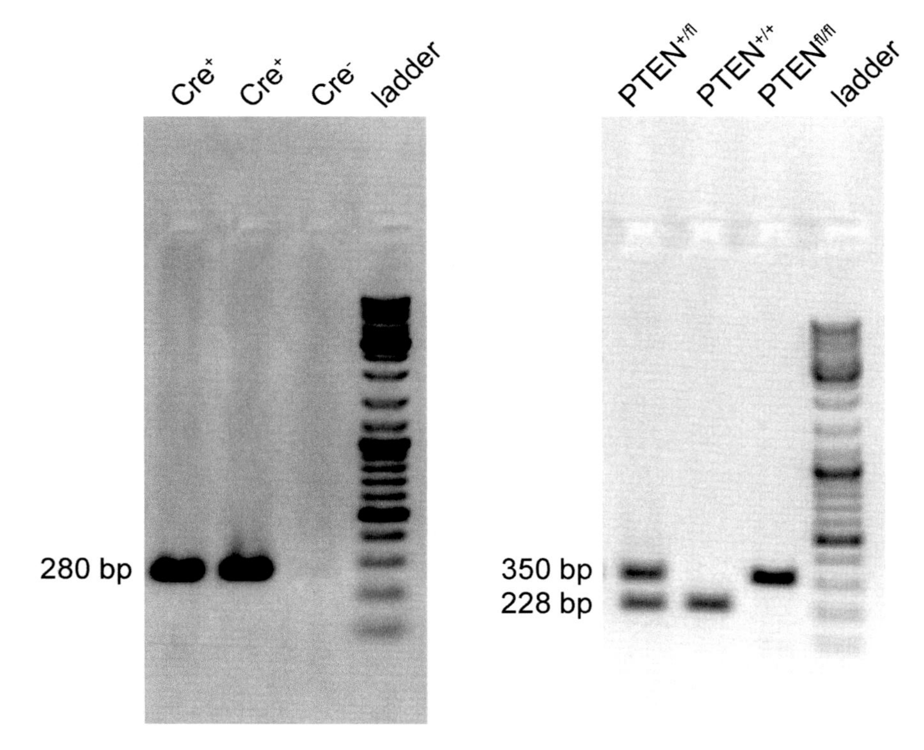

Fig. 1 PCR genotyping for Cre^{+} or Cre^{-} (*left*), and $PTEN^{+/+}$ and/or $PTEN^{fl/fl}$ (*right*) alleles

10. Separate DNA by gel electrophoresis at 130 V for 70 min.
11. Visualize and photograph bands (Fig. 1).

3.3 Blood Glucose Measurement

1. For random blood glucose measurement, test mice at a consistent time in the morning, typically 8:00 am (*see* **Note 10**).
2. For fasting blood glucose measurement, fast mice by placing them in a clean cage 14–16 h prior (for example, fast mice at 5:00 pm) and measure glucose at a consistent time in the morning, typically 8:00 am (*see* **Note 11**).
3. Insert glucometer strip into glucometer.
4. Place mouse in restraint.
5. Grasp tail firmly and nick with scalpel and massage or milk tail as needed to generate a red blood drop (*see* **Note 12**, Fig. 2).
6. Feed blood drop into glucometer strip, ensuring that blood completely fills the visible window of the glucometer strip (*see* **Note 12**).
7. Record final glucometer measurement.

3.4 Glucose Tolerance Test (GTT)

1. Fast mice 14–16 h prior (for example, at 5:00 pm), by placing animals in a clean cage with free access to water (*see* **Note 13**).

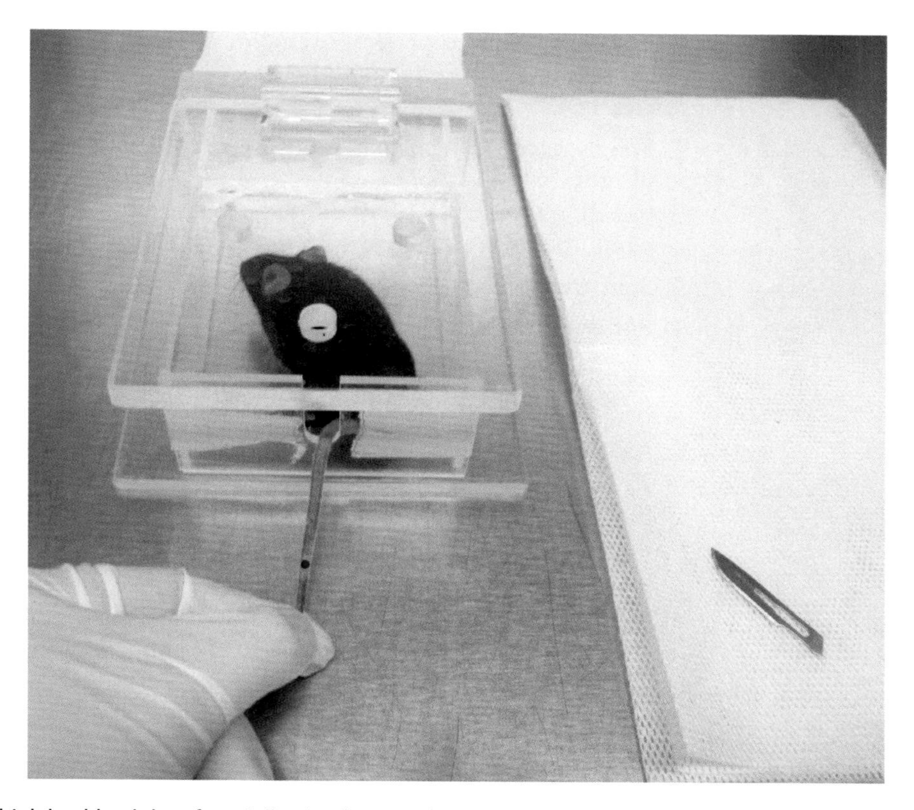

Fig. 2 Obtaining blood drop from tail vein of mouse in restraint

2. Just prior to GTT, weigh mice, and calculate and prepare the required volume of glucose solution for a dose of 1 g/kg body weight. For example, a 20 g mouse would be administered a dose of 40 μL 0.5 g/mL glucose solution (*see* **Note 14**).
3. Start timer.
4. For each mouse:
   (a) Measure and record fasting blood glucose.
   (b) Administer glucose by i.p. injection using a tuberculin syringe (*see* **Note 15**).
5. Measure blood glucose of each mouse at 15, 30, 45, 60, and 120 min. Usually the same nick can be used, ensuring that any blood clot has been cleaned from the area. Mice are allowed unrestricted activity in their cages between blood glucose measurements.
6. At the conclusion of the GTT, restore food and water to mice.
7. Data is usually represented as the average of absolute glucose measurements for each group, and can be analyzed by a two-way ANOVA test. Area under the curve can also be calculated for each mouse and averages for each group compared using a student's *t* test.

3.5 Insulin Tolerance Test (ITT)

1. Fast mice 4 h prior (for example, at 8:00 am), by placing animals in a clean cage with free access to water (*see* **Notes 13** and **16**).
2. Weigh mice, and calculate and prepare insulin dose based on weight in grams. For example, to administer a dose of 1 U/kg to a 20 g mouse, 40 μL of diluted insulin solution (0.5 U/mL) would be used (*see* **Note 17**).
3. Start timer.
4. For each mouse:
   (a) Measure and record blood glucose.
   (b) Administer insulin by i.p. injection.
5. Measure blood glucose of each mouse at 15, 30, 45, 60, and 120 min. Usually the same nick can be used, ensuring that any blood clot has been cleaned from the area. Mice are allowed unrestricted activity in their cages between blood glucose measurements.
6. If mice show signs of severe hypoglycemia, such as unresponsiveness or seizures, terminate ITT by administering 150 μL glucose solution by i.p. injection.
7. At the conclusion of the ITT, restore food and water to mice.
8. Data is usually represented as percent decrease from baseline glucose. The average for each group is usually shown, and can be analyzed by a two-way ANOVA test.

3.6 Glucose-Stimulated Insulin Secretion (GSIS)

1. Fast mice 14–16 h prior (for example, at 5:00 pm), by placing animals in a clean cage (*see* **Note 13**).
2. Weigh mice and calculate the required volume of glucose solution for a dose of 3 g/kg body weight. For example, a 20 g mouse would be administered a dose of 120 μL 0.5 g/mL glucose solution.
3. Place mouse in restraint and obtain approximately 25 μL blood from a tail vein nick in blood collection tubes for measurement of baseline insulin level (*see* **Note 18**).
4. Administer glucose by i.p. injection to mouse and start timer.
5. At 2 and 30 min after injection, obtain approximately 25 μL blood for measurement of insulin.
6. At the conclusion of the GSIS, restore food and water to mice.
7. Ensure that blood collection tubes are properly closed and centrifuge at 2400 × *g* for 10 min. Transfer supernatant to a 0.5 mL Eppendorf tube. Serum can be stored frozen at −80 °C until ready to process.
8. Perform ELISA as per the manufacturer's instructions for wide range assay (*see* **Note 19**). First perform serial dilutions of insulin standard solution provided.

9. Assemble microplate and pipette 95 μL diluent per well.
10. Pipette 5 μL standard solution or mouse serum per well.
11. Incubate plate at 4 °C for 2 h.
12. Wash each well five times with 300 μL wash buffer.
13. Pipette 100 μL anti-insulin enzyme conjugate per well.
14. Incubate plate at room temperature for 30 min.
15. Wash plate seven times with wash buffer.
16. Pipette 100 μL enzyme substrate per well.
17. Incubate plate in the dark at room temperature for 40 min.
18. Add 100 μL stop solution per well.
19. Measure A450 and subtract A630 values within 30 min.
20. Calculate standard curve and insulin concentrations.
21. Data is usually represented as percent of baseline serum insulin and can be analyzed using a student's *t* test.

3.7 Measurement of Glucose Uptake in Muscle

1. Prepare 25 mL flasks with 3 mL KRB buffer supplemented with 8 mM glucose, 32 mM mannitol, and 0.1 % BSA. Warm to 30 °C in a shaking water bath and keep KRB buffer oxygenated by gassing continuously with 95 % O_2-5 % CO_2.
2. Place kimwipes in bell jar and add approximately 1 mL isoflurane. Place mouse in jar to anesthetize and sacrifice mouse by cervical dislocation.
3. Quickly and carefully dissect the soleus and extensor digitorum longus muscles (EDL). Cut muscle into 20–30 mg strips (*see* **Note 20**, Fig. 3).
4. Incubate muscle strips in prepared 25 mL flasks with oxygenated KRB buffer at 30 °C for 30 min.
5. Add insulin to experimental 25 mL flasks for a final concentration of 2 mU/mL and incubate for an additional 30 min (*see* **Note 7**). Do not add insulin to control flasks.
6. Wash muscles in 3 mL KRB buffer with 40 mM mannitol and 0.1 % BSA with or without insulin (if present in previous step) for 10 min.
7. Incubate muscles in 3 mL KRB buffer with 8 mM 2-[3H]DG (2.25 μCi/mL), 32 mM [14C]mannitol (0.3 μCi/mL), 2 mM sodium pyruvate, and 0.1 % BSA for 20 min.
8. Blot muscles at 4 °C and freeze immediately at −80 °C until ready to process.
9. Boil muscles for 10 min in 1 mL water.
10. Transfer extract to an ice bath, vortex, and centrifuge at 1000 × *g*.
11. Measure radioactivity in three 200 μL aliquots of supernatant.
12. Correct for extracellular trapping using [14C]mannitol counts.

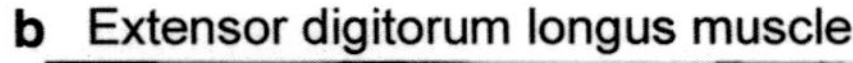

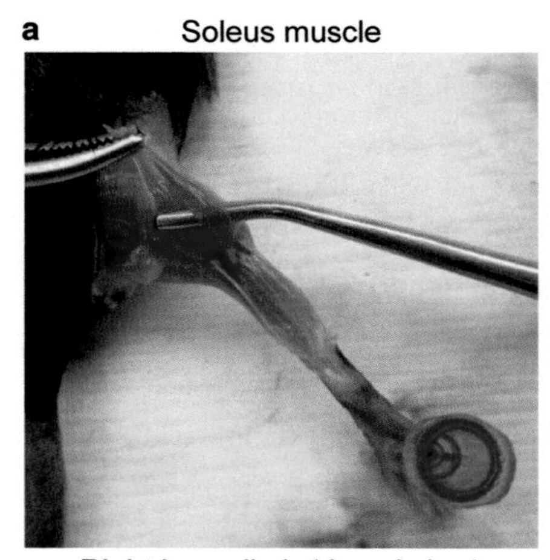

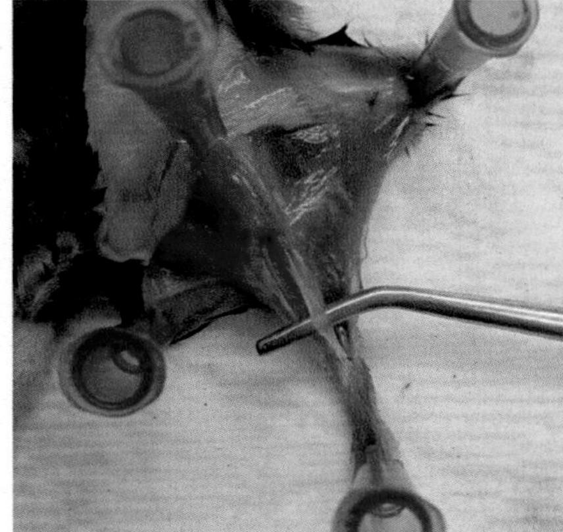

Fig. 3 Mouse muscles typically used to study glucose uptake (for illustrative purposes). Soleus muscle (isolated by probe) shown deep to gastrocnemius muscle in right lower limb (dorsal view) (**a**). Extensor digitorum longus (EDL) muscle (isolated by probe) shown deep to tibialis anterior muscle in left lower limb (ventral view) (**b**)

3.8 Multiple Low-Dose Streptozotocin Model of Type 1 Diabetes (See Note 21)

1. Prepare STZ immediately prior to injection. Using a 3 mL syringe and 18 G needle, draw up 2.5 mL 0.1 M citrate buffer and inject into 50 mg STZ vial. Draw solution up and down to fully dissolve powder (*see* **Note 22**).
2. Using a fresh 3 mL syringe and 18 G needle, add 2.5 mL PBS to STZ vial and draw up and down to mix. Keep STZ on ice until ready to inject.
3. Weigh mice and calculate STZ dose of 40 mg/kg for each mouse. For example, a 20 g mouse would be administered 80 μL of STZ solution.
4. Inject mice i.p. on 5 consecutive days.
5. Monitor glucose prior to first injection and weekly.

3.9 High Fat Diet Feeding Model of Obesity-Associated Insulin Resistance

1. At 6–8 weeks of age, change mouse diet to high fat diet (HFD, 60 % calories from fat) (*see* **Note 23**).
2. Monitor mice every 1 or 2 weeks for body weight and random blood glucose. Fasting blood glucose can also be monitored at regular intervals.
3. HFD feeding is typically continued 8–12 weeks when significant weight gain and increase in blood glucose are usually seen. Earlier time points can be used to examine short-term HFD feeding prior to weight gain, or mice can be fed HFD 6 or more months to investigate chronic obesity [11, 34].

4 Notes

1. Animal care should be conducted in accordance with local animal care guidelines and principles of the National Centre for Replacement, Refinement and Reduction of Animals in Research (NC3R).
2. $Pten^{fl/fl}$ mice have been generated by multiple groups. We use the strain developed by Suzuki et al. and describe our genotyping protocols for this strain [9]. A different mouse strain and methods are also available from Jackson Laboratories, Bar Harbor, Maine (http://www.jax.org/index.html) [35].
3. Commercially available glucometers and glucometer strips are available from a variety of companies, for example, Lifescan, Abbott Diabetes Care, Bayer, and Accu-check. We have generally found widely available instruments for human use to be effective for blood glucose measurements from mice. Models specifically calibrated for rodents are also available, and may have a larger range. Considerations for choosing a meter might include availability, cost of glucometer strips, sample volume, battery life, test time, and result range. Generally, measurements are expected to have an accuracy within 0.8 mmol/L or 15–20 % of laboratory measurements. Please consult the manufacturer's information for further instructions, but most newer models no longer require calibration or strip coding. Spread of blood-borne infections through shared glucometer use is a possibility in humans [36]; if infection is a possibility in the rodents being tested, individual glucometers, specific glucometer models or modified blood sampling technique may be advisable.
4. Warm PBS to 65 °C and dissolve 20 g glucose in 40 mL PBS. Filter with a syringe filter.
5. Dilute 50 μL regular insulin (100 U/mL) in 10 mL sterile PBS. Rapid insulin (insulin lispro or aspart) may be more readily available, and can also be used; however, all experiments where data will be compared or aggregated should be performed with the same insulin.
6. One kit includes one 96-well plate. With nine standards, if samples are run in duplicate, this allows for 39 samples or 13 mice undergoing GSIS. Individual samples from mice not undergoing GSIS can also be tested for random or fasting serum insulin levels.
7. Prepare a 1:100 dilution of regular insulin (100 U/mL) in KRB or PBS buffer. For example, add 10 μL insulin to 990 mL buffer to prepare a 1 U/μL solution. 6 μL can then be added to 3 mL to produce a final concentration of 2 mU/mL.
8. Ideally, sibling pairs should be bred to maintain genetic homogeneity. Sibling $RIP\text{-}cre^{+}$ $Pten^{fl/+}$ pairs can be bred further

generations to increase genetic homogeneity. Mouse lines are typically backcrossed every ten generations to maintain genetic background. We find that a minimum of two breeding pairs should be maintained to continue a line, and four or more breeding pairs used to generate mice for experiments. Mice can be trio or harem bred, in keeping with institutional guidelines. For example, three females can be housed with one male, and two females separated into a cage together once pregnant to avoid overcrowding and share nursing responsibilities.

9. RIP-cre$^-$ mice can also be characterized for completeness, but we do not use these mice as controls because RIP-cre$^+$ mice may have abnormalities of glucose homeostasis that could mistakenly be attributed to effects of the gene knocked out [37].
10. Experiments involving blood glucose measurements should be performed at a consistent time of the day, because circadian rhythms can alter glucose homeostasis [38]. Random or fed blood glucose measurements should be performed early in the morning, as this follows the usual period of nocturnal feeding and activity for mice.
11. Because of the nocturnal feeding and activity patterns of rodents, an overnight fast can represent a significant stress. For this reason, some groups suggest a shorter period of fasting. We also suggest performing random blood glucose measurements prior to experiments or tests involving overnight fasts or ITT, and suggest a period of 4–7 days to allow animals to recover from stress. A clean cage is recommended for fasting to eliminate food particles on the cage floor and bedding.
12. Ensure that blood drop is a deep red. Measurements from clear exudate occasionally obtained with superficial nicks can contribute to inaccurate measurements. Similarly, insufficient blood in the glucometer strip can result in inaccurate blood glucose measurement in older meters.
13. For GTT and ITT, in experienced hands, 15 mice can be tested with one set of supplies (approximately 1 min per mouse per time point), but much smaller numbers should be used initially. For GSIS, mice should be tested individually, particularly for the 0- and 2-min time points.
14. In animals with significant differences in body composition, determining glucose dose by total body weight may be considered less useful as dose is based on combined lean and fat mass. In such cases, dosing can be fixed or calculated by lean body mass, as determined by DEXA, MRI, or other methods capable of determining whole-body composition [39]. To facilitate rapid identification of mice, they can be marked on the tail or other visible area with a felt marker. Mice can also be housed individually for the duration of the experiment; however, this

may affect stress and activity levels if mice are not accustomed to individual housing.

15. Glucose can be administered by i.p. injection or oral gavage, and with practice, both can be performed quickly and with minimal stress to the animal. When glucose is administered orally, the rise in blood glucose will typically be lower due to the incretin effect [40].
16. For an ITT, an overnight fast may increase the incidence of severe and symptomatic hypoglycemia in animals. For that reason, a shorter fast, typically 4–6 h, is performed.
17. For ITT, an insulin dose should be used that is sufficient to induce a significant decrease in blood glucose, but minimizes severe hypoglycemia. Typically, 0.75–1.0 U/kg in adult mice or 1.5 U/kg in particularly obese, insulin-resistant *ob/ob* or *db/db* mice can be used. Lower doses, for example 0.5 U/kg, may be required for younger or particularly insulin-sensitive animals. The same dose of insulin must be used in all animals in groups that will be compared.
18. Alternately, blood can be collected from the saphenous vein.
19. We typically use the protocol for a wide range assay: 0.1–12.8 mg/mL.
20. Soleus and EDL muscles are typically studied because they are composed of predominantly slow type 1 muscle fibers, and fast type 2A, 2B fibers, respectively.
21. A number of streptozotocin protocols have been described [27]. We have found an MLDS protocol to be nonlethal and suitable for studies of type 1 diabetes pathogenesis.
22. STZ is cytotoxic and appropriate safety precautions including gloves and mask should be used. We treat wild-type and knockout mice simultaneously with STZ from the same lot, as differences between lots can occur.
23. Control wild-type and knockout mice fed normal chow diet should be monitored using the same experimental protocols.

Acknowledgements

This work was supported by Canadian Institutes of Health Research (CIHR) Operating Grants MOP-81148 and MOP-93707 to M.W. M.W. is supported by Canada Research Chair in Signal Transduction in Diabetes Pathogenesis. C.T.L. is supported by the Eliot Phillipson Clinician Scientist Training Program, Banting and Best Diabetes Centre Postdoctoral Fellowship, Canadian Diabetes Association (CDA), and Canadian Society of Endocrinology and Metabolism Dr. Fernand Labrie Fellowship Research Award.

References

1. American Diabetes Association (2014) Diagnosis and classification of diabetes mellitus. Diabetes Care 37(Suppl 1):S81–S90
2. Canadian Diabetes Association Clinical Practice Guidelines Expert Committee, Goldenberg R, Punthakee Z (2013) Definition, classification and diagnosis of diabetes, prediabetes and metabolic syndrome. Can J Diabetes 37 (Suppl 1):S8–S11
3. Stambolic V, Suzuki A, de la Pompa JL et al (1998) Negative regulation of PKB/Akt-dependent cell survival by the tumor suppressor PTEN. Cell 95:29–39
4. Steck PA, Pershouse MA, Jasser SA et al (1997) Identification of a candidate tumour suppressor gene, MMAC1, at chromosome 10q23.3 that is mutated in multiple advanced cancers. Nat Genet 15:356–362
5. Cho H, Mu J, Kim JK et al (2001) Insulin resistance and a diabetes mellitus-like syndrome in mice lacking the protein kinase Akt2 (PKB beta). Science 292:1728–1731
6. Jiang ZY, Zhou QL, Coleman KA, Chouinard M, Boese Q, Czech MP (2003) Insulin signaling through Akt/protein kinase B analyzed by small interfering RNA-mediated gene silencing. Proc Natl Acad Sci U S A 100:7569–7574
7. Katome T, Obata T, Matsushima R et al (2003) Use of RNA interference-mediated gene silencing and adenoviral overexpression to elucidate the roles of AKT/protein kinase B isoforms in insulin actions. J Biol Chem 278:28312–28323
8. Wang Q, Somwar R, Bilan PJ et al (1999) Protein kinase B/Akt participates in GLUT4 translocation by insulin in L6 myoblasts. Mol Cell Biol 19:4008–4018
9. Suzuki A, de la Pompa JL, Stambolic V et al (1998) High cancer susceptibility and embryonic lethality associated with mutation of the PTEN tumor suppressor gene in mice. Curr Biol 8:1169–1178
10. Nguyen KT, Tajmir P, Lin CH et al (2006) Essential role of Pten in body size determination and pancreatic beta-cell homeostasis in vivo. Mol Cell Biol 26:4511–4518
11. Wang L, Liu Y, Yan Lu S et al (2010) Deletion of Pten in pancreatic ß-cells protects against deficient ß-cell mass and function in mouse models of type 2 diabetes. Diabetes 59: 3117–3126
12. Wang L, Opland D, Tsai S et al (2014) Pten deletion in RIP-Cre neurons protects against type 2 diabetes by activating the anti-inflammatory reflex. Nat Med 20:484–492
13. Stiles B, Wang Y, Stahl A et al (2004) Liver-specific deletion of negative regulator Pten results in fatty liver and insulin hypersensitivity [corrected]. Proc Natl Acad Sci U S A 101:2082–2087
14. Horie Y, Suzuki A, Kataoka E et al (2004) Hepatocyte-specific Pten deficiency results in steatohepatitis and hepatocellular carcinomas. J Clin Invest 113:1774–1783
15. Wijesekara N, Konrad D, Eweida M et al (2005) Muscle-specific Pten deletion protects against insulin resistance and diabetes. Mol Cell Biol 25:1135–1145
16. Kurlawalla-Martinez C, Stiles B, Wang Y, Devaskar SU, Kahn BB, Wu H (2005) Insulin hypersensitivity and resistance to streptozotocin-induced diabetes in mice lacking PTEN in adipose tissue. Mol Cell Biol 25:2498–2510
17. Butler M, McKay RA, Popoff IJ et al (2002) Specific inhibition of PTEN expression reverses hyperglycemia in diabetic mice. Diabetes 51:1028–1034
18. Ishihara H, Sasaoka T, Kagawa S et al (2003) Association of the polymorphisms in the 5′-untranslated region of PTEN gene with type 2 diabetes in a Japanese population. FEBS Lett 554:450–454
19. Hingorani SR, Petricoin EF, Maitra A et al (2003) Preinvasive and invasive ductal pancreatic cancer and its early detection in the mouse. Cancer Cell 4:437–450
20. Postic C, Shiota M, Niswender KD et al (1999) Dual roles for glucokinase in glucose homeostasis as determined by liver and pancreatic beta cell-specific gene knock-outs using Cre recombinase. J Biol Chem 274:305–315
21. Stanger BZ, Stiles B, Lauwers GY et al (2005) Pten constrains centroacinar cell expansion and malignant transformation in the pancreas. Cancer Cell 8:185–195
22. Herrera PL (2000) Adult insulin- and glucagon-producing cells differentiate from two independent cell lineages. Development 127:2317–2322
23. Wang L, Luk CT, Cai EP et al (2015) PTEN deletion in pancreatic α cells protects against high fat diet-induced hyperglucagonemia and insulin resistance. Diabetes 64:147–157
24. Brüning JC, Michael MD, Winnay JN et al (1998) A muscle-specific insulin receptor knockout exhibits features of the metabolic syndrome of NIDDM without altering glucose tolerance. Mol Cell 2:559–569
25. Eguchi J, Wang X, Yu S et al (2011) Transcriptional control of adipose lipid handling by IRF4. Cell Metab 13:249–259

26. He W, Barak Y, Hevener A, et al. (2003) Adipose-specific peroxisome proliferator-activated receptor gamma knockout causes insulin resistance in fat and liver but not in muscle. Proc Natl Acad Sci U S A 100: 15712-15717
27. Joost H-G, Al-Hasani H, Schurmann A (2012) Animal models in diabetes research, Methods in molecular biology. Humana Press, New York, p 325
28. Friedel RH, Wurst W, Wefers B, Kühn R (2011) Generating conditional knockout mice. Methods Mol Biol 693:205–231
29. Magnuson MA, Osipovich AB (2013) Pancreas-specific Cre driver lines and considerations for their prudent use. Cell Metab 18:9–20
30. Ayala JE, Bracy DP, Malabanan C et al (2011) Hyperinsulinemic-euglycemic clamps in conscious, unrestrained mice. J Vis Exp. doi:10.3791/3188
31. Kim JK (2009) Hyperinsulinemic-euglycemic clamp to assess insulin sensitivity in vivo. Methods Mol Biol 560:221–238
32. O'Dowd JF (2009) The isolation and purification of rodent pancreatic islets of Langerhans. Methods Mol Biol 560:37–42
33. Stocker C (2009) Type 2 diabetes methods and protocols, Methods in molecular biology. Humana Press, New York, London
34. Luk CT, Shi SY, Choi D, Cai EP, Schroer SA, Woo M (2013) In vivo knockdown of adipocyte erythropoietin receptor does not alter glucose or energy homeostasis. Endocrinology 154:3652–3659
35. Groszer M, Erickson R, Scripture-Adams DD et al (2001) Negative regulation of neural stem/progenitor cell proliferation by the Pten tumor suppressor gene in vivo. Science 294:2186–2189
36. (CDC) CfDCaP (2005) Transmission of hepatitis B virus among persons undergoing blood glucose monitoring in long-term-care facilities—Mississippi, North Carolina, and Los Angeles County, California, 2003-2004. MMWR Morb Mortal Wkly Rep 54:220–223
37. Lee JY, Ristow M, Lin X, White MF, Magnuson MA, Hennighausen L (2006) RIP-Cre revisited, evidence for impairments of pancreatic beta-cell function. J Biol Chem 281: 2649–2653
38. Kohsaka A, Bass J (2007) A sense of time: how molecular clocks organize metabolism. Trends Endocrinol Metab 18:4–11
39. Shi SY, Luk CT, Brunt JJ et al (2014) Adipocyte-specific deficiency of Janus kinase (JAK) 2 in mice impairs lipolysis and increases body weight, and leads to insulin resistance with ageing. Diabetologia 57:1016–1026
40. Drucker DJ (2013) Incretin action in the pancreas: potential promise, possible perils, and pathological pitfalls. Diabetes 62:3316–3323

Part III

Methods to Study PTEN Regulation

Chapter 8

Rapid Detection of Dynamic PTEN Regulation in Living Cells Using Intramolecular BRET

Stanislas Misticone, Evelyne Lima-Fernandes, and Mark G.H. Scott

Abstract

Tumor suppressor PTEN phosphatase acts to inhibit the PI3K/AKT pathway and thus regulates cell proliferation, survival, and migration. Dysregulation of PTEN function is observed in a wide range of cancers. In addition to alterations of the *PTEN* gene, repression of PTEN function can also occur at the protein level through changes in PTEN conformation, localization, activity, and stability. The ability to follow switches in PTEN conformation in live cells provides a rapid approach to study changes in PTEN function and may provide a basis to screen pharmacological agents aimed at enhancing or reestablishing PTEN-dependent signaling pathways that have gone awry in cancer. Here, we describe methods to use an intramolecular bioluminescent resonance energy transfer (BRET)-based biosensor that reports dynamic signal-dependent changes in PTEN conformational rearrangement and function.

Key words PTEN, PI3K/AKT, Tumor suppressor, Phosphatase, Biosensor, Bioluminescence resonance energy transfer, Conformational change, Posttranslational modifications

1 Introduction

Tumor suppressor PTEN (phosphatase and tensin homolog deleted on chromosome ten) is a lipid phosphatase that dephosphorylates phosphatidyl (3, 4, 5) trisphosphate and therefore negatively regulates the pro-survival oncogenic PI3K/AKT pathway. In addition to its role in regulation of the PI3K/AKT pathway, PTEN also displays lipid phosphatase-independent functions including the inhibition of cell migration, regulation of p53 levels and activity, and control of genomic stability [1–4]. While the *PTEN* gene is often targeted by mutation, PTEN function can also be repressed in human cancers at the posttranslational level in the absence of any *PTEN* gene alteration [5]. PTEN is tightly controlled by multiple posttranslational modifications (including phosphorylation, ubiquitination, SUMOylation, and acetylation) and interacting partners that combined regulate PTEN plasticity by

Leonardo Salmena and Vuk Stambolic (eds.), *PTEN: Methods and Protocols*, Methods in Molecular Biology, vol. 1388, DOI 10.1007/978-1-4939-3299-3_8, © Springer Science+Business Media New York 2016

impacting conformation, localization, activity, and stability [3, 6, 7]. Perturbation of these different mechanisms can therefore lead to the downregulation of PTEN function in a variety of different ways.

PTEN is comprised of a catalytic domain, a C2 domain, and a C-terminal tail that contains a key regulatory serine-threonine residue cluster [6]. Phosphorylation of these residues by casein kinase 2 (CK2) is thought to promote a "closed" form due to intramolecular interaction of the C-terminal tail with the catalytic and C2 domains of PTEN. This results in enhanced conformational compaction, reduced catalytic activity, and decreased plasma membrane targeting [8–13]. Mutation of the serine-threonine residue cluster in the C-terminal tail disrupts the intramolecular interaction promoting an "open" form of PTEN that displays enhanced activity and plasma membrane targeting. In T-cell acute lymphoblastic leukemia (T-ALL) increased phosphorylation of the serine-threonine cluster in the C-terminal tail leads to the inactivation of PTEN, resulting in enhanced PI3K/AKT signaling [14]. Pharmacological inhibition of CK2 reestablishes PTEN function and inhibits PI3K/AKT signaling in T-ALL cells.

The capacity to monitor PTEN conformation in a direct, rapid, and sensitive manner in live cells is useful to investigate dynamic PTEN regulation. Tools enabling this type of approach may also help form the basis for drug discovery platforms to identify and develop small molecules with the aim of enhancing or reestablishing PTEN-dependent signaling pathways. We recently developed an intramolecular bioluminescence resonance energy transfer (BRET)-based biosensor that is capable of reporting signal-dependent PTEN conformational changes in living cells [15]. In this chapter we provide detailed protocols of how to use the PTEN intramolecular BRET-based biosensor and describe different facets in which it may be employed to investigate cellular regulation.

2 Materials

2.1 Cell Lines

1. Human Embryonic Kidney 293-T (HEK 293 T) cell line.
2. HeLa human cervical cancer cell line.
3. PC-3 human prostate cancer cell line.

2.2 Plasmids

1. BRET plasmids: Rluc-PTEN control vector, Rluc-PTEN-YFP biosensor, Rluc-PTEN-A4-YFP biosensor with Ser380, Thr382, Thr383, and Ser385 phosphorylation sites mutated to alanine [15].
2. Plasmids encoding PTEN regulating proteins: HA-thromboxane A2 receptor (TPαR) and Myc-RhoA-Q63L [15].

2.3 Cell Culture and Transfection

1. Dulbecco's modified Eagle's medium (DMEM), 4.5 g/l glucose, 4 mM glutamine, and 1 mM pyruvate, supplemented with 10 % (v/v) fetal bovine serum, 100 U/ml penicillin, and 0.1 mg/ml streptomycin (Life Technologies).
2. Roswell Park Memorial Institute 1640 Medium (RPMI), GlutaMAX™, 2 g/l glucose, supplemented with 10 % (v/v) fetal bovine serum, 100 U/ml penicillin, and 0.1 mg/ml streptomycin (Life Technologies).
3. Trypsin-EDTA (0.05 %) solution.
4. PBS-EDTA solution: Phosphate-buffered saline (PBS; 1×) without $CaCl_2$ and $MgCl_2$ (Life Technologies) containing 2 mM EDTA.
5. 6- and 12-well plates for cell culture (Falcon, Corning), 96-well sterile white microplates (Perkin Elmer).
6. Poly-L-ornithine (1.5 mg/ml in dH_2O for a 50× stock) used to coat 96-white well plates for adherent cell experiments.
7. Transfection reagents: GeneJuice (Novagen), FuGENE HD Transfection Reagent (Promega).
8. Chemical reagents: U46619 TPαR agonist (Cayman Chemicals), TBB, and CX-4945 CK2 inhibitors (Santa Cruz).

2.4 BRET-Ratio Measurements

1. Coelenterazine-h (Interchim) is dissolved in 100 % ethanol at 2 mM (400× stock solution) and stored at −20 °C in opaque microcentrifuge tubes (*see* **Note 1**).
2. Hanks' balanced salt solution (HBSS; 1×) containing $CaCl_2$ and $MgCl_2$ (Life Technologies).
3. DMEM or RPMI without phenol red.
4. PBS (1×) containing $CaCl_2$ and $MgCl_2$ (Life Technologies).
5. White 96-well plates (Perkin Elmer OptiplateTM-96HB or equivalent) for BRET measurements.
6. Multi-mode microplate reader: Mithras LB 940 (Berthold) or equivalent with BRET 1 filter sets 480 ± 10 nm and 540 ± 20 nm.

3 Methods

3.1 Background

To analyze conformational change of PTEN in live cells, the luminescent BRET-donor (Renilla luciferase, Rluc) is fused to the amino-terminus of PTEN and the fluorescent BRET-acceptor (yellow fluorescent protein, YFP) at the carboxy-terminus, using standard subcloning procedures, to create an Rluc-PTEN-YFP fusion (Fig. 1a and [15]). A control vector encoding Rluc-PTEN and lacking the acceptor YFP is also used to determine the background signal (Fig. 1a). Changes in intramolecular BRET of the Rluc-PTEN-YFP biosensor, depending on the relative orientation

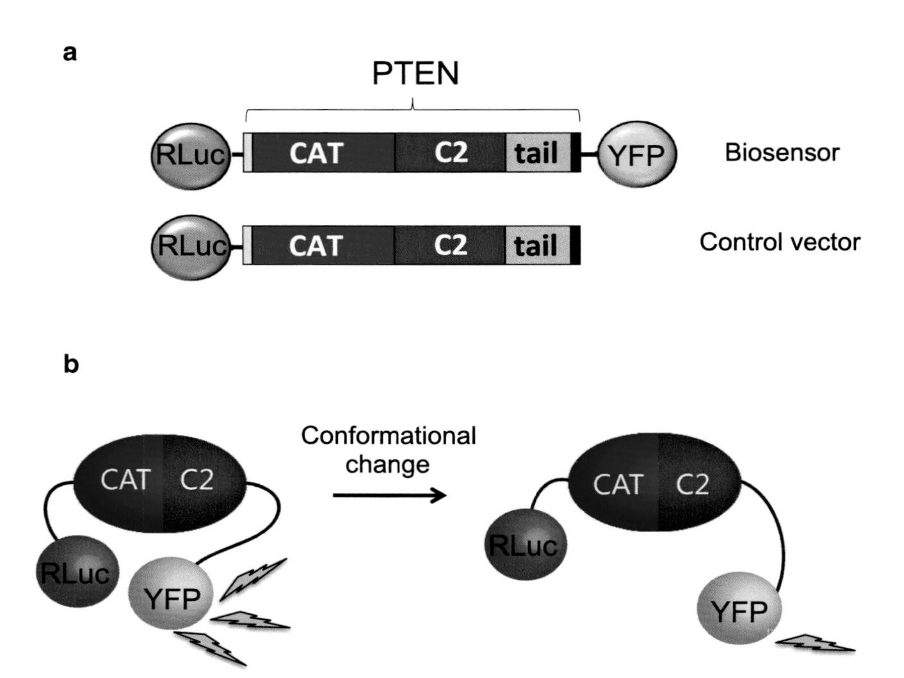

Fig. 1 Schematic representations of the PTEN biosensor. (**a**) Human PTEN is subcloned between Renilla luciferase (Rluc) and yellow fluorescent protein (YFP) to create the genetically encoded Rluc-PTEN-YFP conformational biosensor. An Rluc-PTEN control vector is also used to establish the background BRET signal in the absence of the energy acceptor YFP. (**b**) The diagram depicts how conformational changes in PTEN may induce changes in BRET between Rluc and YFP, although the real orientations of the donor/acceptor pair are not known

of donor and acceptor proteins in the fusion, allow the detection of conformational rearrangement of PTEN in live cells (Fig. 1b). We have previously validated that PTEN within the Rluc-PTEN-YFP fusion maintains functionality: it displays intact lipid phosphatase activity, inhibits cellular pAKT levels, demonstrates expected subcellular localization in live cells, and binds known interaction partners [15]. The Rluc-PTEN-YFP system we describe here is a BRET1 assay but there are now also new-generation BRET2 and BRET3 assays that use modified donor/acceptor pairs (*see* **Note 2**). BRET can be performed in a large variety of intact adherent or suspension mammalian cell lines. For the PTEN biosensor we have tested mouse embryonic fibroblasts, HEK-293 T, HeLa, PC-3, MCF-7, and SKBR3 cell lines. The light emitted by the YFP acceptor over that emitted by the Rluc donor determines the BRET ratio. This is acquired using BRET readers that can perform readings at both wavelengths rapidly and repetitively in wells of 96-well plates. Below, we detail protocols to perform experiments and calculate BRET ratios. In addition we provide details on how to exploit the biosensor to monitor PTEN conformation change in (1) a structure-function mode, (2) experiments using pharmacological inhibitors and PTEN modulators, and (3) real-time kinetics.

3.2 Cell Preparation and Basal BRET Measurements on Adherent Cells

1. 24 h prior to transfection cells are seeded into 6- or 12-well plates to achieve 50–70 % confluence on the day of transfection (*see* **Note 3**). The number of cells seeded per well to achieve this will evidently depend on each cell type, based on cell size and growth rate—in the case of HEK-293 T we seed $100–200 \times 10^3$ cells per well of a 6-well plate in 2 ml of complete DMEM.
2. 24 h after seeding the cells are transfected with 100–200 ng of Rluc-PTEN-YFP or Rluc-PTEN per well of a 6-well plate using either Gene Juice (for HEK 293 T or HeLa cells) or Fugene HD (for PC-3, MCF-7, SKBR3, or MEF cells) according to the manufacturer's instructions. Again the amount transfected should be determined for each new experimental setup.
3. Poly-L-ornithine is used to coat 96-white well plates for adherent cell experiments (*see* **Note 4**). The plates can be prepared in advance and should be incubated for at least 2 h at 37 °C. Prior to cell seeding the plates are washed once with PBS (containing $CaCl_2/MgCl_2$) or complete medium.
4. 24 h post-transfection, cells in each well of the 6-well plate are washed with PBS, detached in either trypsin-EDTA or PBS-EDTA solution at 37 °C, and resuspended in 2 ml of complete medium. The cells are then distributed into the poly-l-ornithine-coated 96-well plates (100 μl per well) and incubated at 37 °C for a further 24 h.
5. The following day, cells are washed with PBS containing $CaCl_2/MgCl_2$. The cells are then overlayed with 90 μl of HBSS or serum-free media without phenol red (*see* **Note 5**) and then 10 μl of a freshly prepared solution of 50 μM coelenterazine-h diluted in HBSS or serum-free media is added to each well giving a final concentration of 5 μM. The samples are then protected from light.
6. After a 3-min incubation at room temperature with coelenterazine-h, BRET readings are initiated using a multi-mode microplate reader that permits sequential integration of light output from Rluc and YFP detected using two filter settings (480 ± 10 nm and 540 ± 20 nm—*see* **Note 6**).

3.3 BRET Measurements on Cells in Suspension

1. Cells are seeded and transfected as described in Subheading 3.2.
2. 24–48 h post-transfection cells are detached from the 6-well plates with PBS-EDTA at 37 °C, transferred to a microcentrifuge tube, and collected by centrifugation at $300 \times g$ for 5 min at room temperature.
3. Cell pellets are resuspended in 500 μl HBSS and 90 μl of the cell suspensions are distributed into wells of a white 96-well plate.
4. As in Subheading 3.2 the BRET measurements are initiated after a 3-min incubation with a freshly prepared coelenterazine solution.

3.4 Calculation of Basal BRET Ratio Obtained with the PTEN Biosensor

The BRET ratio is calculated as the fluorescence signal emitted by YFP (filter 540 ± 20 nm) over the Rluc signal (filter 480 ± 10 nm). Each well is measured for 1 s for each channel and the data (YFP signal, Rluc signal and YFP/Rluc ratio) automatically integrated into BRET reader software (Mikrowin). The wells are read multiple times (generally between three and six times) to generate mean values and the data is exported to a spreadsheet for analysis. Specific BRET ratios (=Net BRET) are then calculated by subtracting the background BRET value, obtained with Rluc-PTEN control vector lacking YFP acceptor, from the mean BRET ratios obtained with Rluc-PTEN-YFP. Values are then converted into mBRET units by multiplying by 1000 to obtain whole numbers, thus circumventing the need to deal with decimal numbers. A spreadsheet with the raw data obtained (Rluc signal, YFP signal, and YFP/Rluc BRET ratio) and the subsequent data analysis in an experiment performed in PC-3 cells is shown in Fig. 2.

3.5 Measuring BRET with Increasing Amounts of PTEN Biosensor

As the donor and acceptor molecule are in the same unimolecular Rluc-PTEN-YFP fusion, the BRET ratio would be predicted to remain stable over a range of different concentrations. An experiment to test this is described below:

1. PC-3 cells are seeded and transfected as described in Subheading 3.2. As different amounts of Rluc-PTEN or Rluc-PTEN-YFP are transfected in this experiment (1–1000 ng) to compare BRET ratios obtained at different expression levels, the total amount of DNA transfected should be completed with "empty" vector such as pcDNA3.1 to the maximal DNA amount.
2. Cells are prepared and BRET measurements initiated as described in Subheading 3.3. BRET ratios obtained can then be plotted against Rluc values (light units) obtained to assess the signal over a range of concentrations (Fig. 3). We have found that the BRET ratio remains stable over a 100-fold range of Rluc values indicating a constant acceptor-to-donor ratio. *See* **Note 7** for evaluating biosensor levels compared to endogenous PTEN.

3.6 Assessing PTEN Conformational Change in a Structure-Function Mode

As mentioned in the introduction, phosphorylation of PTEN on its C-terminal regulatory cluster (Ser 380, Thr 382, Thr 383, Ser 385) promotes a "closed" predominantly cytoplasmic form (Fig. 4a, b). Dephosphorylation of these residues favors an "open" form of PTEN that displays increased plasma membrane targeting (Fig. 4a–d). The biosensor can be used in a structure-function mode to determine whether mutation of amino acids results in conformational change. An experiment to test this is described below:

1. Cells (HEK-293 T, PC-3, or HeLa) are seeded and transfected as described in Subheading 3.2 with either WT Rluc-PTEN-

a

	Rluc-PTEN				Rluc-PTEN-YFP			
Rluc Signal	Well A1	Well B1	Well C1	Well D1	Well E1	Well F1	Well G1	Well H1
Reading 1	111587	121883	118100	121157	120569	121317	123728	124301
Reading 2	107663	116766	114005	116789	117573	117386	121061	119960
Reading 3	102219	111930	109169	110776	113222	113912	116579	115128
Reading 4	97484	107383	104879	106298	108740	108915	111969	109712
YFP Signal								
Reading 5	81962	89006	86580	89443	124664	125834	127811	128525
Reading 6	78429	85379	83125	85067	121844	121914	125436	124067
Reading 7	75005	82212	79622	81674	118205	118697	121142	118638
Reading 8	71659	78546	76963	78156	112449	114380	116228	113934
BRET ratio								
Reading 9	0.7345	0.7303	0.7331	0.7382	1.0340	1.0372	1.0330	1.0340
Reading 10	0.7285	0.7312	0.7291	0.7284	1.0363	1.0386	1.0361	1.0342
Reading 11	0.7338	0.7345	0.7293	0.7373	1.0440	1.0420	1.0391	1.0305
Reading 12	0.7351	0.7315	0.7338	0.7353	1.0341	1.0502	1.0380	1.0385
Mean BRET ratio	0.7330	0.7319	0.7314	0.7348	1.0371	1.0420	1.0366	1.0343
Background BRET ratio		0.7327						
Specific BRET ratio					0.3044	0.3093	0.3038	0.3016
Specific mBRET					304.4	309.3	303.8	301.6
Mean Specific mBRET							305	

b

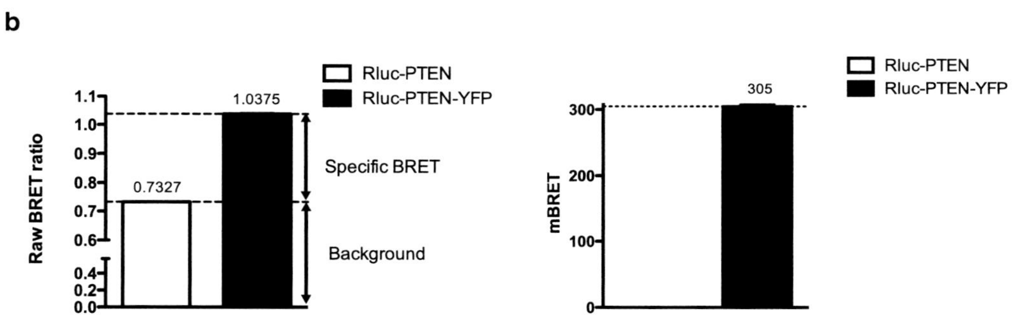

Fig. 2 Calculation of specific BRET ratio and mBRET values obtained with the PTEN biosensor in PC-3 cells. (**a**) Raw BRET data acquisition and analysis. PC-3 cells growing in 6-well plates were transfected with either Rluc-PTEN control vector or Rluc-PTEN-YFP (3-wells transfected for each construct). Cells from each transfection were distributed into four wells of a white 96-well plate and BRET ratios measured four times each for each well. For simplicity, only readings from cells from one 6-well for each construct split into four 96-wells are shown. Wells A1–D1 represent readings obtained from Rluc-PTEN and wells E1–H1 represent those obtained with Rluc-PTEN-YFP. The Rluc values obtained are shown in the Rluc signal rows (Readings 1–4), and those for YFP values in the YFP signal rows (Readings 5–8). The BRET values are shown in the BRET ratio rows (Readings 9–12). The mean BRET ratio is calculated from rows 9–12 and the background determined from the average of these values in wells A1–D1 expressing Rluc-PTEN donor only. Specific BRET ratios are then calculated by subtracting the background BRET value from the mean BRET ratios obtained with Rluc-PTEN-YFP in wells E1–H1. These values are then multiplied by 1000 to obtain mBRET units to circumvent the need to deal with decimal numbers. (**b**) The raw BRET measurements are shown in the *left panel* indicating the background BRET obtained with Rluc-PTEN (*white bar*) and the specific net BRET obtained with Rluc-PTEN-YFP (*filled bar*). The *right panel* shows the specific mBRET values obtained with Rluc-PTEN-YFP (305 mBRET) with Rluc-PTEN being set to 0 mBRET

YFP or the mutant PTEN biosensor Rluc-PTEN(A4)-YFP that has the phosphorylation cluster serine/threonine residues mutated to alanine.

2. 24–48 h after transfection, cells are prepared and BRET measurements initiated as described in Subheading 3.3. BRET measurements are shown for HEK cells in Fig. 4e–g. Figure 4e shows the mBRET units obtained with the WT and A4 biosensor. There was a marked reduction in the BRET signal obtained with the A4 form of PTEN compared to WT PTEN, indicating conformational change. The data can be expressed as ΔmBRET versus the WT PTEN biosensor, which is therefore set to zero, and the change obtained with the mutant is expressed relative to this (Fig. 4f). The data can also be represented as %ΔmBRET where the change is represented as a percentage difference compared to that obtained with WT sig-

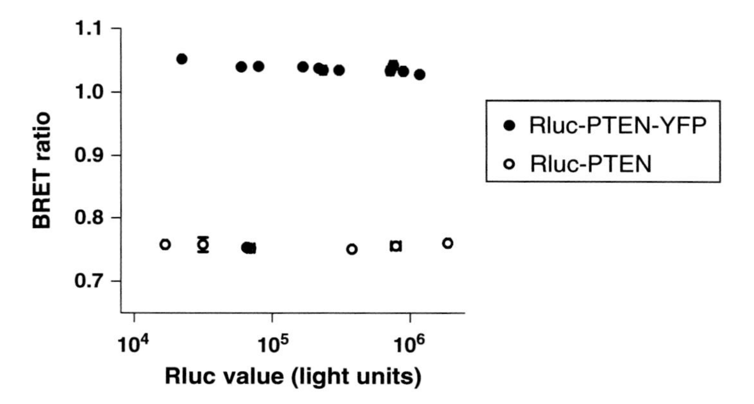

Fig. 3 Stable intramolecular BRET obtained with increasing amounts of PTEN biosensor. PC-3 cells growing in 6-well plates were transfected with increasing amounts of either Rluc-PTEN control vector or Rluc-PTEN-YFP (1–1000 ng). Twenty-four hours later, cells from each transfection were distributed into four wells of a white 96-well plate and BRET ratios measured. The BRET signal is constant over a 100-fold range of Rluc light values due to the constant acceptor-to-donor ratio in the unimolecular biosensor

Fig. 4 (continued) antibody that recognizes phospho-PTEN (Ser380/Thr382/Thr383) in addition to total PTEN expression. (**d**) Live HEK cells expressing WT or A4 forms of PTEN biosensor were directly imaged in IBIDI µ-slide multi-well dishes (Biovalley). The *insets* show magnified regions of the *boxed area* and demonstrate membrane-targeting of PTEN-A4. Scale bar is 10 µm. (**e**) BRET values obtained with the WT and A4 forms of PTEN biosensor in HEK cells, expressed as mBRET. (**f**) The BRET values obtained in (**e**) are expressed as ΔmBRET to show the shift in BRET compared to WT signal, which is set to zero. (**g**) The BRET values obtained in (**e**) can also be expressed as the % change compared to WT signal (Δ%mBRET). (**h**) Δ%mBRET change obtained with the A4 PTEN biosensor in PC-3 cells. (**i**) Δ%mBRET change obtained with the A4 PTEN biosensor in HeLa cells

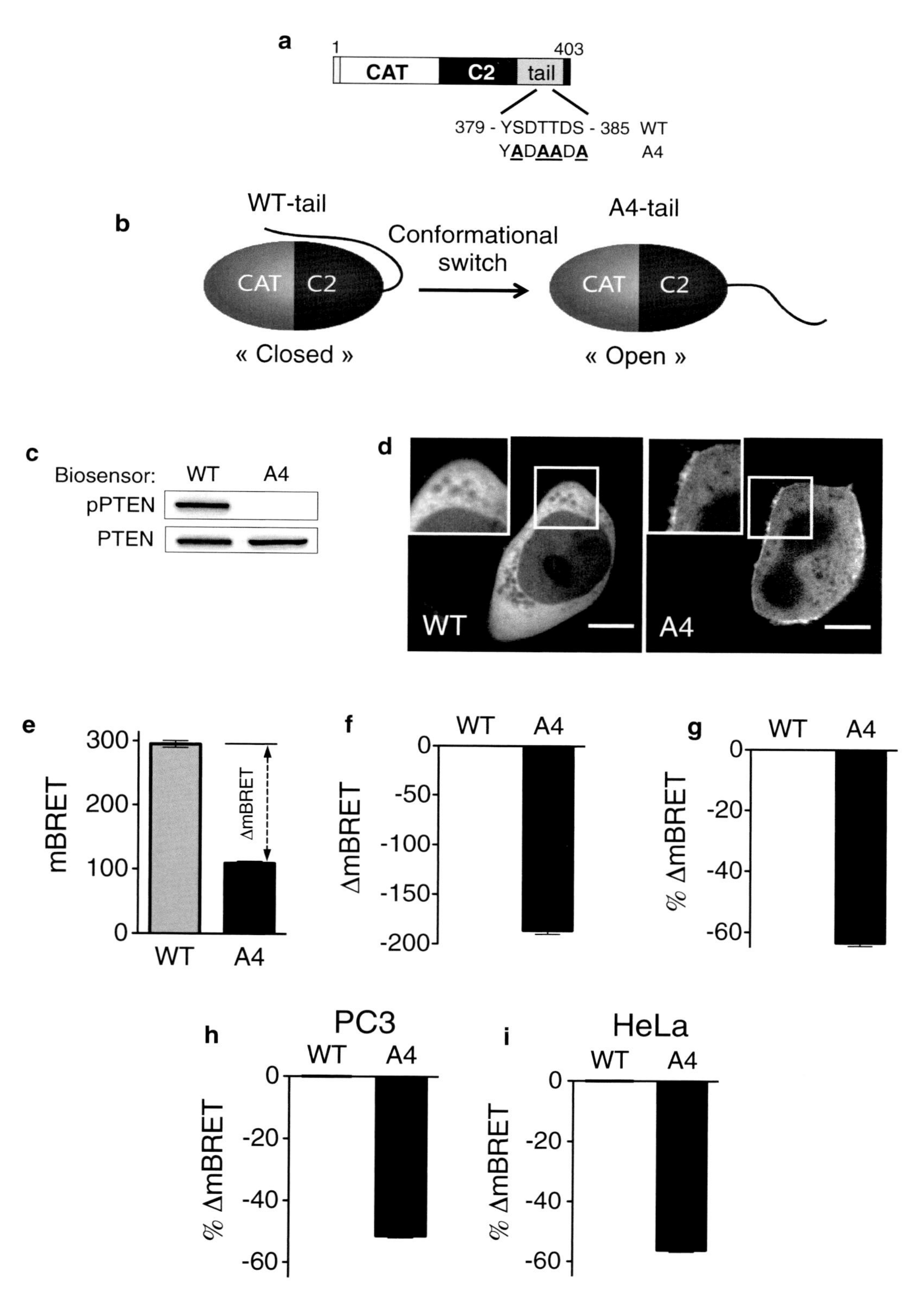

Fig. 4 Detection of PTEN conformational change in live cells following mutation of key regulatory residues in the C-terminal tail. (**a**) Diagram showing key regulatory phosphorylation sites in the C-terminal tail of PTEN at Ser380, Thr382, Thr383, and Ser385. The A4 mutant has all of these phosphorylation sites mutated to alanine. (**b**) Proposed model of the conformational switch that occurs in the A4 form of PTEN. (**c**) A western blot with an

nal (Fig. 4g). Similar data were obtained using PC-3 (Fig. 4h) and HeLa (Fig. 4i) cells. Following changes in BRET signal, secondary tests can be performed to assess potential functional changes (*see* **Note 8**). While these functional tests can be applied to the structure-function mode they are evidently also applicable to the other experimental modes in which the biosensor can be used, described below in Subheadings 3.7–3.9.

3.7 Assessing PTEN Conformational Change by Pharmacological Inhibitors That Impact PTEN Phosphorylation

CK2 overexpression has been documented in different types of cancer and CK2 inhibitors have been developed for clinical use [16, 17]. CK2 is implicated in the phosphorylation of the C-terminal regulatory phosphorylation cluster of PTEN. This leads to phospho-dependent inhibition of PTEN and increased activation of the PI3K/AKT pathway. CK2 inhibitors have been shown to reestablish PTEN signaling and restrain the PI3K/AKT pathway. The experiment described below demonstrates how the PTEN biosensor can be used to report changes in its phosphorylation status in the presence of pharmacological inhibitors.

1. PC-3 cells are seeded and transfected as described in Subheading 3.2.
2. 48 h after transfection cells are treated with the CK2 inhibitors tetrabromobenzotriazole (TBB; 50 μM) or CX-4945 (10 μM) for 2 h (0.1 % DMSO is used as vehicle control). The cells are then prepared for BRET readings as described in either Subheading 3.2 or 3.3. To ensure that pharmacological agents do not affect the spectral properties of YFP, raw YFP measurements can be taken in the presence of vehicle or pharmacological agent (*see* **Note 9**). In this experiment the PTEN phosphorylation status can be assessed in parallel by subsequent western blotting using an anti-phospho Ser380/Thr382/Thr383 antibody (Cell Signaling). Both CK2 inhibitors induce a decrease in BRET signal (Fig. 5). The effect was more pronounced with CX-4945 and this was associated with a greater decrease in phospho Ser380/Thr382/Thr383 compared to that obtained with TBB (Fig. 5).

3.8 Assessing PTEN Conformational Change by Cellular PTEN Regulators

The biosensor can also be used to test the effect of PTEN regulators on conformation. For this the PTEN partner/regulator is simply co-expressed with the biosensor. The experiment described below uses an active form of RhoA (RhoA-Q63L) known to increase PTEN activity.

1. HEK 293 T cells are seeded and transfected as described in Subheading 3.2 with 5 ng biosensor and either 250 ng empty Myc-RK5 vector or Myc-RhoA-Q63L (*see* **Note 10** for DNA ratios of partner/regulator to biosensor).
2. The cells are then prepared for BRET readings as described in Subheading 3.3. In this experiment the levels of cellular pAKT

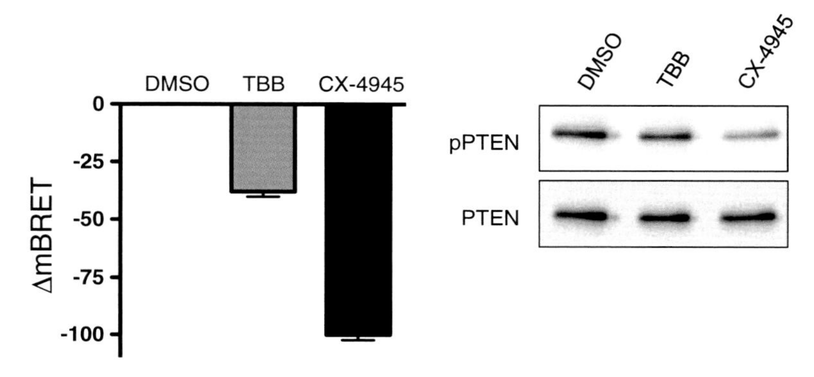

Fig. 5 Detection of PTEN conformational change in live cells following pharmacological modulation of PTEN regulatory pathways. (**a**) ΔmBRET values obtained following a 2-h incubation with the CK2 inhibitors TBB (50 μM) or CX-4945 (10 μM) versus DMSO vehicle control in PC-3 cells expressing WT PTEN biosensor. (**b**) Western blot showing phosphorylation of the Ser380/Thr382/Thr383 cluster in the presence of DMSO, TBB, or CX-4945 in PC-3 cells expressing WT PTEN biosensor

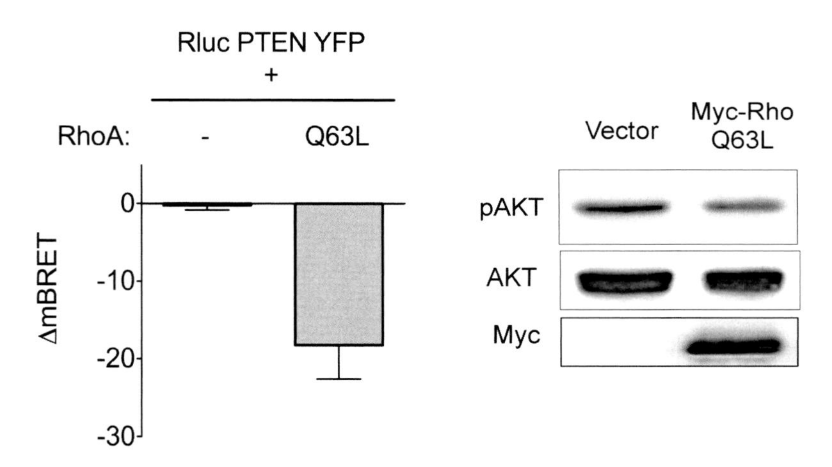

Fig. 6 Detection of signal-dependent PTEN conformational change in live cells. (**a**) ΔmBRET values obtained with the WT PTEN biosensor in HEK cells transfected with either empty vector or Myc-RhoA-Q63L. (**b**) HEK cells transfected with empty vector or RhoA-Q63L were assessed for pAKT levels by western blotting

can be assessed in parallel by subsequent western blotting using an anti-phospho Ser 473 antibody (Cell Signaling). Active RhoA-Q63L induces a decrease in BRET signal and this is associated with a decrease in pAKT levels (Fig. 6).

3.9 Assessing Real-Time Changes in PTEN Conformation Following Cell Surface Receptor Activation

BRET can also be used to follow conformational change of PTEN in real time in living cells. For technical reasons the length of these experiments when using coelenterazine-h is limited to around 20 min (*see* **Note 11**). The experiment described below uses the G protein-coupled receptor, TPαR, which activates PTEN [15, 18].

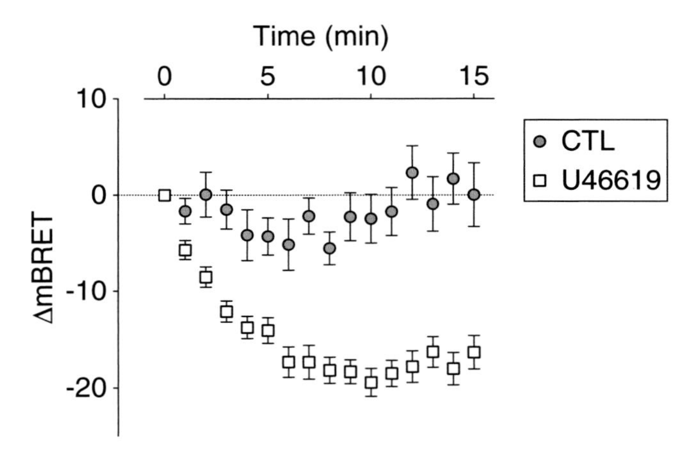

Fig. 7 Real-time analysis of cell surface receptor-induced PTEN conformational change in live cells. Changes in BRET induced by addition of the agonist U46619 for the TPαR. Cells were co-transfected with TPαR and biosensor, and subsequently transferred to poly-L-ornithine-coated plates. BRET readings were acquired every minute after addition of 1 μM U46619 or vehicle control (0.03 % methyl acetate) over a period of 15 min

1. HEK 293 T cells are seeded and transfected as described in Subheading 3.2 with 5 ng biosensor and 250 ng of the TPαR. *See* **Note 10** for DNA ratios of partner/regulator to biosensor.
2. The cells growing in poly-L-ornithine-coated 96-well white plates are serum-starved overnight and then the media is replaced with 80 μl DMEM without phenol red per well. 10 μl of a freshly prepared solution of coelenterazine-h diluted in serum-free media is added to each well, and then 10 μl of an appropriate intermediate stock solution of U46619 or of methyl acetate (vehicle control) is added to each well.
3. BRET readings are then immediately initiated and repeated over the course of the experiment to observe the evolution over time. The BRET ratios obtained are then plotted against time as shown in Fig. 7.

4 Notes

1. Coelenterazine-h is sensitive to light and undergoes slow oxidation. Working solutions are prepared at 50 μM in HBSS or in media without phenol red, immediately before performing experiments. Coelenterazine-h should be protected from light during prolonged storage periods by using opaque tubes. The experiments described here with the PTEN biosensor in a BRET1 assay employed coelenterazine-h as substrate. Other

Rluc substrates are available to obtain either increased luminescence (three- to fivefold increase in peak signal with ViviRen™ from Promega) or extremely stable luminescence over a 24-h period (but with 10–25 times lower signal) using EnduRen™ (Promega). To reduce degradation rates, both of these substrates carry protecting groups that are cleaved by esterases inside cells.

2. Changes in donor and acceptor pairs, as well as different substrates, can be used to generate increased sensitivity BRET1, BRET2, or BRET3 assays. BRET 1 is still the most widely performed assay using Rluc as the energy donor and the enhanced yellow fluorescent protein (YFP) as acceptor. Rluc variants, such as Rluc8, and the YFP variant YPet have been developed that demonstrate increased BRET signals using this BRET couple compared to the standard Rluc/YFP pair. A luciferase subunit (MW: 19 kDa) from the deep sea shrimp *Oplophorus gracilirostris* was recently described and developed by Promega. This new luciferase, Nanoluc™, uses furimazine as substrate and displays around 150-fold greater specific activity than Rluc. The maximal emission peak of Nanoluc™ is 465 nm and it is therefore compatible as donor in BRET1 assays. The increased brightness obtained with Nanoluc™ allows very low expression levels and it is therefore ideally adapted to monitoring protein fusions at "endogenous" cellular levels. BRET2 assays use the substrate coelenterazine 400a with a modified form of Rluc donor called Rluc2, and GFP2, a GFP mutant that can be excited at 400 nm, as the energy acceptor. BRET2 provides the advantage of demonstrating increased spectral resolution between donor and acceptor peaks and therefore displays lower background signal. Incubation of Rluc2 with coelenterazine 400a causes Rluc2 to emit light at 400 nm, promoting the excitation of GFP2, which, subsequently, emits light at 510 nm. Filter sets used for BRET2 are 400 ± 10 nm and 515 ± 10 nm. BRET3 assays use coelenterazine-h with Rluc8 as donor and mOrange (a mutant form of DsRed2) as acceptor that can be excited from 514 to 550 nm and that has maximal emission at 564 nm. This is therefore the BRET configuration that gives the most red-shifted light output.

3. BRET assays should be performed on "fresh" cells that have not been in culture for more than 6 weeks. Routine testing for mycoplasma (e.g., MycoAlert™ Mycoplasma Detection Kit, Lonza) should also be carried out to ensure that cells are free of contamination.

4. Poly-L-ornithine used to coat plates is prepared in 1 ml aliquots at a concentration of 1.5 mg/ml (50×) in sterile dH_2O. Five 96-well plates can therefore be coated at a time using one 50×

aliquot diluted to 1× in dH_2O (100 μl/well of 30 μg/ml poly-L-ornithine) and the plates can be kept at 37 °C for up to a week before use. Prior to cell seeding, the poly-L-ornithine is removed, wells washed once in PBS (containing $CaCl_2$/$MgCl_2$) or complete medium, which is then also removed, and the wells left to air-dry under the hood.

5. When performing BRET on attached cells, the media should be removed by washing once with PBS (containing $CaCl_2$/$MgCl_2$) to remove phenol red, as this quenches luciferase signal.
6. Previously characterized optimal BRET1 filter sets used in the experiments described here are 480 ± 10 nm and 540 ± 20 nm, which were found to provide enhanced BRET readings compared to 485 + 10 nm and 530 ± 12.5 nm filter sets [19].
7. To measure BRET at "physiological" levels of biosensor expression, cells can be transfected with a range of plasmid concentrations, subsequently lysed, and assessed by western blotting with an anti-PTEN antibody (Cell Signaling or Cascade). The level of biosensor can be compared to endogenous PTEN in the same cells or in comparison with a PTEN-expressing line if the cells used for the BRET study are PTEN-null [15]. When decreasing the amount of biosensor it is important to check that the Rluc and YFP stay in the linear range of detection of the BRET reader.
8. Following changes in BRET signal, secondary tests can be performed to assess potential functional changes. The YFP in the fusion enables direct visualization of PTEN subcellular localization in live cells using fluorescence microscopy. We have found that setting cells up in parallel in IBIDI μ-slide multiwell dishes (Biovalley) provides a convenient way to directly visualize PTEN under multiple test conditions (Fig. 4d). Following cell lysis, the biosensor can also be immunoprecipitated using an anti-GFP antibody (Roche) and phosphatase activity against PI(3,4,5)P3 tested in vitro. The same lysates can also be assessed for changes in cellular pAKT levels.
9. Control readings can be performed in experiments using pharmacological agents to check that raw YFP emissions are not altered, which could indirectly affect BRET. This is done simply by measuring the raw fluorescence emitted at 535 nm by Rluc-PTEN-YFP following external excitation at 485 nm on samples exposed to either vehicle or test reagent [15]. Rluc-PTEN is used to assess background fluorescence from the cells.
10. When assessing the effect of a protein partner/regulator or cell surface receptor on PTEN conformation we have found that DNA ratios of test protein:biosensor of 10:1 to 50:1 work best [15]. Using this approach we have shown the expression of CK2 causes an increase in BRET signal and that expression of

β-arrestin 2 promotes a decrease in BRET, as with active RhoA [15]. A subset of partners that bind PTEN via its C-terminal PDZ-binding sequence may not lend themselves to this approach due to the C-terminal fusion of YFP in the biosensor.

11. As Rluc light output declines with time, kinetic experiments using real-time measurements are limited to a time frame of around 20–25 min. For experiments investigating longer time points "reverse kinetics" can be set up, stimulating cells for the longest time point first and then working down to t = 0 min, when all time points can be measured simultaneously for BRET. Alternatively, Enduren™ substrate that provides stable luminescence over long time periods (*see* **Note 1**) could be used.

Acknowledgements

We thank J. Paradis and C. Boularan for critical reading of the manuscript. This work was supported by the Fondation ARC pour la Recherche sur le Cancer, Ligue Contre le Cancer (comité de l'Oise), Fondation pour la Recherche Biomedicale ("Equipe FRM"), CNRS, and INSERM. The laboratory is part of the multidisciplinary "Who am I?" LABEX and S.M. is supported by a LABEX doctoral fellowship. E.L.F. was funded by a doctoral fellowship from the Fondation ARC pour la Recherche sur le Cancer and subsequently a postdoctoral fellowship from the Fondation Nationale pour la Recherche Luxembourg and Marie Curie Actions.

References

1. Salmena L, Carracedo A, Pandolfi PP (2008) Tenets of PTEN tumor suppression. Cell 133(3):403–414
2. Song MS, Salmena L, Pandolfi PP (2012) The functions and regulation of the PTEN tumour suppressor. Nat Rev Mol Cell Biol 13(5): 283–296
3. Worby CA, Dixon JE (2014) PTEN. Ann Rev Biochem 83:641–669
4. Lima-Fernandes E, Enslen H, Camand E, Kotelevets L, Boularan C, Achour L, Benmerah A, Gibson LC, Baillie GS, Pitcher JA, Chastre E, Etienne-Manneville S, Marullo S, Scott MG (2011) Distinct functional outputs of PTEN signalling are controlled by dynamic association with beta-arrestins. EMBO J 30(13): 2557–2568
5. Leslie NR, Foti M (2011) Non-genomic loss of PTEN function in cancer: not in my genes. Trends Pharmacol Sci 32(3):131–140
6. Hopkins BD, Hodakoski C, Barrows D, Mense SM, Parsons RE (2014) PTEN function: the long and the short of it. Trends Biochem Sci 39(4):183–190
7. Naguib A, Trotman LC (2013) PTEN plasticity: how the taming of a lethal gene can go too far. Trends Cell Biol 23(8):374–379
8. Torres J, Pulido R (2001) The tumor suppressor PTEN is phosphorylated by the protein kinase CK2 at its C terminus. Implications for PTEN stability to proteasome-mediated degradation. J Biol Chem 276(2):993–998
9. Vazquez F, Ramaswamy S, Nakamura N, Sellers WR (2000) Phosphorylation of the PTEN tail regulates protein stability and function. Mol Cell Biol 20(14):5010–5018
10. Rahdar M, Inoue T, Meyer T, Zhang J, Vazquez F, Devreotes PN (2009) A phosphorylation-dependent intramolecular interaction regulates the membrane association and activity of the

tumor suppressor PTEN. Proc Natl Acad Sci U S A 106(2):480–485

11. Das S, Dixon JE, Cho W (2003) Membrane-binding and activation mechanism of PTEN. Proc Natl Acad Sci U S A 100(13):7491–7496
12. Bolduc D, Rahdar M, Tu-Sekine B, Sivakumaren SC, Raben D, Amzel LM, Devreotes P, Gabelli SB, Cole P (2013) Phosphorylation-mediated PTEN conformational closure and deactivation revealed with protein semisynthesis. Elife 2:e00691
13. Odriozola L, Singh G, Hoang T, Chan AM (2007) Regulation of PTEN activity by its carboxyl-terminal autoinhibitory domain. J Biol Chem 282(32):23306–23315
14. Silva A, Yunes JA, Cardoso BA, Martins LR, Jotta PY, Abecasis M, Nowill AE, Leslie NR, Cardoso AA, Barata JT (2008) PTEN post-translational inactivation and hyperactivation of the PI3K/Akt pathway sustain primary T cell leukemia viability. J Clin Invest 118(11): 3762–3774
15. Lima-Fernandes E, Misticone S, Boularan C, Paradis JS, Enslen H, Roux PP, Bouvier M, Baillie GS, Marullo S, Scott MG (2014) A biosensor to monitor dynamic regulation and function of tumour suppressor PTEN in living cells. Nat Commun 5:4431
16. Barata JT (2011) The impact of PTEN regulation by CK2 on PI3K-dependent signaling and leukemia cell survival. Adv Enzyme Regul 51(1):37–49
17. Martins LR, Lucio P, Melao A, Antunes I, Cardoso BA, Stansfield R, Bertilaccio MT, Ghia P, Drygin D, Silva MG, Barata JT (2014) Activity of the clinical-stage CK2-specific inhibitor CX-4945 against chronic lymphocytic leukemia. Leukemia 28(1):179–182
18. Song P, Zhang M, Wang S, Xu J, Choi HC, Zou MH (2009) Thromboxane A2 receptor activates a Rho-associated kinase/LKB1/PTEN pathway to attenuate endothelium insulin signaling. J Biol Chem 284(25): 17120–17128
19. Achour L, Kamal M, Jockers R, Marullo S (2011) Using quantitative BRET to assess G protein-coupled receptor homo- and heterodimerization. Methods Mol Biol 756:183–200

Chapter 9

Methods for the Identification of PTEN-Targeting MicroRNAs

Andrea Tuccoli, Marianna Vitiello, Andrea Marranci, Francesco Russo, and Laura Poliseno

Abstract

The identification of *PTEN*-targeting microRNAs usually starts from an in silico bioinformatic prediction and then requires a careful experimental validation that exploits both heterologous and endogenous systems. Here we describe the methods used to carry on these analyses and experiments, examining pitfalls and alternatives for each step. Moreover, we give an overview of the latest high-throughput microRNA target identification techniques which offer a more comprehensive view of the microRNAs that can bind a fundamental tumor suppressor such as *PTEN*.

Key words Posttranscriptional regulation, PTEN, MicroRNAs, MicroRNA responsive element, 3′ Untranslated region

1 Introduction

MicroRNAs (miRNAs) are a class of ~22 nt long single-stranded RNA molecules that have come into light as negative regulators of gene expression at the posttranscriptional level [1, 2]. A growing body of evidence demonstrates that miRNAs are involved in numerous biological and pathological processes. Among the latter, microRNAs play a fundamental role in both cancer development and progression, acting either as oncosuppressor or oncogenic molecules by regulating key genes involved in cellular proliferation and/or survival [3]. One such gene is phosphatase and tensin homolog deleted on chromosome 10 (PTEN), which exerts its tumor-suppressive function by inhibiting the AKT/PI3K signaling pathway [4].

PTEN regulation by microRNAs has been extensively studied and PTEN-targeting microRNAs are the focus of dozens of publications (Table 1). This chapter is intended as a resource that can guide through the identification and validation of new

Leonardo Salmena and Vuk Stambolic (eds.), *PTEN: Methods and Protocols*, Methods in Molecular Biology, vol. 1388, DOI 10.1007/978-1-4939-3299-3_9, © Springer Science+Business Media New York 2016

Table 1
Validated Pr£/V-microRNA interactions

miRNA	Validation technique	Reference (PMID)	Algorithms	
hsa-miR-17-5p	Luc, WB, qPCR	20008935-20227518-21283765-23418359	mW, mT4, mR, P, PITA, Rh, TS	7
hsa-miR-18a-5p	Luc, qPCR, WB	20008935	mW, mT4, PITA, Rh	4
hsa-miR-19a-3p	Luc, qPCR, WB	14697198-18460397-21853360	mW, mT4, mR, mDB, P, PITA, Rh, TS	8
hsa-miR-19b-3p	Luc, qPCR, WB	20008935	mW, mT4, mR, mDB, P, PITA, Rh, TS	8
hsa-miR-20a-5p	Luc, qPCR, WB	20008935-2128376	mW, mT4, mR, P, PITA, Rh, TS	7
hsa-miR-21-5p	GFP, NB, qPCR, WB, ASO, ICC, Luc, Ma	16762633-17681183-18850008-19072831-19167416-19253296-19641183-19672202-19906824-20048743-20216554-20223231-20797623-20813833-20827319-20978511-21544242-22267008-22761812-22770403-22879939-22956424-23226804	PITA, Rh	2
hsa-miR-23a-3p	Luc, qPCR, WB	23019365	mW, mT4, mR, P, PITA, Rh, TS	7
hsa-miR-23b-3p	Luc	23189187	mW, mT4, mR, P, PITA, Rh, TS	7
hsa-miR-26a-5p	GFP, Luc, WB	19487573-20080666-20216554	mW, mT4, mR, mDB, P, PITA, Rh, TS	8
hsa-miR-29a-3p	Luc	21573166-2342637	mW, mT4, mR, mDB, P, PITA, Rh, TS	8
hsa-miR-29b-3p	WB, Ma	21359530	mW, mT4, mR, mDB, P, PITA, Rh, TS	8
hsa-miR-93-5p	GFP, qPCR, WB	22465665	mW, mT4, mR, P, PITA, Rh, TS	7
hsa-miR-103a-3p	WB	22593189	mW, mT4, mR, PITA, Rh, TS	6
hsa-miR-106b-5p	Luc	21283765	mW, mT4, mR, P, PITA, Rh, TS	7
hsa-miR-107	WB	22593189	mW, mT4, mR, PITA, Rh, TS	6

hsa-miR-141-3p	Luc, WB	20053927	mW, mT4, mR, P, PITA, Rh, TS	7
hsa-miR-144-3p	IP, Luc, qPCR, WB	23125220	mW, mT4, mR, mDB, P, PITA, Rh, TS	8
hsa-miR-214-3p	Luc, Ma, qPCR, WB	18199536-20548023-21228352	mW, mT4, mR, mirbridge, PITA, Rh, TS	7
hsa-miR-216a-5p	Luc, WB	20216554-2347158	mW, mT4, mR, P, PITA, Rh, TS	7
hsa-miR-217	Luc, WB	20216554-2347158	mW, mT4, P, PITA, Rh, TS	6
hsa-miR-221-3p	FACS, Flow, Luc, NB, WB, IHC, qPCR	19962668-20618998-23372675	mW, mT4, PITA, Rh	4
hsa-miR-222-3p	FACS, Flow, Luc, NB, WB, IHC, qPCR	19962668-20618998-23028614	mW, mT4, PITA, Rh	4
hsa-miR-494-3p	Flow, Luc, Ma, qPCR, WB	20006626-2254493	mW, mT4, mR, mDB, P, PITA, Rh, TS	8
hsa-miR-519a-3p	Luc, qPCR, WB	22262409	mW, mT4, mR, mDB, PITA, Rh, TS	7
hsa-miR-519c-3p	Luc, qPCR, WB	22262409	mW, mT4, mR, mDB, PITA, Rh, TS	7
hsa-miR-519d-3p	Luc, WB	22262409	mW, mT4, mR, P, PITA, Rh, TS	7

PTEN-microRNA interactions that have been validated using different methodologies, according to miRTarBase. In addition to the list of techniques that have been used in order to validate the microRNA-PTEN mRNA binding, the name of the algorithms that predict each interaction is reported (on the far right, the total number of predictors is shown). Techniques legend: *Luc* Luciferase assay; *WB* western blot; *qPCR* real-time PCR; *GFP* GFP reporter assay; *NB* northern blot; *ASO* antisense oligonucleotide assay; *ICC* immunocytochemistry; *Ma* microarray; *IP* immunoprecipitation; *FACS* fluorescence-activated cell sorting; *Flow* flow cytometry; *IHC* immunohistochemistry. Algorithms legend: *mW* miRWalk; *mT4* Diana microT v4; *mR* miRanda; *P* PicTar; *Rh* RNAhybrid; *TS* TargetScan; *mDB* miRDB

PTEN-targeting miRNAs, using both bioinformatic-based approaches and some newly developed and unbiased high-throughput techniques. In particular, we focus on the more problematic aspects of these processes.

2 Materials

Here we provide a comprehensive list of the materials used in our lab. Unless otherwise stated, all experiments are to be performed following the manufacturer's instructions.

2.1 Molecular Biology

1. pGL3-control Vector (Promega).
2. pRL-TK Vector (Promega).
3. Restriction enzymes (Thermo Scientific).
4. T4 DNA ligase (Thermo Scientific).
5. Alkaline Phosphatase, Calf Intestinal (New England Biolabs).
6. Phusion High-Fidelity DNA Polymerase (Thermo Scientific).
7. QIAquick Gel Extraction Kit (Qiagen).
8. QIAquick PCR Purification Kit (Qiagen).
9. QIAquick Spin Miniprep Kit (Qiagen).
10. PureLink HiPure Plasmid Filter Maxiprep Kit (Life Technologies).
11. Dual-Glo® Luciferase Assay System (Promega).
12. Trizol (Life Technologies).
13. DNase I, amplification grade (Life Technologies).
14. miScript II RT kit (Qiagen).
15. iScript Reverse Transcription Supermix (Bio-rad).
16. Sso Advanced Universal SYBR Green Supermix (Bio-rad).
17. PTEN antibody #9559 (Cell Signaling Technology).
18. pAKT antibody #9271 (Cell Signaling Technology).
19. AKT antibody #9272 (Cell Signaling Technology).
20. QuikChange II Site-directed Mutagenesis Kit (Agilent Technologies)

2.2 Cellular Biology

1. Tissue culture plastic ware (BD).
2. Cell media and supplements (Euroclone).
3. Lipofectamine 2000 (Life Technologies).
4. Opti-mem I Reduced Serum medium (Life Technologies).

2.3 Oligonucleotides

1. Negative control miRNA mimic (sense 5′-UUCUCCGAACGUGUCACGUTT-3′; antisense 5′-ACGUGACACGUUCGGAGAATT-3′) (Genepharma design).

2. GAPDH real-time primers (F 5′-CGCTCTCTGCTCCT CCTGTT-3′; R 5′-CCATGGTGTCTGAGCGATGT-3′).
3. PBGD real-time primers (F 5′-TCCAAGCGGAGCCATG TCTG-3′; R 5′-AGAATCTTGTCCCCTGTGGTGGA-3′).
4. SDHA real-time primers (F 5′-CCACTCGCTATTGCACA CC-3′; R 5′-CACTCCCCATTCTCCATCA-3′).
5. PTEN real-time primers (F 5′-GTTTACCGGCAGCATCA AAT-3′; R 5′-CCCCCACTTTAGTGCACAGT-3′) [15].

3 Methods

3.1 Prediction of PTEN-Targeting MicroRNAs

When approaching the identification of PTEN-targeting microRNAs, an in silico analysis can be performed in order to predict which microRNAs can interact with its 3′UTR. Several algorithms provide lists of predicted miRNAs for a given target and specify the location of the microRNA responsive elements (MREs) along the target mRNA; many are based on sequence complementarity to target sites with emphasis on perfect base-pairing in the seed region (nucleotides 2–7 at the 5′end of the microRNA) and sequence conservation (e.g., TargetScan [5]). Others are based on calculations of mRNA secondary structures and energetically favorable hybridization between the miRNA and the target mRNA (e.g., RNAhybrid [6]). miRanda [7] instead computes optimal sequence complementarity between a set of miRNAs and a given mRNA using a weighted dynamic programming algorithm.

In addition to the identification of MREs located in the 3′UTR of the gene of interest, some algorithms allow for the identification of MREs located in the 5′UTR (e.g., miRWalk [8]) or in the coding sequence (e.g., Diana-microT-CDS [9]); moreover, a few prediction algorithms allow to insert a sequence of interest to search for possible MREs (e.g., RNAhybrid [10]).

There are comprehensive databases that collect both predicted and experimentally verified miRNA-target interactions taken from several resources. One such database is miRWalk [8], which reports binding sites resulting from 12 existing miRNA-target prediction programs. It also documents experimentally verified miRNA-target interactions collected from existing resources via an automated text-mining search. An additional example of a database that collects experimentally validated miRNA targets is miRTarBase [11] (Table 1). More information about the predicted miRNAs can be found by querying specific databases: miRBase collects information regarding sequence annotation, genomic organization, precursor sequences, and literature citations [12]; smiRNAdb is a resource for microRNA expression information [13]; Magia2 reports information based on sequence and gene expression data [14]. In Table 2 we report the most current algorithms for target prediction developed.

Table 2
Most common algorithms for microRNA target prediction

Algorithm	Website	Reference (PMID)
DIANA-microT	http://diana.imis.athena-innovation.gr/DianaTools/index.php?r=microtv4/index	21551220
DIANA-microT-CDS	http://diana.imis.athena-innovation.gr/DianaTools/index.php?r=microT_CDS/index	22285563
miRanda	http://www.microrna.org/microrna/getDownloads.do	20799968
mirBridge	http://www.ncbi.nlm.nih.gov/pubmed/?term=20385095	20385095
miRDB	http://mDB.org/mDB/download.html	18048393
miRMap	http://mirmap.ezlab.org/downloads/mirmap201301e	23034802
miRNAMap	ftp://mirnamap.mbc.nctu.edu.tw/miRNAMap2	18029362
miRWalk	http://www.umm.uni-heidelberg.de/apps/zmf/mW/	21605702
PicTar	http://dorina.mdc-berlin.de/	25416797
PITA	http://genie.weizmann.ac.il/pubs/mir07/mir07_exe.html	17893677
RNA22	https://cm.jefferson.edu/rna22v2/	16990141
RNAhybrid	http://bibiserv.techfak.uni-bielefeld.de/Rh/dl_pre-page.html	15383676
TargetScan	http://www.TS.org/cgi-bin/TS/data_download.cgi?db=vert_61	15652477

MicroRNA-target prediction tools are prone to errors and tend to return a large number of false positive targets (that is, wrong miRNA-mRNA interactions). To reduce the number of false-positive interactions (and thus improve the target prediction), the integration of mRNA and miRNA expression can be used. Target mRNA levels are in fact expected to be anti-correlated with the targeting microRNA levels. However, a thorough experimental validation is still required to sustain the bioinformatic data.

3.2 Validation of Predicted MREs by Luciferase Assays

Luciferase reporter assays are a rapid, sensitive, and quantitative method used to test the direct interaction between a certain microRNA and its predicted target. The 3′UTR of the gene of interest, such as *PTEN*, is cloned into the multiple cloning site located downstream of the Luciferase coding sequence, after the stop codon. In this way, the binding of one or more miRNAs to the Luciferase mRNA leads to its translational inhibition/degradation, assessed by directly measuring the Luciferase signal through a luminometer. In general, a reduced Luciferase activity for the target mRNA compared to a control indicates that the endogenous or exogenous miRNA is capable of binding the target sequence.

All Luciferase-based validation experiments require first and foremost to clone *PTEN* 3′UTR into an appropriate Luciferase vector.

3.2.1 Vector Selection

Several plasmids that encode for Firefly Luciferase can be chosen. Any plasmid containing suitable restriction enzyme sites downstream the Luciferase stop codon can in principle be used. As an example, the pGL3-Control Vector (Promega) can be chosen: in this case, the selected *PTEN* 3′UTR region can be cloned into the XbaI site (*see* **Note 1**). A control vector that expresses the Renilla Luciferase, such as the pRL-TK Vector (Promega), will also be needed. The Renilla Luciferase is used as a normalization control for the Firefly Luciferase encoded by the pGL3-Control Vector; in this way, the effects caused by variations in transfection efficiency can be eliminated and more reproducible results can be achieved (*see* **Note 2**).

3.2.2 Fragment Selection

We suggest to clone the entire *PTEN* 3′UTR (see Ensembl transcript number ENST00000371953) regardless of the location of the predicted MRE in order to preserve the secondary and tertiary structure of the 3′UTR itself. It is also possible to clone just a fragment of the *PTEN* 3′UTR; in our experience, when cloning a short fragment it is advisable to include ~200 bp upstream and downstream the sequence of interest, to preserve the genomic context in which a certain MRE is located (*see* **Note 3**).

3.2.3 Primer Design

When designing the primers, it must be taken into account the fact that *PTEN* 3′UTR has a region of high homology with the 3′UTR of its processed pseudogene *PTENP1* [15]. By aligning the *PTEN* 3′UTR to that of *PTENP1*, it is possible to check the specificity of the primers and to avoid *PTENP1* cross-amplification. If the fragment of interest resides in the highly homolog region, the primers should be moved to a less homolog region that still encompasses the sequence of interest (at least one of the two primers should have a considerable amount of mismatches with *PTENP1* 3′UTR for the PCR to be specific). In alternative, the entire *PTEN* 3′UTR can be amplified specifically (in that case the reverse primer will be located into the low-homology region, *see* ref. 15) and subsequently a nested PCR with the primer pair of interest can be performed on that fragment. Finally, the appropriate enzyme sequences for cloning should be added at the 5′ end of both the forward and the reverse primer.

3.2.4 Cloning

Regardless of the Firefly Luciferase plasmid chosen, the *PTEN* 3′UTR should be cloned downstream of the Luciferase stop codon. In this way, at the functional level, the *PTEN* 3′UTR will be the Firefly Luciferase 3′UTR, and any miRNA binding will exert an effect on the Firefly Luciferase expression. The selected *PTEN*

3′UTR can be amplified from any human genomic DNA using the primer pair designed as above and a proofreading DNA polymerase. A standard cloning procedure with appropriate restriction enzymes can be followed; for suggested materials, *see* Subheading 2. At the end of the procedure, screen the obtained clones by appropriate enzyme digestion and sequencing.

3.2.5 Over-expression of a Predicted MicroRNA

Since the Firefly Luciferase expression is dependent on microRNAs that might bind to the *PTEN* 3′UTR, the over-expression of functional microRNAs should lead to a decreased Firefly/Renilla Luciferase activity ratio. The microRNA can be over-expressed in the cell line of interest by transfecting an appropriate miRNA mimic. A “microRNA mimic” is a short-interfering RNA designed to function as a particular microRNA; it is a double-stranded RNA molecule in which one of the two strands perfectly resembles the mature microRNA of interest. The other strand is perfectly complementary to the miRNA, although mutations can be introduced to favor the incorporation of the miRNA-like strand into RISC (*see* **Note 4**). Moreover, a “UU” overhang is added on both strands at the 3′ end. The microRNA mimic designed as such is a “ready-to-transfect” molecule that will be transiently present into the cells. A negative control mimic will be needed as well; we have successfully used the negative control provided by GenePharma (*see* Subheading 2.3 for sequence).

Many cell lines can be used at this step; it is important to achieve a fair transfection efficiency, and hence to choose a cell line that can be transfected easily. It is advisable to choose a cell line that expresses low amounts of the microRNA of interest, so that the endogenous microRNA will contribute only marginally to the downregulation of the Firefly Luciferase expression. In this respect, the HCT116 Dicer −/− cell line is very convenient, since it lacks the majority of mature microRNAs (we grow them in McCoy’s medium, supplemented with 10 % FBS, glutamine, and Pen-Strep).

3.2.6 Transfection

Cells can be transfected with any commercially available transfectant. To achieve the maximum transfection efficiency while avoiding toxicity, it might be necessary to try different transfection settings in regard to multiple parameters (e.g., number of cells seeded, amount of transfectant). A fluorescent oligo can be used to monitor the transfection efficiency, either by using a fluorescence microscope or by cytofluorometry.

In our lab we have successfully used Lipofectamine 2000 (Life Technologies), following this protocol:

1. The day before transfection, seed 7.5×10^4 HCT116 Dicer −/− cells in each well of a 24-well plate, in order to reach 50–60 % confluence the day after. Seed triplicate wells for every microRNA to be tested, plus the negative control mimic.

2. 24 h later, for each well prepare two mixes: (a) 4 μl of Lipofectamine 2000 + 50 μl of complete McCoy's medium. (b) 50 ng of pGL3-PTEN 3′UTR Vector + 10 ng of pRL-TK vector + miRNA mimic (the final concentration will be 15 nM; e.g., 1 μl of 7.5 μM miRNA mimic (*see* **Note 5**)) + 50 μl of complete McCoy's medium.
3. Mix A and B by carefully pipetting up and down and wait for 20 min.
4. In the meantime, remove the medium from the seeded cells and add 400 μl of complete McCoy's medium to each well (*see* **Note 6**).
5. Add the mix to the cells.
6. After 6 h, replace the medium and continue incubation.

3.2.7 Luciferase Activity Measurement

24 h after transfection, use the Dual-Glo® Luciferase Assay System (Promega) to perform the Luciferase assay, following the manufacturer's protocol.

1. Transfer 20 μl of cellular lysate into a 96-well white plate, making duplicates for each sample.
2. Read the Firefly and Renilla Luciferase luminescence on an appropriate instrument (we use the GloMax Microplate Luminometer (Promega)).
3. Calculate the ratio between Firefly and Renilla Luciferase activity for every sample; data can be presented normalizing the Luciferase ratio of the miRNA mimic on the Luciferase ratio of the negative control mimic (in this way, the negative control value will be set at 1). A functional microRNA (that is, a microRNA that is capable of inhibiting *PTEN* expression by binding to its 3′UTR) will determine a lower Firefly/Renilla ratio when compared to the negative control mimic (that is, a Luciferase activity ratio < 1).

3.2.8 Predicted MicroRNA Inhibition

The Firefly/Renilla Luciferase activity ratio can also be used to measure the effects of the inhibition of the microRNA under study. The steps to achieve the inhibition are similar to those mentioned for the over-expression, with some important differences:

1. Several techniques and molecules can be used to inhibit the activity of a microRNA of interest; in our lab, we have successfully used Exiqon LNA inhibitors. These molecules are synthetically modified oligonucleotides designed to perfectly pair to a microRNA of interest and block its pairing with the target; they are "ready-to-transfect" molecules that will be transiently present into the cells. A negative control LNA (designed by Exiqon) is used as a control.

2. Contrary to the over-expression experiments, in this case a cell line that expresses the microRNA of interest at high level should be chosen (*see* **Note 7**).
3. Transfection and Luciferase activity measurement: follow the protocol described for the over-expression of the microRNA of interest (*see* Subheadings 3.2.6 and 3.2.7). To achieve full microRNA inhibition, it is advisable to test different doses of the LNA. A functional microRNA will determine a higher Firefly/Renilla ratio when cells are transfected with the corresponding LNA compared to those transfected with the negative control LNA.

3.2.9 MRE Mutagenesis

Most microRNA target prediction algorithms rely on the common consensus that the microRNA seed is the most important region as far as the microRNA-target interaction is concerned (Table 2). Indeed, it has been shown that even single-nucleotide mutations in the microRNA seed (as well as mutations in the seed match, that is, the sequence on the target that pairs to the microRNA seed) can disrupt the miRNA-target interaction [16]. Thus, a functional MRE must retain microRNA seed-seed match pairing to be functional.

The mutagenesis of the seed match sequence of a predicted MRE cloned into the Luciferase vectors discussed in Subheading 3.2.1 is used as a means to further validate predicted targets; the general idea is that mutations in the seed match of a functional MRE will abolish the effects on the Firefly/Renilla activity ratio measured over-expressing or inhibiting the predicted microRNA.

When approaching the mutagenesis of MREs (either already validated with the assays described above or not) and the consequent experiments, a few key points have to be considered:

1. To exert the strongest effect on the MRE, mutate the seed match sequence, that is, the sequence on the target that perfectly pairs the microRNA seed (bases 2–7 on the miRNA 5′ end) (*see* **Note 8**). In most cases, three point mutations should suffice to disrupt a fully functional MRE; G:U wobble formation should be avoided, since such pairings have been shown to be at least partially functional [16]. MRE mutagenesis can be performed by using commercially available kits, such as the QuikChange II Site-directed Mutagenesis Kit (Agilent Technologies). It consists of a PCR performed on a plasmid that contains the fragment of interest (that is, the MRE or the whole 3′UTR) by using primers bearing the desired mutations. Despite the primer-target sequence mismatches, the PCR efficiently amplifies the sequence of interest, generating a product that contains the desired mutations. The kit manual gives precise directions on how to design the PCR primers to be used on the plasmid, as well as how to carry on the subsequent

steps. The entire fragment-containing plasmid is amplified in its mutated form; the PCR is not free of errors, despite the proofreading ability of the polymerase provided with the kit. Therefore, accurate sequencing of the resulting mutated fragment is strongly suggested in order to double-check the absence of undesired mutations that might influence the results. Moreover, to avoid sequencing the entire plasmid, it is advisable to sub-clone the mutated fragment into a non-PCR-amplified digested copy of the same backbone (using a standard cloning protocol).

2. Mutated MRE analysis: The same assays described in Subheadings 3.2.5 (over-expression of the predicted microRNA) and 3.2.8 (inhibition of the predicted microRNA) can be performed on the mutated plasmid. As discussed in Subheading 3.2.7, a functional MRE will determine a lower Firefly/Renilla activity ratio when a plasmid containing its wild-type sequence is co-transfected with a microRNA mimic compared to the same plasmid co-transfected with a negative control mimic; the mutated version of the same MRE should instead show a comparable Firefly/Renilla activity ratio in the cells co-transfected with the specific miRNA mimic and in those co-transfected with the negative control mimic. In other words, the mutagenesis should abolish the difference seen using the wild-type construct.
3. The same reasoning holds true for the inhibition assay: the functional MRE should determine a higher Firefly/Renilla activity ratio upon specific LNA co-transfection compared to the negative control LNA, while a mutated MRE should not. In addition, in a cell line that expresses appropriate amounts of the microRNA of interest, it is possible to directly compare the Firefly/Renilla activity ratio among the wild-type plasmid and the mutated one; a functional MRE will determine a higher Firefly/Renilla ratio when mutated rather than in its wild-type form, since in the latter case the Firefly Luciferase expression is inhibited by the endogenous microRNA (*see* **Note 9**).

3.3 Validation of Predicted PTEN-Targeting MicroRNAs by Measurement of PTEN mRNA and Protein Levels

The Luciferase assays described above are useful tools to validate the physical microRNA-target interaction, but they rely on a heterologous system in which *PTEN* 3′UTR is cloned downstream a reporter gene. The end point of PTEN-regulating microRNA identification is to measure their effects on the endogenous PTEN, measurable at the mRNA and protein levels. Moreover, a useful readout of microRNA regulation of PTEN is the phosphorylation of AKT, which is inversely correlated with PTEN activity.

The following key steps have to be undertaken in order to perform these validation experiments.

3.3.1 Measurement of PTEN mRNA Levels

Most microRNAs are capable of inhibiting gene expression by acting both at the RNA and at the protein level [17]. Thus, it is possible to measure the activity of a microRNA by measuring the target mRNA expression.

In general terms, the over-expression of a predicted microRNA (or a microRNA that has been already validated by means of Luciferase assays) should result in a lower level of target mRNA, as measured by real-time PCR. On the contrary, the microRNA inhibition should result in higher mRNA target levels (*see* **Note 10**).

A cell line with suitable *PTEN* levels should be chosen, that is, *PTEN* mRNA should be amplifiable in a range that is well within the upper and lower range limits of the real-time machine to be used. Real-time Ct values higher than 30 (that is, low *PTEN* mRNA levels) should be avoided especially when over-expressing the microRNA, since low PTEN expression will diminish the effect of further PTEN inhibition. On the contrary, real-time Ct values lower than 20 (that is, very high *PTEN* mRNA levels) should be avoided when inhibiting the microRNA; moreover, as discussed in Subheading 3.2.8, the microRNA inhibition has to be carried out in a cell line that expresses appropriate levels of the microRNA of interest. The DU 145 prostate cancer cell line has been successfully used to measure the effects of PTEN-targeting miRNAs on *PTEN* mRNA, both using miRNA mimics and miRNA inhibitors [15]; we grow them in Minimum Essential Medium Eagle (EMEM) supplemented with 10 % FBS, glutamine, and Pen-Strep. A PTEN-targeting siRNA should be used as positive control, since it is capable of targeting PTEN, decreasing both its mRNA and protein levels. We had success in using a siPTEN purchased from Dharmacon [15].

A standard transfection protocol is used to deliver either the microRNA mimic or the LNA inhibitor into the cells (for details on mimics and inhibitors, refer to Subheadings 3.2.5 and 3.2.8, respectively):

1. The day before transfection, seed 1.5×10^5 DU 145 cells in each well of a 12-well plate, in complete EMEM medium; seed one well for each microRNA to be tested, plus two wells for controls (one for the negative control mimic or LNA and the other for the siPTEN).
2. 24 h later, for each well prepare two mixes: (a) 6 μl of Lipofectamine 2000 + 100 μl of Opti-mem medium. (b) miRNA mimic (or LNA inhibitor) in order to reach a final concentration of 60 nM (e.g., 3 μl of 20 μM microRNA mimic or LNA inhibitor) + 100 μl of complete Opti-mem medium.
3. Mix A and B by carefully pipetting up and down and wait for 20 min.
4. In the meantime, remove the medium from the seeded cells and add 800 μl of Opti-mem medium to each well.

5. Add the mix to the cells.
6. After 6 h, replace the medium with EMEM complete medium and continue incubation.

3.3.2 Real-Time PCR

To perform the real-time PCR, 24 h after transfection (*see* **Note 11**), collect the cells and proceed with the following steps:

1. Extract RNA from the cell pellet using Trizol (Life Technologies), following the manufacturer's instructions. Add 1 μl of 20 mg/ml nuclease-free glycogen to each sample after recovering the cleared supernatant and before adding the isopropanol, to increase RNA yield. At the end of the procedure, resuspend the RNA pellet in no more than 30 μl of nuclease-free water, in order to avoid excessive dilution of the RNA.
2. Quantify the RNA using a spectrophotometer (we use the Nanodrop Lite from Thermo Scientific). Take note of the 260/280 ratio, which ideally should be comprised within the 1.8–2.1 range.
3. To get rid of possible genomic DNA contaminations, treat the RNA with DNase (*see* **Note 12**). We usually treat 1 μg of total RNA with DNase I, Amplification Grade (Life Technologies), following the manufacturer's instructions.
4. Retrotranscribe the RNA with a suitable kit; we use the iScript Reverse Transcription Supermix (Bio-rad), using 500 ng of DNase-treated RNA in a 20 μl reaction.
5. Perform the real-time PCR; we use the Sso Advanced SYBR Green Supermix (Bio-rad), with 1 μl of a 1:5 dilution of the cDNA for every 15 μl reaction and 1 μl of 10 μM forward and reverse primer. We use three housekeeping genes as reference (GAPDH, PBGD, and SDHA, see primer sequences in Subheading 2.3); perform each reaction in duplicate or triplicate. We run the reaction on a CFX96 Touch Real-time System (Bio-rad). Data within each sample is normalized to the three housekeeping genes using the CFX manager software (Bio-rad) and each sample is then normalized on the negative control mimic sample, which is set to 1. The siPTEN is expected to exert the highest decrease in *PTEN* mRNA levels; the functional microRNAs are those that will show a *PTEN* mRNA level < 1 and that will be statistically significant by *t*-test analysis (we usually analyze the results of three independent experiments) (*see* **Note 13**).

3.3.3 Measurement of PTEN Protein Levels

PTEN protein levels are affected by PTEN-regulating microRNAs [18]; a functional microRNA should lower the PTEN protein level when over-expressed and increase it when inhibited.

In regard to the cell line and negative/positive control choices, as well as for the transfection protocol, please refer to Subheading 3.3.1 above.

1. To measure PTEN protein levels, collect cells 48 h after transfection (*see* **Note 11**) and extract proteins by cell lysis with an appropriate buffer.
2. Follow a standard western blot protocol [18] using a PTEN antibody (we have successfully used the PTEN #9559 antibody from Cell Signaling Technology; we dilute it 1:1000). We suggest using a 10 % SDS–polyacrylamide gel and loading ~30 μg of protein extract. As a loading control, strip the membrane by placing it in warm water (70–80 °C) for 30 min and then incubate it with a control antibody, such as one for tubulin or actin. Similarly to the *PTEN* mRNA assay, the siPTEN is expected to exert the highest decrease in PTEN protein levels; the functional microRNAs are those that will show a PTEN protein level lower than that of the negative control mimic when over-expressed; on the contrary, they should show a higher PTEN protein level when inhibited with LNAs, compared to the negative control LNA.
3. To further explore the effect of PTEN-regulating miRNAs, the western blot membrane can be stripped and incubated to measure pAKT levels (we use the pAKT (Ser^{473}) #9271 antibody from Cell Signaling Technology, diluted 1:1000). In fact, PTEN levels are inversely correlated with pAKT levels [4]. A higher pAKT level should be observed when over-expressing the microRNA, while a lower pAKT level should be obtained in the samples in which the microRNA has been inhibited.
4. For an additional and elegant loading control, the membrane can be stripped again and incubated with a total AKT antibody (such as the AKT #9272 from Cell Signaling Technology, diluted 1:1000).

3.4 Validation of Predicted PTEN-Targeting MicroRNAs by Functional Assays

Functional effects of the downregulation of PTEN by a signature of miRNAs can be analyzed exploiting cellular assays such as growth curves, colony formation efficiency, and soft agar, each of which tests the tumorigenic features of the cells. Growth curves allow monitoring of cell proliferation over time; on the other hand, the colony efficiency and soft agar assays test the ability of tumor cells to form colonies in 2-D and 3-D, respectively.

Since PTEN has a tumor-suppressive role, the over-expression of a PTEN-targeting miRNA should lead to a faster growth (as measured by growth curve assay) and a higher number of colonies in 2-D and 3-D (as measured by colony efficiency and soft agar assays, respectively). The opposite result is expected by transfecting the cells with a microRNA inhibitor (slower growth and less colonies formed).

3.4.1 Transfecting Cells for Functional Assays

A single transfection can be performed for all three cellular assays; after transfection, the cells are in fact detached and seeded for growth curve, colony efficiency, and soft agar assays:

1. The day before transfection, seed 2×10^5 DU 145 cells in a 6-well plate, to reach 80–90 % confluence the day after. Seed a well for the negative control mimic (or control LNA inhibitor) and one for the positive control (siPTEN).
2. 24 h later, for each well prepare two mixes: (a) 10 μl of Lipofectamine 2000 + 250 μl Opti-mem medium. (b) miRNA mimic or LNA inhibitor (at 60 nM final concentration; e.g., 6 μl of 20 μM miRNA mimic/LNA inhibitor) + 250 μl of Opti-mem medium.
3. Mix A and B by carefully pipetting up and down and wait for 20 min.
4. In the meantime, replace cell medium with 1.5 ml of Opti-mem.
5. Add the mix to the cells.
6. After 6 h, detach cells and seed them according to the paragraphs below, depending on which assay is to be performed (*see* **Note 14**).

3.4.2 Growth Curve Assay

1. The day of the transfection, prepare as many 12-well plates as needed, considering three wells per sample for each time point (usually T1 to T4). Consider also that it is crucial for samples belonging to different time points to be seeded on different multi-well plates, since they will be fixed separately.
2. After detaching the cells, make from each sample a dilution of 3×10^3 cells/ml. Make a total volume of ~14 ml.
3. Seed 1 ml (3×10^3 cells) per well in triplicate, and do so for each time point (T1 to T4, *see* **Note 15**).
4. Starting the day after the transfection, fix a time point every other day: first, aspirate medium and wash once with PBS. Then, add 4 % PFA (usually 1 ml for each well is more than enough) and incubate for 10 min at room temperature. Lastly, remove PFA and store in PBS at 4 °C, carefully sealing the plate with parafilm to prevent evaporation. Repeat the same process for each of the four time points.
5. Once all time points have been fixed, stain all plates together: first, aspirate the PBS and add 1 ml of crystal violet solution (0.1 % crystal violet, 20 % methanol, in water). Leave the plates rocking on a plate at 20–50 rpm for 15 min; then, aspirate the crystal violet solution and wash with abundant tap water. Let the plates dry afterwards (usually overnight, uncapped and upside down).

6. Destain the plates by adding ~0.5 ml of 10 % acetic acid per well and leaving them rocking at 20–50 rpm for 15 min. Afterwards, transfer 150 μl of each sample in duplicate to a 96-well plate. Read the absorbance at 590 nm on a spectrophotometer (*see* **Note 16**).
7. Each sample should be normalized on its T1; the data can be showed as variation of cell percentage over time. The results from three independent experiments can be subjected to statistical analysis.
8. As stated above, the expected cell growth should be higher compared to the control when transfecting the miRNA mimic, and lower than the control when using the LNA inhibitor. The highest growth should be visible in the siPTEN-transfected sample.

3.4.3 Colony Efficiency Assay

1. The day of the transfection, prepare three 60 mm plates for each sample.
2. Dilute cells to 40 cells/ml (make ~17 ml for each sample).
3. Seed 5 ml (corresponding to 200 cells) in each 60 mm plate, in triplicate. Cells should be distributed evenly to avoid the growth of clusters of nearby colonies.
4. Place the cells in the incubator and do not disturb them before 7 days from the seeding.
5. Take a look at the cells after ~7 days; when round colonies are visible (usually after 7–10 days, *see* **Note 17**), fix and stain the cells by adding 2.5 ml of a crystal violet solution (0.2 % crystal violet, 4 % formaldehyde, in water) directly into the plates. Let the plates stand at room temperature for ~90 min.
6. Remove the solution and wash with abundant tap water.
7. Count the colonies and normalize the data on the number of colonies obtained with the negative control. The average colony number ± SEM for each sample can be expressed on a bar graph as fold change variation compared to the control sample. The highest number of colonies should be observed in the siPTEN sample; the miRNA mimic-transfected samples should show more colonies that the negative control mimic sample, while the samples transfected with the LNA inhibitor should yield less colonies than the LNA control sample.

3.4.4 Soft Agar Assay

In all the following steps, try to work in conditions as sterile as possible:

1. The day before transfection, prepare the soft agar base directly in 6-well plates (*see* **Note 18**); make three wells for each sample. First, for each 6-well plate to be used (scale-up if more are

needed) prepare a 50 ml tube with 18.5 ml of non-complete EMEM and warm it to 37 °C by placing it in a water bath (this is crucial, since a medium not fully warmed will cause the melted agar to solidify very rapidly, making it impossible to dispense properly). Then, prepare 250 ml (*see* **Note 19**) of a 6 % agar solution in PBS 1× by melting the agar/PBS mixture in a microwave oven, being careful not to cap the bottle tightly. Stop the microwave and shake the mixture every minute or so, to facilitate melting.

2. Add 2 ml of 6 % agarose solution to the warm EMEM (if 18.5 ml were prepared; if more, scale-up the agarose volume accordingly). Mix quickly using a 25 ml pipette; work rapidly, to avoid agar solidification.
3. Using the same 25 ml pipette, dispense 3 ml of EMEM/agar solution in each well, making triplicates.
4. Incubate the plate overnight in the cell incubator at 37 °C to allow agar solidification.
5. The day of the transfection, make a 4×10^4 cell dilution (7.2 ml of total volume) in a 50 ml tube.
6. Equilibrate the suspension at 37 °C in a water bath.
7. Prepare a 3 % agarose solution in PBS 1× by melting the agarose in a microwave.
8. Add 1 ml of the agarose solution to the cell suspension. Mix gently, but quickly.
9. Add 2 ml of the mixture to each well containing the agarose base prepared the day before.
10. Shake the plate delicately to obtain a homogeneous cell distribution.
11. Incubate the plates on a flat surface at 4 °C for 1 h to rapidly solidify the agar.
12. Place in a cell incubator at 37 °C.
13. After 10–15 days, count five microscope fields with a 10× objective for each well. Normalize the data on the number of colonies obtained in the negative control (*see* **Note 20**). The average colony number ± SEM of each sample can be normalized on that of the negative control sample; data can be represented on a bar graph as fold change variations compared to the control sample. The highest number of colonies formed should be observed in the positive control sample (siPTEN); the samples transfected with the miRNA mimic should show more colonies than the negative control mimic sample; instead, the samples transfected with the LNA inhibitor should have less colonies than the LNA control sample.

3.5 Considerations on MicroRNA Cooperation

A single microRNA can potentially bind many targets [5]; on the other hand, a single target can be bound by more than one microRNA at the same time. Keeping this in mind, we suggest using mixes of microRNAs in order to measure their cumulative effect on PTEN; all the assays described in the previous chapter can benefit of the "microRNA-mix effect," although particular attention must be paid to the following points.

3.5.1 Cell Line

As stated above, the cell line of interest should express low levels of the microRNA of interest when the over-expression with the miRNA mimic is performed, while it should express medium-to-high levels of the same miRNA when the inhibition with the LNA is performed. When working with multiple microRNAs, it clearly becomes harder to find a cell line that expresses a suitable amount of all of them. For the over-expression experiments, we still suggest using the HCT116 Dicer −/− cell line; for the inhibition experiments, a cell line that expresses the microRNA signature of interest must be chosen, by carefully determining the expression of every microRNA to be inhibited.

3.5.2 Transfection

Follow the same transfection protocols described above, depending on the assay chosen; the only modification is that multiple mimics (or inhibitors) will have to be added to the "B" mix. We recommend not exceeding 100 nM final concentration of mimic (or inhibitor) mix, to avoid cell toxicity. Depending on the cell line, the tolerance to high doses of transfected oligos varies, so we recommend performing a dose–response curve using different doses of the control mimic/LNA inhibitor and monitoring cell survival.

3.5.3 Luciferase Assays

To test multiple miRNAs by means of Luciferase assays, all their predicted MREs need to be cloned downstream of Firefly Luciferase. We strongly recommend against multiple cloning of stand-alone MREs, while cloning of the entire *PTEN* fragment containing all the MREs of interest is preferred. This should leave intact the distance and relative position of the predicted MREs, allowing for better testing of a possible cooperation between the microRNAs that bind to them.

When performing a mutagenesis assay, sequentially mutate the single MREs on the *PTEN* fragment of interest, carefully sequencing the resulting constructs at every step. As suggested in Subheading 3.2.9, sub-clone the final mutated fragment into a non-PCR-amplified digested copy of the same backbone.

3.5.4 Analysis of the Results

The activity of multiple microRNAs on PTEN should be stronger than that of single microRNAs. Compared to a single microRNA over-expression, the use of a mix of functional microRNAs should determine:

1. Lower Firefly/Renilla Luciferase activity ratio in the Luciferase assay, when using the wild-type construct.
2. Lower *PTEN* mRNA level as measured by real-time PCR.
3. Lower PTEN protein level as measured by western blot, as well as higher pAKT levels.
4. Higher cell growth in the growth curve assay.
5. Higher number of colonies in 2-D and 3-D, when performing colony efficiency assay and soft agar assay, respectively.

Not all the microRNAs that bind *PTEN* mRNA will necessarily have the same effect on PTEN expression and function in all cell lines and/or conditions, so the possibility that some miRNAs will not enhance the effects seen on PTEN when added to a mix must be taken into account. Regardless, the study of multiple microRNAs at the same time is strongly suggested and constitutes a rationale for using more high-throughput techniques that can give a complete picture of all the microRNAs binding to PTEN.

3.6 High-Throughput Methods for the Identification of PTEN-Targeting MicroRNAs

The experimental methods to identify miRNA-mRNA interactions described above are often applied to just one microRNA at a time. These techniques are unable to reveal all the miRNAs that can be associated with the *PTEN* transcript, because they miss interactions that for whatever reason do not "follow the rules" and thus are not predicted by algorithms. Moreover, as discussed in Subheading 3.1, computational analyses of miRNA-mRNA binding often yield high false-positive rates (25–70 %), even when merging the results of several algorithms [19].

Recently, two new affinity techniques, the MS2-TRAP [20] and a biotinylated antisense oligonucleotide capture affinity technique [21], have overcome these limitations, giving the chance to study all the microRNAs that bind to a specific mRNA. Both methods are based on the affinity-precipitation of the transcript of interest and its bound miRNAs using conjugated beads (see details below). After RNA extraction, the bound miRNAs can be identified by next-generation sequencing (NGS).

3.6.1 MS2-Tagged RNA Affinity Purification

The MS2-tagged RNA affinity purification (MS2-TRAP) method (also known as pull-down assay or RNA immunoprecipitation (RIP) assay) is based on the addition of several MS2 RNA tags (usually 12–24 copies) to the 3′UTR of interest; these tags are 19-nucleotide sequences with a hairpin-loop structure that are bound with high affinity and specificity by the MS2-binding protein [22]. For a detailed protocol, *see* ref. 20; in synthesis, these are the main passages (Fig. 1):

1. The 3′UTR of interest is cloned into a suitable plasmid that bears 12–24 MS2 RNA tags; Yoon and colleagues suggest cloning the 3′UTR upstream the MS2 RNA tags, to avoid the possible translation of the tags.

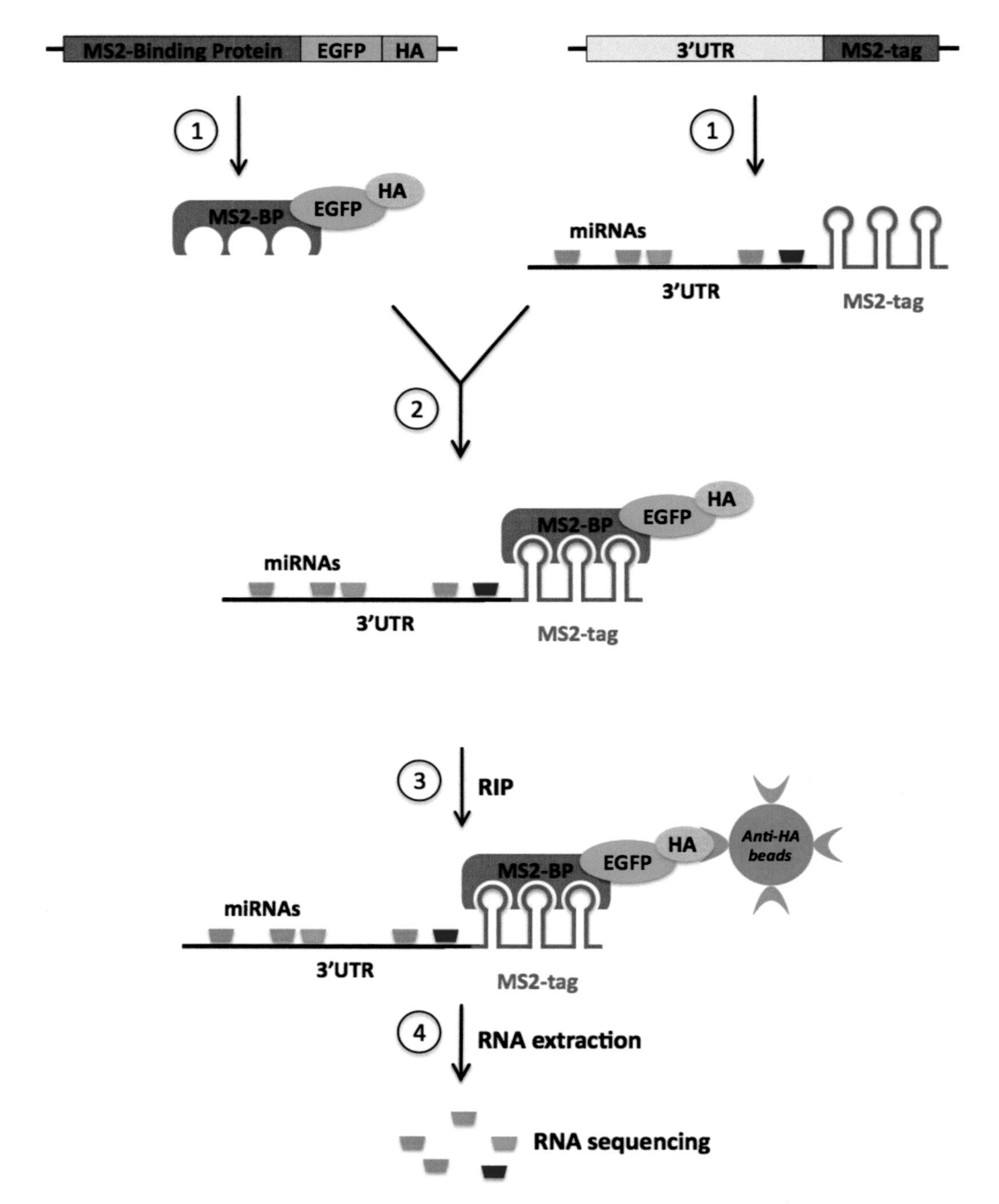

Fig. 1 MS2-tagged RNA affinity purification. (1) The vectors expressing the MS2-binding protein/EGFP-HA chimerical protein and the 3′UTR-MS2 RNA tag transcript are co-transfected in mammalian cells. (2) Inside the cells, the MS2-binding protein binds the MS2-hairpin motifs, and the "3′UTR-MS2 RNA tags :: MS2-binding protein-EGFP-HA" complex is formed. (3) The "3′UTR-MS2 RNA tags :: MS2-binding protein-EGFP-HA" complexes are immunoprecipitated using anti-HA beads. RIP: RNA immunoprecipitation. (4) After some washes, the beads are recovered and treated with DNase and proteinase K. The RNA is subsequently extracted for further analysis

2. The plasmid that expresses the "3′UTR of interest-MS2 RNA tags" transcript is transfected into a cell line of choice along with the plasmid that expresses the MS2-binding protein. Usually, the MS2-binding protein is in frame with a fluorescent reporter used to assess transfection efficiency (e.g., EGFP) and a tag used for immunoprecipitation (e.g., HA).
3. Inside the cells, the MS2 binding protein binds the "3′UTR-MS2 RNA tags" transcript, forming a stable complex.
4. After 24–48 h from transfection, cells are harvested and then lysed with a buffer that maintains the ribonucleoproteins (RNPs) intact.
5. The resulting lysate is used in an affinity precipitation using the anti-EGFP or the anti-HA antibody conjugated with sepharose beads; since the MS2-binding protein is fused with the tag that is used for the immunoprecipitation and at the same time binds to the RNA of interest (through the MS2-binding tags), the RNA of interest is enriched along with the beads/HA/MS2 protein complexes.
6. After some washes, the complexes associated to the beads are eluted and recovered. Subsequently, they are treated with DNase and proteinase K.
7. Finally, RNA is extracted and can be then tested by NGS: the microRNAs bound to the RNA of interest are identified as those that exhibit enrichment compared to the control sample (e.g., a sample immunoprecipitated with a control antibody, such as mouse IgG).

This method has been recently adapted to *PTEN* in two different works [23, 24]. Tay et al. used an MS2-tagged *PTEN* 3′UTR to confirm that some of the published *PTEN*-targeting miRNAs are indeed bound to *PTEN* mRNA in the DU 145 cell line [23]. Karreth et al. used the same assay as a tool to prove that *ZEB2* can act as a *PTEN* ceRNA via miRNA sequestration [24]. The authors in fact demonstrated that a pool of miRNAs can bind to both *PTEN* and *ZEB2* 3′UTRs. They also performed the RIP of *ZEB2* 3′UTR after PTEN silencing and vice versa (the RIP of *PTEN* 3′UTR after ZEB2 silencing); in both silencing conditions they observed an increased binding of miR-92a to the non-silenced transcript due to the absence of the other competitive transcript.

3.6.2 RNA Affinity Purification Using a Biotinylated Antisense Oligonucleotide

Hassan et al. (2013) have developed a technique that can be used to perform a high-throughput identification of PTEN microRNAs; in this assay, a short (20 nt) biotinylated antisense oligonucleotide complementary to the target mRNA is used for the affinity separation of the miRNAs bound to the target mRNA. The advantage of this new method when compared to the MS2-TRAP is the possibility to directly assess the binding of endogenous miRNAs without perturbing the system by over-expressing the target RNA. Moreover, since it does not rely on fusion constructs, this method preserves the entire transcript structure, with all its secondary and tertiary motifs.

The biotinylated antisense oligonucleotide capture affinity technique consists of the following steps (Fig. 2) (for a more detailed protocol, *see* ref. 21):

1. The secondary structure of the target mRNA is predicted using bioinformatic tools such as M-Fold, in order to identify an exposed single-strand region of the secondary structure that can be easily accessed by a biotinylated probe that is around 20 nt long.
2. After trypsinization, cells are treated with formaldehyde to cross-link the mRNA:miRNA:RISC ternary complexes. The cross-linking reaction is then stopped by adding glycine and the cells are collected and lysed.
3. The supernatant containing the mRNA:miRNA:RISC ternary complexes is treated with DNase and incubated with streptavidin beads conjugated with the previously designed biotinylated probe. The affinity purification is then carried on.
4. The biotin complexes are eluted from the beads and recovered.
5. The cross-linking is reversed by heating the sample and the RNPs are digested using proteinase K.
6. At this stage, the RNA can be extracted and used for further studies, such as the identification of microRNAs by NGS; those microRNAs that will result enriched compared to a sample treated with a control biotinylated probe should be considered as target-binding and should be subjected to additional validation studies.

4 Notes

1. A single-enzyme cloning protocol might prove tricky, since a higher number of colonies containing a self-ligated vector is to be expected (resulting in higher background, hence in higher false-positive rates); moreover, the fragment of interest could be inserted in both orientations, therefore a careful clone screening is required.

   As an alternative, a multi-cloning site can be cloned into the XbaI site, facilitating the subsequent cloning of *PTEN*-related fragments of interest (*see* ref. 18). Commercially available vectors, such as the widely used pMIR-REPORT (Promega), already bear a multi-cloning site downstream Luciferase, after the stop codon.

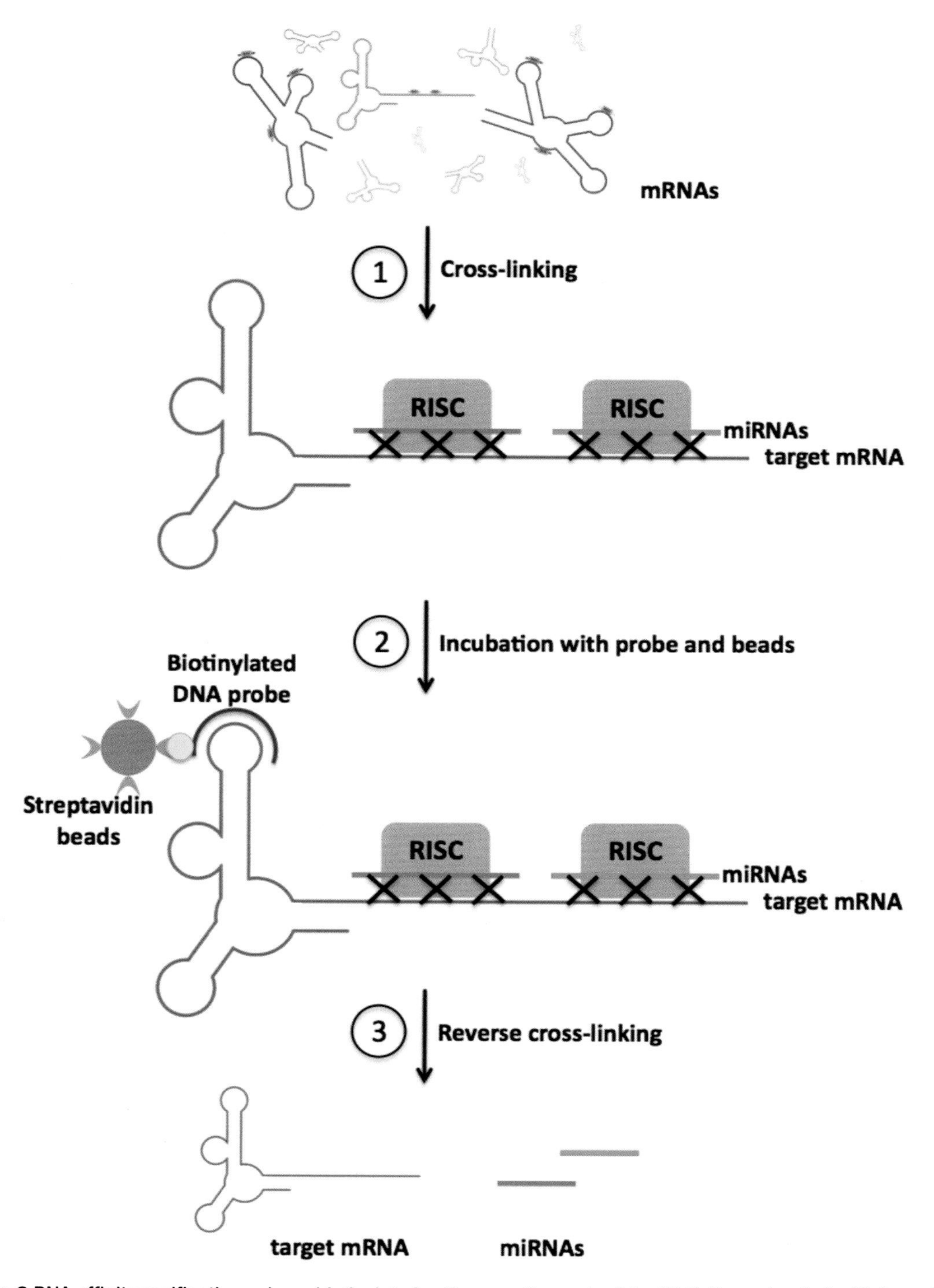

Fig. 2 RNA affinity purification using a biotinylated antisense oligonucleotide. (1) Cells are treated with formaldehyde to cross-link the miRNA:RISC complexes with the target mRNA and then lysed. (2) The lysate is treated with DNase and incubated with streptavidin beads bound with the biotinylated DNA probe of choice. (3) After some washes, the biotin complexes are eluted from the beads and recovered. Subsequently, the formaldehyde cross-linking is reversed and the samples are treated with proteinase K. At this point, RNA can be extracted and subjected to further analyses

2. Vectors such as the pMIR-GLO (Promega) encode for both Firefly and Renilla Luciferase, eliminating the need of co-transfecting a second plasmid.

   The psiCHECK-2 Vector (Promega) is another option for concomitant expression of both Firefly and Renilla Luciferase. In this case it is the Renilla Luciferase that is influenced by the microRNA expression, while the Firefly serves as a normalization control.

3. MicroRNAs are capable of binding MREs located in the 5′UTR and ORF regions of mRNAs [25, 26]. In order to study 5′UTR or ORF-targeting miRNAs, it is the 5′UTR or ORF sequence of *PTEN* that needs to be cloned downstream the Luciferase. To this end, the same protocol used for the 3′UTR can be followed. However, if analyzing the PTEN ORF, it is advisable to either mutate the ATG start codon or to clone just the region of interest (that is, the MRE of interest). In this way, possible secondary effects caused by PTEN protein over-expression (e.g., feedback loops) can be avoided.

4. The thermodynamic stability of the microRNA mimic ends determines which strand will be preferentially loaded into RISC [27]. The strand with lower 5′ end stability will in fact be favored in most cases.

   To ensure that the strand with the same sequence as the microRNA of interest is preferentially loaded into RISC (we call it the "sense strand"), it is advisable to decrease the thermodynamic stability of its 5′ end by inserting mutations in the 3′ end of the complementary strand (the "antisense strand").

   In spite of this, since strand selection by RISC is not a strict process and depends on a number of factors, many of which are still unknown [27], both strands might still be incorporated. We therefore suggest verifying the proper "sense strand" over-expression by real-time PCR, following a microRNA-oriented protocol (e.g., retrotranscription of the transfected cells RNA with miScript II RT Kit (Qiagen) followed by real-time PCR with appropriate primers).

   As an alternative, several companies provide ready-to-use microRNA mimics that should ensure the over-expression of the strand of interest (e.g., Dharmacon's miRIDIAN microRNA mimics).

5. A dose–response curve might be required to optimize the amount of both the plasmids and the miRNA mimic to be used.

6. HTC116 Dicer −/− cells are kept in their growth medium for the entire duration of the experiment because in our experience they do not tolerate Opti-mem medium. We indeed noticed a very high cell mortality rate when using Opti-mem, which brought us to switch to McCoy's complete medium.

Most cell lines, such as DU 145, are instead transfected in Opti-mem.

7. Finding a cell line that expresses a certain microRNA of interest might prove to be a daunting task. For "hard-to-find" microRNAs, we advise to over-express the miRNA by using a mimic, as described in Subheading 3.2.5, and concomitantly transfect an LNA microRNA inhibitor designed for the same miRNA or a control LNA. If the Firefly/Renilla ratio is higher in the mimic/specific LNA transfection compared to the mimic/control LNA transfection, then the effect seen on the Luciferase is indeed dependent on the microRNA.

   When co-transfecting a miRNA mimic and a LNA, we have successfully used the same amounts that are used for the transfection of either the mimic or the LNA alone (that is, 15 nM of mimic and 15 nM of LNA).

8. If the predicted MRE does not perfectly pair with the microRNA seed, we still suggest to mutate those bases that do pair. Since extensive pairing between the microRNA 3′end and the target might compensate for mismatches in the microRNA seed-seed match pairing [28], we also suggest to insert additional mutations in that region.

9. In the experiments performed with the mutated MREs, the amount of miRNA mimic to be transfected has to be thoroughly assessed by means of a dose–response curve. Excessive concentration of the miRNA of interest might in fact generate off-target effects that can abolish differences between wild-type and mutated MREs.

10. Depending on the relative microRNA-target expression levels, it is not always possible to measure opposite effects with the over-expression versus inhibition of the microRNA using the same cell line. As an example, if the microRNA is expressed in large excess compared to the target, its over-expression will not probably lead to a further decrease in the target mRNA/protein levels, since all the MREs on the target are already occupied by the endogenous miRNA. Analogously, very low levels of endogenous microRNA are not suitable for inhibition studies, since the microRNA targets are most likely all de-repressed.

    In summary, it is crucial to perform the desired experiment in the right cell line, usually one that expresses high levels of miRNA for the inhibition experiments and one that expresses low levels of miRNA for the over-expression experiments.

11. A time-course experiment might be needed to establish the time point at which the mRNA or protein decrease/increase is most prominent. It should also be taken into account the fact that the time point of maximum decrease obtained with the

miRNA mimic might not necessarily coincide with the time point of maximum increase with the LNA inhibitor. Moreover, most likely the time point for maximum decrease/increase of the mRNA will be earlier than that in which the maximum decrease/increase of protein level is observed.

12. Ideally, all real-time primers should be designed in order to yield a PCR product on cDNA but not on genomic DNA, that is, primers should be placed in different exons. Since for some genes this is not possible and since it is hard to predict which genes will be needed to be amplified in the future starting from a given cDNA sample, we suggest treating the RNA with DNase by default.
13. Although the majority of the microRNAs are known to cause a decrease in the target mRNA levels [17], this does not apply to all of them. If a decreased/increased mRNA level is not detected with the mimic/inhibitor transfection, it is still worth trying measuring the protein level.
14. Since most of the cells will be used to seed the three assays, it is advisable to transfect two wells for each sample in parallel. After detaching the cells, these two wells can be merged in order to have enough cells to (1) perform the functional assays described; (2) make a pellet of around 3×10^5 cells to use for RNA extraction, in order to confirm the increase/decrease in *PTEN* mRNA levels. In alternative, proteins can be extracted to confirm the increase/decrease of PTEN protein levels.
15. We usually fix the cells at four different time points, every other day starting from the day after the transfection. Additional time points can be programmed, but the cells should not be over-confluent by the last time point (eventually adjust the number of cell seeded).
16. If the number of cells is very high, the solution will be extremely dark, and the instrument might return an error message. Make sure that the absorbance level falls within the instrument acceptable range; otherwise, dilute the samples accordingly using the acetic acid solution.
17. To establish when to fix the cells, use the negative control as reference: the colonies diameter should more or less fill the field seen with a 10× objective when cells are observed under a microscope. Avoid over-growing the colonies to the point that they start fusing with each other.
18. The 6 % agarose base can be stored for future use at 4 °C for 2–3 days.
19. We suggest preparing a large amount of 6 % agar solution (that is, 250 ml) to facilitate agarose melting; we indeed have had difficulties in melting smaller volumes.

20. Since the colonies formed in soft agar develop in all three dimensions and might be located at different heights, move the focus of the microscope from one end to the other when looking at a given field. Count a colony as "one" only when it clearly grew in an evident three-dimensional structure. We suggest not considering colonies that are not three-dimensional.

References

1. Bartel DP (2009) MicroRNAs: target recognition and regulatory functions. Cell 136:215–233. doi:10.1016/j.cell.2009.01.002
2. Bartel DP (2004) MicroRNAs: genomics, biogenesis, mechanism, and function. Cell 116:281–297
3. Jansson MD, Lund AH (2012) MicroRNA and cancer. Mol Oncol 6:590–610. doi:10.1016/j.molonc.2012.09.006
4. Song MS, Salmena L, Pandolfi PP (2012) Nature reviews. Mol Cell Biol 13:283–296. doi:10.1038/nrm3330
5. Lewis BP, Burge CB, Bartel DP (2005) Conserved seed pairing, often flanked by adenosines, indicates that thousands of human genes are MicroRNA targets. Cell 120:15–20. doi:10.1016/j.cell.2004.12.035
6. Kruger J, Rehmsmeier M (2006) RNAhybrid: microRNA target prediction easy, fast and flexible. Nucleic Acids Res 34:W451–W454. doi:10.1093/nar/gkl243
7. Betel D, Koppal A, Agius P, Sander C, Leslie C (2010) Comprehensive modeling of microRNA targets predicts functional non-conserved and non-canonical sites. Genome Biol 11:R90. doi:10.1186/gb-2010-11-8-r90
8. Dweep H, Sticht C, Pandey P, Gretz N (2011) miRWalk--database: prediction of possible miRNA binding sites by "walking" the genes of three genomes. J Biomed Inform 44:839–847. doi:10.1016/j.jbi.2011.05.002
9. Reczko M, Maragkakis M, Alexiou P, Grosse I, Hatzigeorgiou AG (2012) Functional microRNA targets in protein coding sequences. Bioinformatics 28:771–776. doi:10.1093/bioinformatics/bts043
10. Rehmsmeier M, Steffen P, Hochsmann M, Giegerich R (2004) Fast and effective prediction of microRNA/target duplexes. RNA 10:1507–1517. doi:10.1261/rna.5248604
11. Hsu SD, Tseng YT, Shrestha S, Lin YL, Khaleel A, Chou CH, Chu CF, Huang HY, Lin CM, Ho SY, Jian TY, Lin FM, Chang TH, Weng SL, Liao KW, Liao IE, Liu CC, Huang HD (2014) miRTarBase update 2014: an information resource for experimentally validated miRNA-target interactions. Nucleic Acids Res 42:D78–D85. doi:10.1093/nar/gkt1266
12. Kozomara A, Griffiths-Jones S (2014) miRBase: annotating high confidence microRNAs using deep sequencing data. Nucleic Acids Res 42:D68–D73. doi:10.1093/nar/gkt1181
13. Landgraf P, Rusu M, Sheridan R, Sewer A, Iovino N, Aravin A, Pfeffer S, Rice A, Kamphorst AO, Landthaler M, Lin C, Socci ND, Hermida L, Fulci V, Chiaretti S, Foa R, Schliwka J, Fuchs U, Novosel A, Muller RU, Schermer B, Bissels U, Inman J, Phan Q, Chien M, Weir DB, Choksi R, De Vita G, Frezzetti D, Trompeter HI, Hornung V, Teng G, Hartmann G, Palkovits M, Di Lauro R, Wernet P, Macino G, Rogler CE, Nagle JW, Ju J, Papavasiliou FN, Benzing T, Lichter P, Tam W, Brownstein MJ, Bosio A, Borkhardt A, Russo JJ, Sander C, Zavolan M, Tuschl T (2007) A mammalian microRNA expression atlas based on small RNA library sequencing. Cell 129:1401–1414. doi:10.1016/j.cell.2007.04.040
14. Bisognin A, Sales G, Coppe A, Bortoluzzi S, Romualdi C (2012) MAGIA^2: from miRNA and genes expression data integrative analysis to microRNA-transcription factor mixed regulatory circuits (2012 update). Nucleic Acids Res 40:W13–W21. doi:10.1093/nar/gks460
15. Poliseno L, Salmena L, Zhang J, Carver B, Haveman WJ, Pandolfi PP (2010) A coding-independent function of gene and pseudogene mRNAs regulates tumour biology. Nature 465:1033–1038. doi:10.1038/nature09144
16. Brodersen P, Voinnet O (2009) Nature reviews. Mol Cell Biol 10:141–148. doi:10.1038/nrm2619
17. Guo H, Ingolia NT, Weissman JS, Bartel DP (2010) Mammalian microRNAs predominantly act to decrease target mRNA levels. Nature 466:835–840. doi:10.1038/nature09267
18. Poliseno L, Salmena L, Riccardi L, Fornari A, Song MS, Hobbs RM, Sportoletti P, Varmeh S, Egia A, Fedele G, Rameh L, Loda M, Pandolfi

PP (2010) Identification of the miR-106b~25 microRNA cluster as a proto-oncogenic PTEN-targeting intron that cooperates with its host gene MCM7 in transformation. Sci Signal 3:ra29. doi:10.1126/scisignal.2000594

19. Sethupathy P, Megraw M, Hatzigeorgiou AG (2006) A guide through present computational approaches for the identification of mammalian microRNA targets. Nat Methods 3:881–886. doi:10.1038/nmeth954
20. Yoon JH, Srikantan S, Gorospe M (2012) MS2-TRAP (MS2-tagged RNA affinity purification): tagging RNA to identify associated miRNAs. Methods 58:81–87. doi:10.1016/j.ymeth.2012.07.004
21. Hassan T, Smith SG, Gaughan K, Oglesby IK, O'Neill S, McElvaney NG, Greene CM (2013) Isolation and identification of cell-specific microRNAs targeting a messenger RNA using a biotinylated anti-sense oligonucleotide capture affinity technique. Nucleic Acids Res 41:e71. doi:10.1093/nar/gks1466
22. Peabody DS (1993) The RNA binding site of bacteriophage MS2 coat protein. EMBO J 12:595–600
23. Tay Y, Kats L, Salmena L, Weiss D, Tan SM, Ala U, Karreth F, Poliseno L, Provero P, Di Cunto F, Lieberman J, Rigoutsos I, Pandolfi PP (2011) Coding-independent regulation of the tumor suppressor PTEN by competing endogenous mRNAs. Cell 147:344–357. doi:10.1016/j.cell.2011.09.029
24. Karreth FA, Tay Y, Perna D, Ala U, Tan SM, Rust AG, DeNicola G, Webster KA, Weiss D, Perez-Mancera PA, Krauthammer M, Halaban R, Provero P, Adams DJ, Tuveson DA, Pandolfi PP (2011) In vivo identification of tumor- suppressive PTEN ceRNAs in an oncogenic BRAF-induced mouse model of melanoma. Cell 147:382–395. doi:10.1016/j.cell.2011.09.032
25. Lytle JR, Yario TA, Steitz JA (2007) Target mRNAs are repressed as efficiently by microRNA-binding sites in the 5' UTR as in the 3' UTR. Proc Natl Acad Sci U S A 104:9667–9672. doi:10.1073/pnas.0703820104
26. Qin W, Shi Y, Zhao B, Yao C, Jin L, Ma J, Jin Y (2010) miR-24 regulates apoptosis by targeting the open reading frame (ORF) region of FAF1 in cancer cells. PLoS One 5:e9429. doi:10.1371/journal.pone.0009429
27. Kim VN, Han J, Siomi MC (2009) Nature reviews. Mol Cell Biol 10:126–139. doi:10.1038/nrm2632
28. Brennecke J, Stark A, Russell RB, Cohen SM (2005) Principles of microRNA-target recognition. PLoS Biol 3:e85. doi:10.1371/journal.pbio.0030085

Chapter 10

Posttranscriptional Regulation of PTEN by Competing Endogenous RNAs

Yvonne Tay and Pier Paolo Pandolfi

Abstract

PTEN expression can be dysregulated in cancers via multiple mechanisms including genomic loss, epigenetic silencing, transcriptional repression, and posttranscriptional regulation by microRNAs. MicroRNAs are short, noncoding RNAs that regulate gene expression by binding to recognition sites on target transcripts. Recent studies have demonstrated that the competition for shared microRNAs between both protein-coding and noncoding transcripts represents an additional facet of gene regulation. Here, we describe in detail an integrated computational and experimental approach to identify and validate these competing endogenous RNA (ceRNA) interactions.

Key words MicroRNA, miRNA, Competing endogenous RNA, ceRNA, PTEN

1 Introduction

The tumor suppressor PTEN (phosphatase and tensin homolog deleted on chromosome 10) is one of the most frequently lost and mutated genes in a wide spectrum of human cancers [1]. *PTEN* is located on chromosome 10q23, a region that is susceptible to loss of heterozygosity [2–4], and disruptions in the *PTEN* locus have been described in multiple sporadic tumors such as brain, breast, and prostate cancers [5, 6]. PTEN encodes a plasma-membrane lipid phosphatase that functions as a negative regulator of the proto-oncogenic and pro-survival PI3K-AKT signaling pathway [6–8]. Additionally, PTEN may also dephosphorylate highly acidic protein and peptide substrates on phosphorylated serine, threonine, and tyrosine residues [9], as well as possess phosphatase-independent nuclear functions such as the maintenance of chromosomal integrity [10]. PTEN has been shown to play critical roles in a plethora of cellular processes including cell survival, proliferation, motility, senescence, and metabolism [5, 6, 8].

Leonardo Salmena and Vuk Stambolic (eds.), *PTEN: Methods and Protocols*, Methods in Molecular Biology, vol. 1388, DOI 10.1007/978-1-4939-3299-3_10,

Genetic deletions or mutations resulting in the heterozygous or homozygous loss of PTEN function are a widespread cause of PTEN dysregulation in human cancers. Indeed, various mouse models have demonstrated that heterozygous and homozygous *PTEN* loss, and even subtle reductions in its expression, can lead to an increase in spontaneous tumors in many tissue types, providing further support for its key role as a haploinsufficient tumor suppressor in some tissues [5, 11–14]. Additional mechanisms which contribute to aberrant PTEN expression in cancers include epigenetic silencing, transcriptional repression, posttranslational modifications, protein-protein interactions, and posttranscriptional regulation by microRNAs [6, 8].

MicroRNAs are a class of small noncoding RNAs that fine-tune gene expression by binding to specific microRNA response elements (MREs) on target transcripts, resulting in transcript degradation or translational inhibition [15–17]. MicroRNAs are critical regulators of diverse biological processes and have been functionally linked to various human cancers. As subtle variations in PTEN expression have critical consequences for cancer susceptibility in vivo, the fine-tuning of its levels by microRNAs may have tremendous functional importance in physiological settings [11, 18, 19]. To date, numerous microRNAs that promote tumorigenesis by directly targeting *PTEN* have been identified [20].

Recent studies have shed light on a previously uncharacterized dimension of microRNA:target interactions. It has been proposed that RNA transcripts which contain MREs for shared microRNAs can communicate with and co-regulate each other by competing for these microRNAs and titrating their availability [21, 22]. Such natural microRNA sponges or competing endogenous RNAs (ceRNAs) may be either protein-coding or noncoding transcripts. Functional ceRNA regulation has been described in diverse species ranging from plants to mice and humans, suggesting that these interactions may form an additional layer of posttranscriptional regulation with tremendous physiological relevance.

PTEN has one of the most extensively characterized ceRNA networks which encompasses both noncoding and protein-coding transcripts. The noncoding *PTEN* pseudogene *PTENP1* was shown to regulate PTEN levels by acting as a decoy for *PTEN-targeting* microRNAs [23]. Protein-coding transcripts including *CNOT6L, RB1, TNKS2, VAPA, VCAN,* and *ZEB2* have been reported to act as PTEN ceRNAs that modulate PTEN expression in a microRNA- and 3′UTR-dependent manner and antagonize downstream PI3K signaling [24–27]. PTEN ceRNAs have been found to possess tumor-suppressive properties, be co-expressed with *PTEN,* and selectively lost in various human cancers [23–27].

The ability to predict and validate ceRNA crosstalk will therefore facilitate the functionalization of previously uncharacterized noncoding and protein-coding transcripts in part based on their ceRNA activity. Here, we present a computational and experimental framework for the identification and validation of PTEN ceRNAs, which can also be applied to the study of ceRNA interactions for any potential transcript of interest.

2 Materials

2.1 Tissue Culture

1. Phosphate-buffered saline, trypsin (from various vendors).
2. Cell-culture-treated plastics such as 10 cm, 15 cm, 6-well, and 12-well plates, serological pipettes (from various vendors).
3. Hemacytometer or automated cell counter, Trypan blue.
4. Cell-culture medium for cell lines of interest (DU145 and HCT116 cells were grown in DMEM supplemented with 10 % fetal calf serum, 100 U/ml penicillin, 0.1 mg/ml streptomycin, and 2 mM L-glutamine, all from Life Technologies).

2.2 Transfection of siRNAs, miRNA Mimics, and Plasmids

1. Transfection reagents such as Dharmafect 1 or Lipofectamine 2000 (available from various vendors).
2. Expression vectors such as psiCHECK-2 for luciferase reporter assays or pcDNA3.1+ for 3′UTR overexpression (available from various vendors).
3. siRNAs, miRNA mimics, and inhibitors (available from various vendors, reconstituted in RNase-free water to 20 pmol/μl).
4. Reduced serum media for transfection such as Opti-MEM (Life Technologies).

2.3 Molecular Cloning

1. PCR amplification kit (available from various vendors).
2. Standard DNA gel electrophoresis system (available from various vendors).
3. Gel extraction and PCR purification kit (available from various vendors).
4. Site-directed mutagenesis kit (available from various vendors).

2.4 RNA Analysis

1. Trizol reagent (Life Technologies).
2. RNA extraction kit (from various vendors).

2.5 Protein Analysis

1. RIPA protein lysis buffer supplemented with proteinase inhibitors (Roche).
2. Standard western blotting system (Life Technologies XCell SureLock Mini-Cell Electrophoresis System with NuPage 4–12 % Bis-Tris gels).

2.6 Luciferase Assays

1. Microplate luminometer (Promega GloMax-96).
2. White-walled clear-bottomed 96-well plates.
3. Dual-Luciferase Reporter Assay System (Promega).

2.7 RNA Immunoprecipitation

1. NT2 buffer; 2.5 ml 1 M Tris pH 7, 1.5 ml 5 M NaCl, 50 μl 1 M $MgCl_2$, 25 μl NP-40, make up to 50 ml with RNase-free water.
2. NT2-crowders; 25 mg Ficoll PM400, 75 mg Ficoll PM70, 2.5 mg dextran sulfate 670 k (Fluka) in 10 ml of NT2.
3. NET2 buffer; 850 μl NT2-crowders, 10 μl 0.1 M DTT, 30 μl 0.5 M EDTA, 5 μl RNase OUT, 5 μl Superase IN.
4. PLB (10×); 1 ml 1 M HEPES pH 7, 2.5 ml 4 M KCl, 0.5 ml 1 M $MgCl_2$, 5 ml 0.5 M EDTA pH 8, 0.5 ml NP-40, 0.5 ml RNase-free water.
5. PLB (1×); 5 ml 10× PLB, 100 μl 1 M DTT, 1 Proteinase Inhibitor tablet (Roche), 62.5 μl RNase OUT, 125 μl Superase IN, make up to 50 ml with RNase-free water.
6. 2× SDS-TE; 200 μl 1 M Tris pH 7.5, 40 μl 0.5 M EDTA pH 8, 2 ml 10 % SDS, make up to 20 ml with RNase-free water.
7. Other reagents; Protein A-Sepharose beads (Sigma), sodium azide, Trizol LS, chloroform, 5 M LiCl, glycogen/GlycoBlue, phenol:chloroform:isoamyl (25:24:1), chloroform:isoamyl (24:1), 3 M sodium acetate.

3 Methods

3.1 Prediction of ceRNA Interactions

To characterize the PTEN ceRNA network in a given setting, it is first necessary to identify putative transcripts that contain microRNA response elements (MREs) for PTEN-targeting microRNAs. Either validated or predicted PTEN-targeting microRNAs can be used effectively to predict PTEN ceRNA interactions (*see* **Note 1**).

3.1.1 Mutually Targeted MRE Enrichment (MuTaME) Using Validated PTEN-Targeting microRNAs

This analysis focuses on the validated PTEN-targeting microRNAs previously implicated in the ceRNA-mediated regulation of PTEN by its pseudogene PTENP1: miRs-17-5p, 19a, 19b, 20a, 20b, 26a, 26b, 93, 106a, and 106b [24]. Utilize a microRNA target prediction algorithm such as rna22 to generate MuTaME scores for the entire human transcriptome, based on several key considerations

1. The number of shared microRNAs between PTEN and candidate ceRNA X:

$$\frac{\#\,\text{microRNAs predicted to target X}}{\#\,\text{microRNAs under consideration}}$$

2. The number of MREs predicted in X for the *i*-th microRNA and the width of the span covered:

$$\frac{\#\,\text{MREs in X for } i\text{-th microRNA}}{\text{Distance between leftmost and rightmost predicted MRE for the } i\text{-th microRNA}}$$

3. How evenly distributed the predicted MREs for the *i*-th microRNA are over the distance spanned in X:

$$\frac{(\text{Distance between leftmost and rightmost predicted MRE for the } i\text{-th microRNA})^2}{(\text{Sum of squared distances between successive MREs of the } i\text{-th microRNA})^2}$$

4. The relationship between the total number of MREs predicted in X and the total number of microRNAs that give rise to these MREs:

$$\frac{(\#\,\text{MREs in X for all considered microRNAs} - \#\,\text{microRNAs predicted to target X} + 1)}{\#\,\text{MREs in X for all considered microRNAs}}$$

Obtain combined MuTaME scores for each candidate transcript X by multiplying these four components. Additional criteria such as requirements for candidate ceRNAs to contain MREs for at least seven of the ten validated PTEN-targeting microRNAs and for all predicted MREs to occur in the candidate's 3′UTR can be set.

3.1.2 Mutually Targeted MRE Enrichment (MuTaME) Using Predicted PTEN-Targeting microRNAs

MuTaME can also be performed successfully with predicted PTEN-targeting microRNAs [25]. In this analysis, utilize a microRNA target prediction algorithm such as TargetScan to predict PTEN-targeting microRNAs, and set a stringent cutoff of at least seven shared microRNAs. Obtain similarity scores based on all microRNAs predicted by TargetScan and the total number of MREs located in their 3′UTRs. This approach is described in detail in [28].

3.2 Coexpression Analysis of Putative PTEN ceRNAs

The ceRNA hypothesis predicts that transcripts within a given ceRNA network should be coregulated (*see* **Note 2**). It is thus critical to ascertain whether putative PTEN ceRNAs are coexpressed with PTEN in various settings such as in human prostate cancer, glioblastoma, and melanoma, malignancies in which PTEN levels are often reduced or lost. In particular, it is especially informative to interrogate datasets in which integrated expression profiling of both mRNAs and microRNAs is available to assess whether coexpression correlation of PTEN and its predicted ceRNAs is affected by changes in expression levels of the shared microRNAs.

3.2.1 Prostate Cancer and Glioblastoma

1. Use a publicly available dataset such as GEO super-series GSE21032 for prostate cancer coexpression analysis [29]. Download the processed and normalized whole-transcript (GSE21034) and microRNA (GSE21036) expression data for human primary and metastatic prostate cancer samples from NCBI for analysis.
2. Use a publicly available dataset such as GEO series GSE15824 to assess coexpression in glioblastoma. Download processed and normalized expression data from NCBI, and link entrez gene IDs to Affymetrix probe sets by Affymetrix annotation na31.
3. First, calculate the average PTEN value among all samples. Divide samples into two subsets according to PTEN expression: a PTEN-down subset consisting of samples with PTEN expression levels lower than the PTEN average value and a PTEN-up subset consisting of samples with PTEN expression levels higher than this average value. Examine the expression levels of putative PTEN ceRNAs in these two datasets, and ascertain statistical significance using the Student's *T*-test. Calculate Pearson correlation coefficients to determine whether PTEN and its putative ceRNAs are significantly co-expressed.

3.2.2 Normal Tissues

1. Coexpression between PTEN and its candidate ceRNAs can also be examined in various normal tissues as well as across species. Use the annotated TS-CoExp Browser (http://www.cbu.mbcunito.it/ts-coexp) to analyze ceRNA expression in multi-tissue, normal tissue-specific human-mouse conserved and human-specific coexpression networks [30]. Coexpression is considered significant if a specific candidate is found in the top ranked 1 % of genes coexpressed with PTEN.

3.2.3 Integrated mRNA and microRNA Expression Analysis

1. Subdivide samples according to low or high expression levels of the ten validated PTEN-targeting microRNAs used in this analysis. Require at least one of these microRNAs to show an expression level lower (or higher) than a standard deviation from its average expression level in all samples; discard samples in which the different levels of expression among microRNAs are not consistent. Calculate Pearson correlation coefficients to determine whether correlation between expression of PTEN and its putative ceRNAs is significantly increased when microRNA expression levels are taken into consideration.

3.3 Effect of Putative ceRNAs on PTEN Levels

A *bona fide* PTEN ceRNA will effectively regulate PTEN levels by sequestering shared PTEN-targeting microRNAs. Investigation of the effect of putative ceRNAs on PTEN levels is thus essential to validate predicted ceRNA interactions.

3.3.1 ceRNA Knockdown

1. Perform knockdown experiments with reverse transfection to maximize transfection efficiency.
2. Prepare the transfection mix by adding 100 pmol of siRNA to 100 μl of Opti-MEM in a first tube and 2 μl of Dharmafect1 to 100 μl of Opti-MEM in a second tube. Perform biological duplicates, and scale the volume of reagents accordingly.
3. After 5–10-min incubation, combine the siRNA and Dharmafect mixes and incubate for an additional 15–20 min. During this second incubation time, trypsinize DU145 cells and prepare suspensions of 100,000 cells per 800 μl.
4. Aliquot 200 μl of the siRNA + Dharmafect1 transfection mix into one well of a 12-well plate, followed by 800 μl of the cell suspension. Mix gently and incubate at 37 °C. Replace growth medium after 4–12 h.

3.3.2 ceRNA Overexpression

1. PCR amplify the 3′UTR of selected ceRNAs from cell line genomic DNA and clone into a mammalian expression vector such as pcDNA3.1+ according to standard protocols (*see* **Note 3**). Very long 3′UTRs may need to be cloned as two separate fragments. Choose a vector with a selectable marker if you are working with cell lines with low transfection efficiency, and alternatively use a retroviral or lentiviral system for overexpression.
2. For plasmid transfection, seed DU145 cells in 12-well plates at a density of 120,000 cells/well (*see* **Note 4**).
3. Twenty-four hours after seeding, prepare the transfection mix by adding 1 μg of plasmid to 100 μl of Opti-MEM in a first tube and 4 μl of Lipofectamine 2000 to 100 μl of Opti-MEM in a second tube. Perform biological duplicates or triplicates, and scale the volume of reagents accordingly.
4. After 5–10-min incubation, combine the plasmid and Lipofectamine mixes and incubate for an additional 15–20 min. Aliquot 200 μl of the siRNA+ Lipofectamine transfection mix into each well. Mix gently and incubate at 37 °C. Replace growth medium after 4–12 h.

3.3.3 Effect on Endogenous PTEN Transcript Levels

1. Forty-eight hours post-transfection, wash cells with ice-cold PBS, aspirate, and add 0.5 ml of Trizol reagent per well. Let stand for 5 min, pipette to mix, and transfer to a 1.5 ml microcentrifuge tube.
2. Add 0.1 ml chloroform, and invert vigorously for 15 s to mix. Let stand for 2 min and centrifuge at maximum speed for 10 min at 4 °C.
3. Transfer 250 μl of the supernatant to a clean microcentrifuge tube, being careful not to disturb the DNA-containing white

interface. Add an equal volume of 70 % EtOH, mix well, and transfer into a column-based RNA extraction kit such as the RNeasy kit from Qiagen.

4. Centrifuge at maximum speed for 30 s, decant the flow-through, and proceed with wash instructions according to the manufacturer's protocol. Elute in 50 μl of RNase-free water and store at −80 °C or proceed to subsequent step.
5. Perform cDNA synthesis from 1 μg of total RNA using a kit such as the High Capacity cDNA Archive kit (Applied Biosystems). Dilute cDNA 10× and use 4.5 μl of diluted cDNA per 10 μl real-time PCR reaction.
6. Determine PTEN expression using either Taqman or SYBR Green probes and the appropriate master mix (*see* **Note 5**). As a positive control for the ceRNA knockdown experiments, determine the expression levels of each candidate ceRNA in both negative control and the corresponding siRNA samples to verify that effective knockdown has been achieved.

3.3.4 Effect on Endogenous PTEN Protein Levels

1. Seventy-two hours post-transfection, wash cells with ice-cold PBS, aspirate, and add 80–100 μl of RIPA lysis buffer supplemented with protease inhibitors. Incubate on ice for 20 min, transfer the lysates to microcentrifuge tubes, and centrifuge at maximum speed for 15 min at 4 °C.
2. Transfer the supernatants to clean microcentrifuge tubes and determine protein concentrations using Bradford dye (Bio-Rad). Keep samples on ice.
3. Size-fractionate 5–10 μg of total protein by SDS-PAGE on 4–12 % Bis-Tris NuPAGE gels using MOPS or MES running buffer (Life Technologies).
4. Transfer to PVDF or nitrocellulose membranes in NuPAGE transfer buffer containing 10 % methanol. Check transfer efficiency using Ponceau S staining and block with 5 % milk in PBST for at least 30 min before probing with specific primary antibodies for PTEN and housekeeping genes such as GAPDH or HSP90 (*see* **Note 6**).

3.3.5 Effect on Exogenous PTEN Reporters

1. Seed DU145 cells in 12-well plates at a density of 120,000 cells/well.
2. Twenty-four hours after seeding, prepare the transfection mix by adding 100 ng of empty psiCHECK-2 or psiCHECK-2+PTEN3′UTR and either 100 pmol of siRNA or 1 μg of plasmid to 100 μl of Opti-MEM in a first tube and 2 μl of Lipofectamine 2000 to 100 μl of Opti-MEM in a second tube. Perform biological duplicates, and scale the volume of reagents accordingly.

3. After 5–10-min incubation, combine the plasmid/siRNA and Lipofectamine mixes and incubate for an additional 15–20 min. Aliquot 200 μl of the plasmid/siRNA+ Lipofectamine transfection mix into each well. Mix gently and incubate at 37 °C. Replace growth medium after 4–12 h.
4. Seventy-two hours post-transfection, wash cells with ice-cold PBS, aspirate, and add 100 μl of 1× lysis buffer (Promega). Place on an orbital shaker for 10 min to dissociate the cell layer. Pipette gently to mix and transfer 20 μl of each lysate into one well of a white-walled 96-well plate. Measure firefly and Renilla luciferase activities with the dual-luciferase reporter system (Promega).

3.4 Reciprocal Effect of PTEN Perturbation on PTEN ceRNA Levels

As mentioned earlier, ceRNA interactions are likely to function in a reciprocal manner (*see* **Note 7**). To investigate this, perform PTEN knockdown and overexpression experiments as described in Subheadings 3.3.1 and 3.3.2, respectively. Determine the effect on endogenous PTEN ceRNA transcript and protein levels and exogenous PTEN ceRNA reporters as described in Subheadings 3.3.3, 3.3.4, and 3.3.5, respectively (*see* **Note 8**).

3.5 Biochemical Analyses of microRNAs Associated with PTEN and Putative ceRNA Transcripts

MicroRNAs have been shown to display context-specific expression patterns. An accurate knowledge of the exact microRNA regulators involved in a given context will facilitate a precise understanding of ceRNA networks in particular settings (*see* Fig. 1).

3.5.1 Bead Preparation: Swelling and Antibody-Binding

1. Resuspend desired amount of Protein A-Sepharose beads in 5–10 volumes of PLB containing 5 % BSA and 0.05 % sodium azide—use 50 μl of swelled bead bed volume in a total of 300 μl PLB per IP. Leave on rotator at 4 °C overnight.
2. Vortex to mix and add 2 μg of antibody per IP. Leave on rotator at 4 °C overnight.
3. Wash with 1 ml of NT2, then vortex, and spin at 5000 ×*g* for 30 s. Repeat 5×. Use NT2-crowders for the last wash. For each IP, resuspend 50 μl of bead bed volume in 850 μl of NET2-crowders.

3.5.2 Harvesting Cytoplasmic Lysates

1. Place cell culture plates on ice, remove media, and wash 4× with ice-cold PBS.
2. Add 500 μl of ice-cold 1× PLB to each well (6-well plate) and leave for 10 min.
3. Scrape, collect, and resuspend in 1.5 ml tubes. Spin at 4000 ×*g* for 10 min at 4 °C and collect supernatant.

3.5.3 Immunoprecipitation

1. Add 100 μl of collected supernatant to each IP bead slurry tube. Keep 100 μl of cell lysate supernatant in PLB at 4 °C for positive ("Total") control. Leave overnight on rotator at 4 °C.
2. Spin beads down at 5000 ×*g* for 30 s. Discard supernatant, and wash beads with 500 μl of ice-cold NT2-crowders 1×, and NT2 5×. Spin beads down. Discard supernatant and resuspend beads in 100 μl of NET2; add 100 μl of 2× SDS-TE. Incubate at 55 °C for 30 min, mixing every few minutes.
3. Spin beads down at 5000 ×*g* for 2 min and collect 200 μl of supernatant containing protein-RNA complexes in fresh tubes.

3.5.4 RNA Extraction and Recovery

1. Add 200 μl of Trizol LS to 200 μl of the supernatant above, and then add 40 μl of chloroform. For "Total" control samples, add 200 μl of Trizol LS to 100 μl of PLB lysate and 100 μl of NET2, and then add 40 μl of chloroform. Spin at maximum speed for 10 min at 4 °C.
2. Carefully transfer 300 μl of aqueous phase containing RNA to fresh 2 ml tubes. Add pre-mixed salt solution to each: 40 μl 5 M NaCl, 40 μl 5 M LiCl, 10 μl 5 μg/μl glycogen (or 4 μl of 15 mg/ml GlycoBlue).
3. Add 1.5 ml of 100 % EtOH per IP, mix. Leave at −80 °C overnight.
4. Spin at maximum speed for 30 min at 4 °C. Wash with 1 ml of 75 % EtOH, spin at maximum speed for 10 min at 4 °C (2×). Discard supernatant and leave to dry. Resuspend pellet in 200 μl H_2O.

3.5.5 Phenol/Chloroform RNA Cleanup

1. Add 200 μl of phenol:chloroform:isoamyl (25:24:1) to 200 μl of the RNA above.
2. Mix and spin at maximum speed for 5 min at RT.
3. Keep supernatant (200 μl, keep volumes consistent), add equal volumes of chloroform:isoamyl (24:1). Mix and spin at maximum speed for 5 min at RT.
4. Keep supernatant (200 μl, keep volumes consistent).
5. To each sample add 20 μl of 3 M sodium acetate, 20 μl of 5 μg/μl glycogen (Ambion; alternatively add 7 μl of 15 mg/ml GlycoBlue), and 500 μl of 100 % EtOH in fresh 2 ml tubes.
6. To precipitate RNA, leave at −80 °C overnight.
7. To recover RNA, spin at maximum speed for 30 min at 4 °C.
8. Wash with 1 ml of 75 % EtOH. Spin at maximum speed for 10 min at 4 °C.
9. Second wash with 1 ml of 75 % EtOH. Spin at maximum speed for 10 min at 4 °C.
10. Discard supernatant and leave to dry in vacuum rotator. Check for pellet.

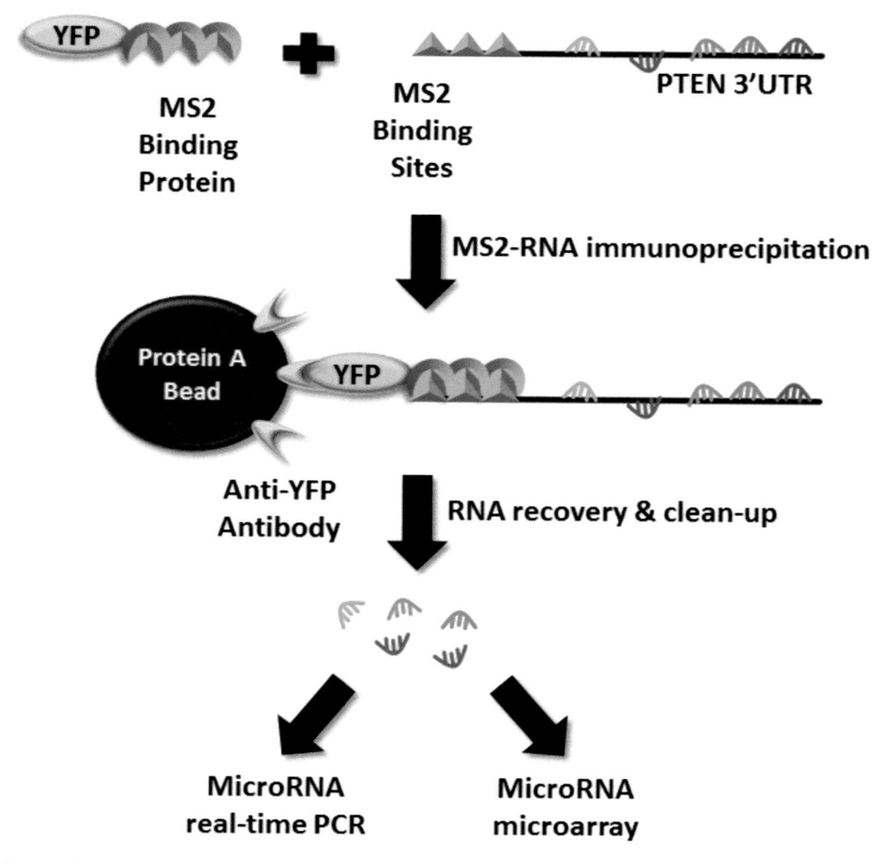

Fig. 1 Schematic outlining the MS2-RNA Immunoprecipitation (RIP) strategy to validate endogenous microRNA:PTEN 3′UTR interactions

11. Resuspend pellet in 20 or 50 μl H_2O. Tap to mix and spin to collect. Quantitate.

3.6 MicroRNA Dependency of ceRNA Regulation

3.6.1 Effect of microRNAs on Endogenous PTEN and PTEN ceRNA Levels

Perform microRNA overexpression and knockdown experiments with reverse transfection to maximize transfection efficiency. Transfect 100 pmol of microRNA mimic or antisense per well of a 12-well plate as described in Subheading 3.3.1, and determine the effect on endogenous PTEN and PTEN ceRNA transcript and protein levels as described in Subheadings 3.3.3 and 3.3.4, respectively.

3.6.2 Effect of microRNAs on Exogenous PTEN and PTEN ceRNA Reporters

1. Clone the 3′UTRs of the PTEN ceRNAs of interest as described in Subheading 3.3.2. Very long 3′UTRs may need to be cloned as two separate fragments.
2. Perform transfections with 100 pmol of the microRNA mimic of interest and 100 ng of empty psiCHECK-2 or psiCHECK-

2+PTEN3′UTR or psiCHECK-2+ceRNA3′UTR per well of a 12-well plate, as described in Subheading 3.3.5.

3.6.3 Effect of Defects in microRNA Activity on ceRNA Interactions

Co-regulation of long noncoding RNAs and mRNAs alone does not always imply *bona fide* ceRNA interactions; for example they may be intermolecular binding partners in Staufen-mediated mRNA decay [31]. To validate putative ceRNA interactions, it is critical to demonstrate the microRNA dependence of this co-regulation.

1. If the putative ceRNA interactions involve only one or two microRNAs, their dependence on the presence of these microRNAs can be assayed using microRNA inhibition as described in Subheading 3.6.1. When microRNA function is inhibited, the observed effect of putative ceRNAs on PTEN should be at least partially abrogated.
2. If the putative ceRNA interactions involve many microRNAs, such that inhibiting all the microRNAs simultaneously is impractical, an alternative approach examining the effect of defective processing of mature microRNAs on ceRNA activity may be used. For example, a Dicer-mutant HCT116 colon cancer cell line has been developed in which gene targeting was used to disrupt a conserved portion of the N-terminal helicase domain in exon 5 of Dicer, an enzyme critical for the cleavage of the pre-microRNA hairpin (*see* **Note 9**) [32]. Determine the effect of putative ceRNA perturbation (as described in Subheadings 3.3.1 and 3.3.2) on PTEN protein and luciferase reporter levels (as described in Subheadings 3.3.4 and 3.3.5, respectively) in both Dicer-mutant and isogenic wild-type HCT116 cells. Transfect HCT116 cells at a density of 130,000 or 160,000 cells per 12 well for siRNA or plasmid transfections, respectively (*see* **Note 10**). *Bona fide* ceRNA interactions should be at least partially abrogated in the Dicer-mutant cells relative to the wild-type cells (*see* **Note 11**).

3.6.4 Mutagenesis of Specific microRNA Response Elements

A complementary method to investigate the microRNA dependency of the putative ceRNA activity involves the mutation of the specific microRNA response elements thought to be involved in the crosstalk (*see* **Note 12**). This can be done using commonly available site-directed mutagenesis kits from various vendors, such as the QuikChange II Site-Directed Mutagenesis Kit.

1. Generate MRE-mutant constructs for selected ceRNAs of interest, using the wild-type ceRNA 3′UTR constructs generated previously in Subheading 3.3.2. Mutagenic oligonucleotide primers should be designed with as few mutations as possible to abrogate microRNA binding while preserving secondary structure. If most of the binding in the microRNA:target

heteroduplex is predicted to occur within the seed region, then two to three mutations in the seed region should be sufficient to significantly attenuate microRNA binding. However, if strong binding is predicted to occur along the rest of the heteroduplex then additional mutations may be required. In this scenario, a multiple mutation kit such as the QuikChange Multi Site-Directed Mutagenesis Kit may be needed to achieve optimal mutation efficiency.

2. If these MREs are critical for the observed cross talk between the putative ceRNAs and PTEN, then the overexpression of these MRE-mutant constructs (as described in Subheading 3.3.2) should have a markedly reduced effect on PTEN levels (as described in Subheadings 3.3.3–3.3.5) as compared to that of their corresponding wild-type counterparts.

4 Notes

1. PTEN is a relatively unique candidate for ceRNA analyses, as a large number of PTEN-targeting microRNAs have been experimentally validated. For most other transcripts which have few or no known microRNA regulators, ceRNA analyses will be limited to those involving predicted targeting microRNAs for both the gene of interest and candidate ceRNAs.
2. Although it has been suggested that microRNA-mediated repression functions largely to destabilize RNAs, this is not always the case [33]. MicroRNAs have been shown to inhibit translation initiation without affecting mRNA stability [34]. Coexpression analyses is thus a useful filter to identify putative ceRNAs, however its use will inevitably limit subsequent analyses to candidates which are destabilized at the transcript level by microRNA targeting.
3. If the MREs of interest are located in the 5′UTR or CDS regions of putative ceRNAs, then the corresponding fragment should be cloned into reporter vectors for overexpression studies.
4. Prior to transfection, check cell seeding carefully under a microscope. Cells should be distributed evenly throughout the well, transfection efficiency will be markedly reduced if the cells are clumped in the center of the well.
5. As mentioned in **Note 2**, microRNA regulation may not always affect transcript stability. It is possible that microRNA and ceRNA regulation will lead to a decrease in PTEN protein levels without a corresponding change in its transcript levels. It is thus essential to always examine the effect of putative ceRNAs on PTEN protein levels and not rely solely on transcript analyses.

6. Care should be taken not to overload the gel or to overexpose the film. ceRNA regulation may only result in a 50 % increase or decrease in protein levels, this difference may be overlooked if the signal is saturated and not within the dynamic range of the film.
7. Due to differences in factors such as expression level and number of MREs, ceRNA regulation may not always be reciprocal. For example, ceRNA A which is highly expressed and contains multiple MREs for the shared microRNA X may be a strong PTEN ceRNA, but PTEN which may be moderately expressed and contain only a single MRE for microRNA X may not efficiently regulate ceRNA A expression. As such, although reciprocity should be investigated, it is not a necessary requirement to validate *bona fide* PTEN ceRNAs.
8. Many predicted ceRNAs may be previously uncharacterized or little known genes, and there may not be good commercially available antibodies for these candidates. These putative interactions will have to be studied using transcript and reporter assays first. Specific antibodies could then be generated for particularly promising candidates.
9. It is important to note that Dicer-mutant cells still contain low levels of many microRNAs and thus low levels of ceRNA cross talk may be possible. Any experiments in these cells must be performed in parallel in the isogenic wild-type HCT116 cell line, and results interpreted relative to those obtained from that line.
10. HCT116 cells take significantly longer to adhere to cell culture surfaces than other cell lines such as DU145 or U2OS. They should be allowed to settle for at least 12 h/overnight before the transfection media is changed to prevent loss of cells.
11. Some microRNAs have been shown to be processed via a Dicer-independent biogenesis pathway [35, 36]. If the microRNAs in question are processed via this pathway then ceRNA cross talk will not be attenuated in the Dicer-mutant cells. It is essential to check the levels of the mature microRNAs being studied in both wild-type and Dicer-mutant cells. If they are not significantly different then an alternative approach, such as using Drosha knockdown, will have to be employed.
12. This approach is more appropriate when ceRNA cross talk is dependent on a single or small number of MREs. If many MREs are involved in the cross talk, then mutagenesis of all the MREs may result in significant changes in secondary structure and topology that may affect other types of posttranscriptional regulation such as binding of RNA binding proteins or Staufen-mediated decay. This may confound the results of such experiments.

Acknowledgements

We thank Shen Mynn Tan for critical reading of the manuscript. Y.T. was supported by a Singapore National Research Foundation Fellowship and a National University of Singapore President's Assistant Professorship. P.P.P. was supported in part by NIH grants 1R01CA170158 and R01 CA82328.

References

1. Hollander MC, Blumenthal GM, Dennis PA (2011) PTEN loss in the continuum of common cancers, rare syndromes and mouse models. Nat Rev Cancer 11(4):289–301
2. Steck PA et al (1997) Identification of a candidate tumour suppressor gene, MMAC1, at chromosome 10q23.3 that is mutated in multiple advanced cancers. Nat Genet 15(4): 356–362
3. Cantley LC, Neel BG (1999) New insights into tumor suppression: PTEN suppresses tumor formation by restraining the phosphoinositide 3-kinase/AKT pathway. Proc Natl Acad Sci U S A 96(8):4240–4245
4. Li J et al (1997) PTEN, a putative protein tyrosine phosphatase gene mutated in human brain, breast, and prostate cancer. Science 275(5308):1943–1947
5. Di Cristofano A, Pandolfi PP (2000) The multiple roles of PTEN in tumor suppression. Cell 100(4):387–390
6. Salmena L, Carracedo A, Pandolfi PP (2008) Tenets of PTEN tumor suppression. Cell 133(3):403–414
7. Maehama T, Dixon JE (1998) The tumor suppressor, PTEN/MMAC1, dephosphorylates the lipid second messenger, phosphatidylinositol 3,4,5-trisphosphate. J Biol Chem 273(22): 13375–13378
8. Song MS, Salmena L, Pandolfi PP (2012) The functions and regulation of the PTEN tumour suppressor. Nat Rev Mol Cell Biol 13(5): 283–296
9. Myers MP et al (1997) P-TEN, the tumor suppressor from human chromosome 10q23, is a dual-specificity phosphatase. Proc Natl Acad Sci U S A 94(17):9052–9057
10. Shen WH et al (2007) Essential role for nuclear PTEN in maintaining chromosomal integrity. Cell 128(1):157–170
11. Trotman LC et al (2003) Pten dose dictates cancer progression in the prostate. PLoS Biol 1(3):E59
12. Di Cristofano A et al (1998) Pten is essential for embryonic development and tumour suppression. Nat Genet 19(4):348–355
13. Suzuki A et al (1998) High cancer susceptibility and embryonic lethality associated with mutation of the PTEN tumor suppressor gene in mice. Curr Biol 8(21):1169–1178
14. Podsypanina K et al (1999) Mutation of Pten/Mmac1 in mice causes neoplasia in multiple organ systems. Proc Natl Acad Sci U S A 96(4):1563–1568
15. Bartel DP, Chen CZ (2004) Micromanagers of gene expression: the potentially widespread influence of metazoan microRNAs. Nat Rev Genet 5(5):396–400
16. Bartel DP (2009) MicroRNAs: target recognition and regulatory functions. Cell 136(2): 215–233
17. Thomas M, Lieberman J, Lal A (2010) Desperately seeking microRNA targets. Nat Struct Mol Biol 17(10):1169–1174
18. Alimonti A et al (2010) Subtle variations in Pten dose determine cancer susceptibility. Nat Genet 42(5):454–458
19. Berger AH, Knudson AG, Pandolfi PP (2011) A continuum model for tumour suppression. Nature 476(7359):163–169
20. Tay Y, Song SJ, Pandolfi PP (2013) The Lilliputians and the giant: an emerging oncogenic microRNA network that suppresses the PTEN tumor suppressor in vivo. Microrna 2(2):127–136
21. Salmena L et al (2011) A ceRNA hypothesis: the Rosetta Stone of a hidden RNA language? Cell 146(3):353–358
22. Tay Y, Rinn J, Pandolfi PP (2014) The multilayered complexity of ceRNA crosstalk and competition. Nature 505(7483):344–352
23. Poliseno L et al (2010) A coding-independent function of gene and pseudogene mRNAs regulates tumour biology. Nature 465(7301): 1033–1038

24. Tay Y et al (2011) Coding-independent regulation of the tumor suppressor PTEN by competing endogenous mRNAs. Cell 147(2):344–357
25. Karreth FA et al (2011) In vivo identification of tumor-suppressive PTEN ceRNAs in an oncogenic BRAF-induced mouse model of melanoma. Cell 147(2):382–395
26. Lee DY et al (2010) Expression of versican 3′-untranslated region modulates endogenous microRNA functions. PLoS One 5(10):e13599
27. Sumazin P et al (2011) An extensive microRNA-mediated network of RNA-RNA interactions regulates established oncogenic pathways in glioblastoma. Cell 147(2):370–381
28. Karreth FA et al (2014) Pseudogenes as competitive endogenous RNAs: target prediction and validation. Methods Mol Biol 1167:199–212
29. Taylor BS et al (2010) Integrative genomic profiling of human prostate cancer. Cancer Cell 18(1):11–22
30. Piro RM et al (2011) An atlas of tissue-specific conserved coexpression for functional annotation and disease gene prediction. Eur J Hum Genet 19(11):1173–1180
31. Park E, Maquat LE (2013) Staufen-mediated mRNA decay. Wiley Interdiscip Rev RNA 4(4):423–435
32. Cummins JM et al (2006) The colorectal microRNAome. Proc Natl Acad Sci U S A 103(10):3687–3692
33. Eichhorn SW et al (2014) mRNA destabilization is the dominant effect of mammalian microRNAs by the time substantial repression ensues. Mol Cell 56(1):104–115
34. Thermann R, Hentze MW (2007) Drosophila miR2 induces pseudo-polysomes and inhibits translation initiation. Nature 447(7146): 875–878
35. Cheloufi S et al (2010) A dicer-independent miRNA biogenesis pathway that requires Ago catalysis. Nature 465(7298):584–589
36. Cifuentes D et al (2010) A novel miRNA processing pathway independent of Dicer requires Argonaute2 catalytic activity. Science 328(5986):1694–1698

Chapter 11

In Cell and In Vitro Assays to Measure PTEN Ubiquitination

Amit Gupta, Helene Maccario, Nisha Kriplani, and Nicholas R. Leslie

Abstract

The lipid and protein tyrosine phosphatase, PTEN, is one of the most frequently mutated tumor suppressors in human cancers and is essential for regulating the oncogenic pro-survival PI3K/AKT signaling pathway. Because of its diverse physiological functions, PTEN has attracted great interest from researchers in multiple research fields. The functional diversity of PTEN demands a collection of delicate regulatory mechanisms, including transcriptional control and posttranslational mechanisms that include ubiquitination. Addition of ubiquitin to PTEN can have several effects on PTEN function, potentially regulating its stability, localization, and activity. In cell and in vitro ubiquitination assays are employed to study the ubiquitination-mediated regulation of PTEN. However, PTEN ubiquitination assays are challenging to perform and the data published from these assays has been of mixed quality. Here we describe protocols to detect PTEN ubiquitination in cultured cells expressing epitope tagged ubiquitin (in cell PTEN ubiquitination assay) and also using purified proteins (in vitro PTEN ubiquitination assay).

Key words Phosphatase, PTEN, Ubiquitin, Phosphoinositide, Cancer, Tumor suppressor

1 Introduction

Protein ubiquitination is a posttranslational modification in which lysine residues in target proteins are modified with the covalent addition of the small protein ubiquitin, a 76 amino acid polypeptide of 8.5 kDa. The addition of ubiquitin can affect the turnover of proteins by increasing their degradation via the proteosome, or affect their cellular location, activity, or interactions with other proteins [1]. As its name suggests, the ubiquitin system appears to contribute to the regulation of most proteins in eukaryotic cells, therefore influencing diverse areas of physiology and pathology [2].

The covalent attachment of ubiquitin to target proteins requires the consecutive action of three enzyme activities: ubiquitin-activating enzyme E1, ubiquitin-conjugating enzyme E2, and ubiquitin ligase E3. The first step involves the activation of ubiquitin by the formation of a thioester bond with the ubiquitin-activating

Leonardo Salmena and Vuk Stambolic (eds.), *PTEN: Methods and Protocols*, Methods in Molecular Biology, vol. 1388, DOI 10.1007/978-1-4939-3299-3_11,

enzyme, E1, and requires ATP. In the second step, E1 delivers the activated ubiquitin to the E2 ubiquitin-conjugating enzyme. Finally, E3 ligases catalyze the transfer of ubiquitin from E2 to a lysine residue in the substrate protein, forming an isopeptide bond between the carboxylic acid group of the terminal amino acid of the ubiquitin protein (glycine 76) and the epsilon amino group of the target lysine [3]. In the ubiquitination cascade, E1 can bind separately with any one of a variety of E2s and E3s in a hierarchical way. There appear to be two genes encoding E1 proteins in the human genome (UBA1 and UBA6), 35 E2s, and over 600 E3 ligases. Unlike E1 and E2, E3 ubiquitin ligases display strong specificity for individual substrate proteins [3]. Protein regulation via ubiquitination is further influenced by deubiquitinases, which are specialized proteases that oppose the role of ubiquitin ligases by removing ubiquitin from substrate proteins.

Further functional diversity can be derived from monoubiquitination and polyubiquitination of single or multiple lysine residues on target proteins (*see* Fig. 1). Monoubiquitination is the addition of a single ubiquitin group to one or more lysines within a target protein, whereas polyubiquitination involves the addition of chains of linked ubiquitin moieties. Since ubiquitin itself contains seven lysines that can participate in further ubiquitin conjugation, polyubiquitin chains can be formed with different linkage topologies, lengths, and functions.

The PTEN tumor suppressor regulates the oncogenic PI3K/AKT signaling pathway by catalyzing degradation of the phosphatidylinositol 3,4,5-trisphosphate (PIP_3) lipid generated by class I PI3K [4]. Even subtle reductions in the expression levels of PTEN appear to enhance rates of tumor formation in some tissues, particularly the breast [5], highlighting the apparent significance of the regulation of PTEN function and stability via ubiquitination. PTEN has been reported to have at least two possible ubiquitination sites, lysine 13 and lysine 289 [6, 7], and NEDD4-1 has been identified as an efficient E3 ubiquitin ligase for PTEN [7]. While PTEN polyubiquitinylation correlates with proteasome-mediated degradation, monoubiquitinylation of PTEN at Lys289 and 13 has been shown to promote the nuclear accumulation of PTEN [7]. Additionally, reports that PTEN activity can be directly regulated by ubiquitination [12] and of regulated PTEN deubiquitination [8, 9] along with the identification of new E3 ligases for PTEN such as XIAP (X-linked inhibitor of apoptosis protein) [10] and WWP2 (WW domain containing E3 ubiquitin protein ligase 2) [11] all add to the apparent complexity of PTEN's regulation by ubiquitination and show that most of the details of this regulation remain to be revealed. Here, we describe in cell and in vitro protocols to study the ubiquitin modification of PTEN.

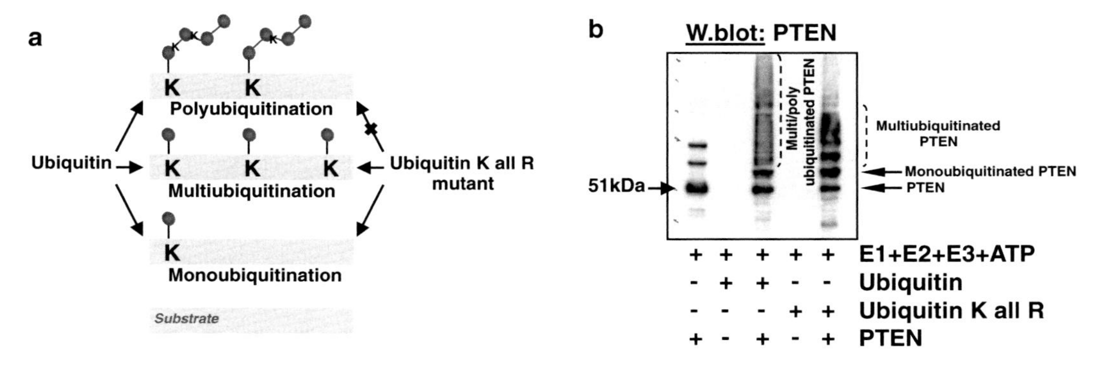

Fig. 1 Detection of PTEN ubiquitination by in vitro PTEN ubiquitination assay. (**a**) A scheme is shown that distinguishes among polyubiquitination, multisite monoubiquitination (here labeled multiubiquitination), and single monoubiquitination of different lysine (K) residues on a ubiquitinated substrate protein. Using a ubiquitin mutant with each lysine residue mutated to arginine (ubiquitin K all R mutant), the formation of polyubiquitin chains is blocked. (**b**) An anti-PTEN Western blot (W.blot) is shown comparing an in vitro ubiquitinated purified PTEN protein sample with a similar reaction using the ubiquitin K all R mutant and with a sham ubiquitinated sample (i.e., reaction in the absence of ubiquitin). Other immunoreactive bands within this negative control sample represent residual GST-tagged PTEN after an incomplete cleavage of this affinity purification tag. Polyubiquitinated, multiubiquitinated, and monoubiquitinated PTEN is illustrated in the figure. Smear of high molecular weight represents polyubiquitinated PTEN. This figure was originally published in Journal of Biological Chemistry. Maccario H., Perera N.M., Gray A., Downes, C.P., and Leslie, N.R. Ubiquitination of PTEN (phosphatase and tensin homolog) inhibits phosphatase activity and is enhanced by membrane targeting and hyperosmotic Stress. *J Biol Chem. 2010;* 285:12620–12628. © the American Society for Biochemistry and Molecular Biology)

2 Materials

All reagents are prepared and stored at room temperature unless indicated otherwise.

2.1 In Vitro PTEN Ubiquitination Assay Components

1. PTEN ubiquitination buffer: 50 mM Tris–HCl pH 7.5, 5 mM $MgCl_2$, 1 mM dithiothreitol (DTT) (store for <24 h, *see* **Note 1**).
2. Purified recombinant human Ubiquitin-activating enzyme, UBE1 (commercially available from, e.g., R&D systems, Minneapolis, MN, USA; Ubiquigent, Dundee, Scotland, UK, or expressed and purified, *see* **Note 2**). Store in small aliquots at −80 °C.
3. Purified recombinant human ubiquitin-conjugating enzyme, UbcH5b (Boston Biochem, Cambridge, MA, USA; Ubiquigent, Dundee, Scotland, UK, or expressed and purified) (*see* **Notes 2** and **3**). Store in small aliquots at −80 °C.
4. Purified recombinant glutathione S-transferase tagged human E3 ubiquitin ligase—NEDD4 (*see* **Note 4**). Store in small aliquots at −80 °C.

5. 10 mM ATP, pH 7.0. Stored in aliquots at −20 °C.
6. Purified recombinant PTEN protein (0.5 mg/ml in buffer containing): 50 mM Tris–HCl pH 7.4, 150 mM NaCl, 1 mM EGTA, 1 mM DTT. Store in small aliquots at −80 °C (*see* **Note 5**).
7. Purified untagged ubiquitin (*see* **Note 2**). Store at −80 °C in small aliquots. Details of ubiquitination can be studied using mutants of the ubiquitin protein, such as K all R Ubiquitin, in which each lysine residue within ubiquitin is mutated to arginine to block the formation of polyubiquitin chains.
8. 4× Lithium dodecyl sulfate (LDS) sample loading buffer.
9. 2-Mercaptoethanol.
10. 4–12 % Precast SDS-polyacrylamide gradient gel (*see* **Note 6**).
11. Polyvinylidene difluoride (PVDF) membrane.
12. Mouse monoclonal Anti-PTEN antibody, A2B1 unconjugated (Santa Cruz Biotechnology, Dallas, TX, USA). Stored at 4 °C.

2.2 In Cell PTEN Ubiquitination Assay Components

1. HEK293T cells or U87MG PTEN-null glioblastoma cells.
2. Mammalian expression vector encoding untagged wild-type PTEN (pcDNA3.1+ PTEN WT) (*see* **Note 2**).
3. Mammalian expression vector encoding FLAG-tagged ubiquitin (pCMV5-Flag-Ub) (*see* **Note 2**).
4. Transfection reagent.
5. Proteasome inhibitor I (PSI). Prepare a 10 mM stock in DMSO. Store aliquots at −20 °C.
6. 1× Phosphate-buffered saline (PBS).
7. Cell lysis buffer: 25 mM Tris–HCl (pH 7.4), 150 mM NaCl, 1 % Nonidet P-40, 1 mM EGTA, 1 mM EDTA, 5 mM sodium pyrophosphate, 10 mM glycerophosphate, and 50 mM sodium fluoride containing freshly added 0.2 mM phenylmethylsulfonyl fluoride, 1 mM benzamidine, 10 μg/ml aprotinin, 10 μg/ml leupeptin, and 10 mM *N*-ethylmaleimide (*see* **Note 7**).
8. Bradford Coomassie Protein Assay Reagent.
9. Protein G sepharose FastFlow beads (Sigma-Aldrich, St. Louis, MO, USA) (*see* **Note 8**).
10. Sepharose 4B-CL beads (Sigma-Aldrich, St. Louis, MO, USA) (*see* **Note 8**).
11. Washing buffer: 25 mM Tris–HCl (pH 7.4), 500 mM NaCl, 1 % Nonidet P-40, 1 mM EGTA, 1 mM EDTA, 5 mM sodium pyrophosphate, 10 mM beta-glycerophosphate, and 50 mM sodium fluoride containing freshly added 0.2 mM phenylmethylsulfonyl fluoride, 1 mM benzamidine, 10 μg/ml aprotinin, 10 μg/ml leupeptin, and 10 mM *N*-ethylmaleimide (*see* **Note 7**).

12. 4× Lithium dodecyl sulfate (LDS) sample loading buffer (Invitrogen, Waltham, MA, USA).
13. 2-Mercaptoethanol (Sigma-Aldrich, St. Louis, MO, USA).
14. 4–12 % Precast SDS-polyacrylamide gel (Invitrogen, Waltham, MA, USA) (*see* **Note 6**).
15. Polyvinylidene difluoride (PVDF) membrane.
16. Mouse monoclonal Anti-PTEN antibody, A2B1 unconjugated (Santa Cruz Biotechnology, Dallas, Texas, USA).
17. Mouse monoclonal Anti-FLAG antibody, M2 (Sigma-Aldrich, St. Louis, MO, USA).
18. Mouse monoclonal Anti-β-Actin antibody (Sigma-Aldrich, St. Louis, MO, USA).

3 Methods

Carry out all procedures at room temperature unless otherwise specified.

3.1 In Vitro PTEN Ubiquitination Assay

1. The reaction is performed in total final volume of 25 μl.
2. For each reaction, prepare a mix containing the following (*see* Table 1):
3. Add 200 ng of purified recombinant PTEN protein diluted in ubiquitination buffer to the above mix.
4. Make the total volume to 25 μl with PTEN ubiquitination buffer. The buffers in which the proteins are stored should be considered and making a 5× stock of PTEN ubiquitination buffer may be beneficial.
5. As controls, prepare similar reactions with one component of the ubiquitination assay omitted: as a minimum, perform the reactions in the absence of ubiquitin or PTEN protein.

Table 1
In vitro PTEN Ubiquitination Mix

Component	Stock solution (mM)	Volume required (μl)	Final concentration (mM)
6His-UBE1	0.0005	2.5	0.00005
GST-UbcH5b	0.05	0.5	0.001
E3 ubiquitin ligase-NEDD4-1	0.025	1.0	0.001
ATP	10.0	5.0	2.0
Ubiquitin (*see* **Note 9**)	1	2.5	0.1

6. Incubate the mixture at 30 °C for 1 h.
7. Terminate the reaction by adding 10 μl 4× LDS sample loading buffer and 4 μl 2-mercaptoethanol (*see* **Note 10**).
8. Heat the mixture at 70 °C for 10 min.
9. Resolve the samples on 4–12 % SDS-PAGE gel.
10. Electroblot the resolved samples onto polyvinylidene difluoride (PVDF) membrane.
11. Detect ubiquitinated PTEN by probing the membrane with anti-PTEN antibody (A2B1): Following electro blotting, dry the PVDF membrane for 30 min and then block the membrane in 5 % (w/v) low-fat milk made up in TBS-T (10 mM Tris–HCl pH 7.6, 150 mM NaCl, 0.1 % Tween 20) for 2 h. Incubate the blocked PVDF membrane overnight at 4 °C with anti-PTEN antibody (200 ng/ml dilution, A2B1 monoclonal) made up in 2 % (w/v) BSA. Wash the membrane over 30 min with three buffer changes of TBS-T and incubate it with HRP-conjugated anti-mouse secondary antibody (1:10,000 dilution) made up in 5 % (w/v) low-fat milk for 1 h. Wash the membrane over 30 min with four buffer changes of TBS-T. Process for HRP detection using chemiluminescent HRP substrate, ECL (Millipore, Waltham, MA, USA). For an example result, *see* Fig. 1.

3.2 In Cell PTEN Ubiquitination Assay

1. Seed U87MG or HEK293T cells in 6 cm dishes in 5 ml recommended medium supplemented with 10 % FBS. Grow cells in a 37 °C incubator with 5 % CO_2.
2. Perform transfections when cells are ~50 % confluent using TransIT-LT1 transfection reagent following the manufacturer's instructions.
3. Co-transect the cells with both 2 μg PTEN-WT plasmid DNA and 2 μg Flag-Ub plasmid DNA per dish (*see* **Note 11**).
4. As negative controls, co-transfect the cells with each vector alone and 2 μg of empty vector DNA (pCMV5 or pcDNA3.1+). Also, include an untransfected control.
5. As positive control to enhance the accumulation of ubiquitinated proteins, 24 h post-transfection treat the cells with 10 μM PSI (or other proteasome inhibitor) for 2 h.
6. Wash the cells twice with 4 ml ice-cold 1× PBS.
7. Immediately lyse cells on ice using 1 ml cell lysis buffer per 6 cm dish. Swirl the dish carefully to let the lysis buffer cover the entire area of grown cells (*see* **Notes 12** and **13**).
8. Collect the cell debris with a cell scraper and transfer the cell lysates into a 1.5 ml microfuge tube on ice. To ensure optimal protein recovery, incubate the lysate on ice for 30 min.

9. Centrifuge the samples at 20,000 × *g* for 20 min at 4 °C to remove any unlysed cells or organelles along with insoluble or aggregated proteins.
10. Transfer the resulting supernatant to a new microfuge tube on ice. Be careful not to disturb the pellet.
11. Option to snap freeze the lysates in liquid nitrogen and store at −80 °C.
12. Measure the protein concentration (e.g., using Bio-rad or Pierce protein assay reagents or similar following the manufacturer's guidelines, in particular keeping concentrations of lysis buffer components, including detergents, below interfering limits (e.g., NP-40 approx 0.125 % final concentration).
13. It is recommended to first ascertain the equal expression of FLAG-tagged ubiquitin in Flag-Ub plasmid DNA-transfected cells before detecting the PTEN ubiquitination. To determine this, immunoblot the lysates with anti-FLAG antibody M2: Prepare samples by mixing the cell lysates containing 10 μg of soluble protein with 4× LDS sample loading buffer containing 10 % 2-mercaptoethanol. Heat the mixture at 70 °C for 10 min. Resolve the samples by SDS-PAGE and analyze the ubiquitinated protein levels by immunoblotting the lysates with Anti-FLAG antibody M2. Similar cellular protein ubiquitination levels in the FLAG-ubiquitin-expressing cells indicates equal expression of FLAG-tagged ubiquitin in Flag-Ub plasmid DNA-transfected cells. Depending upon the experimental design, at this stage researchers may also analyse PTEN expression by immunoblotting to allow the immunopurification of similar quantities of PTEN protein from each sample in subsequent steps.
14. For detecting PTEN ubiquitination, immunoprecipitate PTEN from an equal volume of the diluted cell lysates containing 1 mg of soluble protein using A2B1 antibody by following the below steps:
    (a) Cut the narrow end of a P-200 pipette tip and transfer 10 μl of 50 % protein G sepharose bead slurry to a 1.5 ml microfuge tube (*see* **Note 8**). Wash the beads thrice with 100 μl ice-cold PBS by centrifuging at 1000 × *g* for 1 min.
    (b) Resuspend the beads in 500 μl of ice-cold PBS. Pre-couple the protein G sepharose beads with the antibody by incubating it with 4 μg PTEN A2B1 antibody at 4 °C for 1 h while gently mixing on a rotating platform. Centrifuge the antibody-coupled beads at 2000 × *g* for 1 min at 4 °C and discard the supernatant.
    (c) Simultaneously with the previous steps, incubate the cell lysate with 10 μl of inert sepharose beads (Sepharose 4B-CL)

that have been pre-washed three times in ice-cold PBS (*see* **Note 8**). Incubate lysates and beads for 1 h at 4 °C while gently rotating on a microfuge tube rotator. This step is an additional preclearing step where aggregated proteins and those that bind non-specifically to sepharose are bound to the inert matrix and are removed from the cell extract. Centrifuge it at 2000 × *g* for 1 min at 4 °C and collect the cleared lysate.

(d) Add this pre-cleared cell lysate to the prepared PTEN antibody coupled protein G sepharose beads on ice. Make the final volume of this mixture to 1 ml with ice-cold cell lysis buffer (*see* **Note 14**).

(e) Incubate the cell lysate-bead mixture at 4 °C for 3 h with rotation.

(f) Spin down the beads at 2000 × *g* for 1 min. Aspirate the supernatant. Wash the beads with the washing buffer thrice (*see* **Notes 14–16**).

15. Aspirate the residual wash buffer and add 20 μl of 1× LDS sample loading buffer containing 10 % 2-mercaptoethanol.
16. Heat the samples at 70 °C for 10 min to release the proteins bound to the beads.
17. Resolve the samples on 4–12 % SDS-PAGE gel.
18. Electroblot the resolved samples onto polyvinylidene difluoride (PVDF) membrane.
19. Detect ubiquitinated PTEN with anti-FLAG antibody M2: Following electro blotting, dry the PVDF membrane for 30 min and then block the membrane in 5 % (w/v) low-fat milk made up in TBS-T (10 mM Tris–HCl, pH 7.6, 150 mM NaCl, 0.1 % Tween 20). Wash the membrane over 15 min with three buffer changes of TBS-T and incubate the blocked PVDF membrane overnight at 4 °C with anti-FLAG M2 antibody (200 ng/ml) made up in 2 % (w/v) BSA. Wash the membrane over 30 min with three buffer changes of TBS-T and incubate it with HRP-conjugated anti-mouse secondary antibody (1:10,000 dilution) made up in 5 % (w/v) low-fat milk for 1 h. Remove the unbound antibody by washing the membrane over 30 min with four buffer changes of TBS-T. Detect FLAG-Ubiquitin conjugated proteins bound to HRP-antibody using chemiluminescent HRP substrate, ECL (Millipore, Waltham, MA, USA). If ubiquitinated PTEN is being cleanly detected in PTEN immunoprecipitates, signals should be absent or very low from samples not expressing PTEN and also cells not expressing FLAG-Ub.
20. Strip the blot and reprobe the membrane with anti-PTEN antibody (A2B1) to detect the immunoprecipitated PTEN (e.g., Fig. 2).

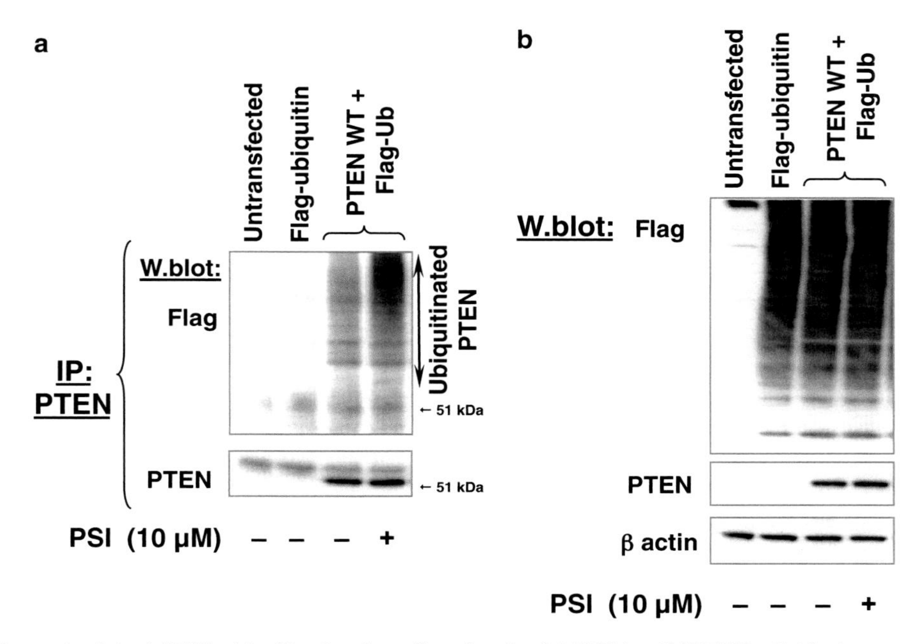

Fig. 2 Assay to detect PTEN ubiquitination in cultured cells. (**a**) PTEN-null U87MG glioblastoma cells were transfected with expression vectors encoding FLAG-ubiquitin (Flag-Ub) either alone or in combination with a vector encoding untagged wild-type PTEN (PTENWT). After 24 h, a sample of cells coexpressing FLAG-ubiquitin and PTEN was then treated with 10 μM PSI for 2 h. Cells were lysed, and PTEN was immunoprecipitated (IP) before Western blotting (W.blot) these immunoprecipitates with antibodies raised against the FLAG epitope and PTEN. A smear of high molecular weight in FLAG western blot represents polyubiquitinated PTEN. Note that this apparent polyubiquitinated PTEN protein accumulated when the cells were treated with a proteasome inhibitor for 2 h, correlating with the inhibition of PTEN protein degradation due to proteasomal inhibition (positive control for the experiment). (**b**) An anti-FLAG western blot of the total cell lysates showing similar cellular protein ubiquitination levels in the FLAG-ubiquitin-expressing cells. This figure was originally published in Journal of Biological Chemistry. Maccario H., Perera N.M., Gray A., Downes, C.P., and Leslie, N.R. Ubiquitination of PTEN (phosphatase and tensin homolog) inhibits phosphatase activity and is enhanced by membrane targeting and hyperosmotic Stress. *J Biol Chem. 2010;* 285:12620–12628. © the American Society for Biochemistry and Molecular Biology)

4 Notes

1. Prepare PTEN ubiquitination buffer fresh each time using frozen aliquots of 1 M DTT in 10 mM sodium acetate (pH 5.2).
2. Expression vector construction/origin for ubiquitin and ubiquitin activating enzyme E1 and described in [12]. Examples are also available from Addgene.
3. Polyubiquitin chain topology is strongly influenced by the specific E2 ligase protein used. However, although there is evidence to support specific E3 ligases in PTEN ubiquitination, the identity of E2 ligase(s) is even less certain. Our in vitro

experiments have used UbcH5b which in this context is likely to act as a promiscuous E2. More physiological in vitro PTEN ubiquitination would benefit from the identification of relevant E2 component(s).

4. GST-NEDD4 is expressed in *E. coli* BL21 cells transformed with pGEX-NEDD4 (construction as described in [12], also available from Addgene). Expression is induced in exponentially growing cultures with 200 μM IPTG at 20 °C for 18 h and purified using GSH-Sepharose 4B (GE Healthcare, Little Chalfont, UK) following the manufacturer's guidelines and published protocols [13]. Purified protein is dialyzed at 4 °C into 50 mM Tris–HCl, 150 mM NaCl, 0.1 % 2-mercaptoethanol, and 10 % glycerol and snap frozen in 20 μl aliquots for storage at −80 °C.
5. Recombinant PTEN protein is expressed in *E. coli* cells as fusion proteins with glutathione S-transferase (GST). GST-PTEN is immobilized on GSH-Sepharose 4B bead slurry and untagged PTEN protein is purified by release from the GST-tag by PreScission protease enzyme. Add 10 % glycerol to the untagged purified PTEN protein aliquots, snap freeze in liquid nitrogen, and store the aliquots at −80 °C.
6. Before loading the samples onto the Precast SDS-polyacrylamide gels rinse the gels with tap water and gel wells at least three times with 1× running buffer.
7. Prepare the *N*-ethylmaleimide solution from fresh solid immediately before use to prevent hydrolysis of maleimide group.
8. Resuspend the protein G sepharose or inert sepharose beads thoroughly before dispensing.
9. To distinguish among polyubiquitinated and multi/monoubiquitinated PTEN include a mutant in which each lysine residue within ubiquitin is mutated to arginine (referred to as "ubiquitin K all R") in the experiment which blocks formation of polyubiquitin chains.
10. If it is required to measure PTEN activity after in vitro ubiquitination then after in vitro ubiquitin assay for 1 h, immunoprecipitate PTEN using anti-PTEN monoclonal antibody A2B1 in 50 mM Tris–HCl (pH 7.5) buffer containing 150 mM NaCl, and 5 mM 2-mercaptoethanol. Wash twice with reaction/wash buffer (25 mM Tris pH 7.5, 150 mM NaCl, 1 mM EGTA, and 2 mM DTT) before performing PTEN activity assays on the immune complexes. Perform control immunoprecipitation and phosphatase assays using sham-ubiquitinated PTEN that has been treated in parallel with one component of the ubiquitination assay omitted.
11. Since efficiency of transductions is usually greater than that of transfections, lentiviruses encoding for PTEN could also be employed to express PTEN and Flag-Ub protein in the cultured cells.

12. It is very important to include deubiquitinating protease inhibitors such as NEM in the cell lysis buffer as well as during immunoprecipitation and washing steps to prevent PTEN deubiquitination during the experimental procedure. This precludes subsequent assay of PTEN from cell lysates as NEM irreversibly alkylates the PTEN active site.
13. Do not include 2-mercaptoethanol in the cell lysis buffer as 2-mercaptoethanol reacts with NEM removing its ability to inhibit deubiquitinases.
14. In order to detect ubiquitination specifically on PTEN and not on non-covalently interacting proteins, we use stringent condition for cell lysis, immunoprecipitation, and washing. We incorporate 0.5 M NaCl in the washing buffer so that nonspecific binding to the beads is reduced and thus likely to be specific to PTEN.
15. Wash the immune complexes at least three times to avoid background signal.
16. Be very careful not to remove any beads during aspiration.

Acknowledgements

We would like to thank the members of the NRL for constructive discussions. This work was funded by the Medical Research Council (grant code G0801865).

References

1. Komander D, Rape M (2012) The ubiquitin code. Ann Rev Biochem 81:203–229
2. Popovic D, Vucic D, Dikic I (2014) Ubiquitination in disease pathogenesis and treatment. Nat Med 20:1242–1253
3. Dikic I, Robertson M (2012) Ubiquitin ligases and beyond. BMC Biol 10:22–24
4. Leslie NR, Batty IH, Maccario H et al (2008) Understanding PTEN regulation: PIP2, polarity and protein stability. Oncogene 27: 5464–5476
5. Alimonti A, Carracedo A, Clohessy JG et al (2010) Subtle variations in Pten dose determine cancer susceptibility. Nat Genet 42: 454–458
6. Trotman LC, Wang X, Alimonti A et al (2007) Ubiquitination regulates PTEN nuclear import and tumor suppression. Cell 128:141–156
7. Wang X, Trotman LC, Koppie T et al (2007) NEDD4-1 is a proto-oncogenic ubiquitin ligase for PTEN. Cell 128:129–139
8. Song MS, Salmena L, Carracedo A et al (2008) The deubiquitinylation and localization of PTEN are regulated by a HAUSP-PML network. Nature 455:813–817
9. Zhang J, Zhang P, Wei Y et al (2013) Deubiquitylation and stabilization of PTEN by USP13. Nat Cell Biol 15:1486–1494
10. Van Themsche C, Leblanc V, Parent S et al (2009) X-linked inhibitor of apoptosis protein (XIAP) regulates PTEN ubiquitination, content, and compartmentalization. J Biol Chem 284:20462–20466
11. Maddika S, Kavela S, Rani N et al (2011) WWP2 is an E3 ubiquitin ligase for PTEN. Nat Cell Biol 13:728–733
12. Maccario H, Perera NM, Gray A et al (2010) Ubiquitination of PTEN (phosphatase and tensin homolog) inhibits phosphatase activity and is enhanced by membrane targeting and hyperosmotic Stress. J Biol Chem 285:12620–12628
13. McConnachie G, Pass I, Walker SM et al (2003) Interfacial kinetic analysis of the tumour suppressor phosphatase, PTEN: evidence for activation by anionic phospholipids. Biochem J 371:947–955

Part IV

Methods to Study *PTEN* Localization

Chapter 12

Assessing PTEN Subcellular Localization

Anabel Gil, José I. López, and Rafael Pulido

Abstract

PTEN subcellular localization is fundamental in the execution of the distinct PTEN biological activities, including not only its $PI(3,4,5)P_3$ phosphatase activity when associated to membranes but also its subcellular compartment-specific interactions with regulatory and effector proteins, including those exerted in the nucleus. As a consequence, PTEN subcellular localization is tightly regulated in vivo by both intrinsic and extrinsic mechanisms. The plasma membrane/nucleus/cytoplasm partitioning of PTEN has been the focus of several studies, both from a mechanistic and from a disease-association point of view. Here, we summarize the current knowledge on PTEN plasma membrane/nucleus/cytoplasm distribution, and present subcellular fractionation, immunofluorescence, and immunohistochemical methods to study the distribution and shuttling of PTEN between these subcellular compartments.

Key words PTEN, Tumor suppression, Subcellular fractionation, Immunofluorescence, Immunohistochemistry

1 Introduction

PTEN (phosphatase and tensin homolog deleted in chromosome 10) is a unique gene whose protein product is a highly relevant tumor suppressor in a wide variety of tissues [1, 2]. A large number of human cancers carry mutations or deletions in the *PTEN* gene, or have dysregulated PTEN protein expression [3, 4], and germline *PTEN* mutations are associated with PHTS (PTEN harmatoma tumor syndrome) [5, 6]. PTEN is a potent negative effector of the pro-oncogenic and pro-survival PI3K/AKT pathway by dephosphorylating the major product of PI3K activity, phosphatidylinositol 3,4,5-trisphosphate [$PI(3,4,5)P_3$] [7]. Consequently, changes in the expression and function of PTEN protein have important consequences in tumor initiation and progression, as well as in the response to anticancer therapies, which is currently being translated to the clinics [8–10]. The $PI(3,4,5)P_3$ phosphatase activity of PTEN takes place in association with the inner face of the plasma membrane, where $PI(3,4,5)P_3$ is generated by PI3K in response to

Leonardo Salmena and Vuk Stambolic (eds.), *PTEN: Methods and Protocols*, Methods in Molecular Biology, vol. 1388, DOI 10.1007/978-1-4939-3299-3_12,

growth and survival factors. Accordingly, PTEN actively shuttles from the cytoplasm to the plasma membrane, which is essential for its tumor suppressor function [11–13]. PTEN is also present in other cell compartments, including mitochondria, endoplasmic reticulum, secretory vesicles, and exosomes [14, 15]. In addition, PTEN is localized in the nucleus in many cell types and tissues [16, 17], and nucleolar PTEN localization has also been described [18].

PTEN nuclear accumulation associates in a variety of cancers with tumor stage or prognosis, although variability in nuclear/cytoplasmic detection exists depending on the anti-PTEN antibodies and the immunohistochemistry (IHC) protocols used [19–22]. Low nuclear PTEN expression has been documented in thyroid carcinomas [23], endocrine pancreatic tumors [24], and melanoma [25], and lack of nuclear PTEN correlated with lower survival or poor clinical outcome in esophageal squamous cell carcinoma [26], gastric cancer [27], prostate cancer [28], and colorectal carcinoma [29, 30]. In addition, frequent PTEN nuclear localization has been observed in breast tumors [31]. Thus, nuclear PTEN appears to be required for tumor suppression in many tissues, although an excess of nuclear accumulation could also have oncogenic consequences by diminishing the pool of PTEN associated to the plasma membrane [32]. Remarkably, catalysis of $PI(3,4,5)P_3$ is not a central function of nuclear PTEN [33], and most of the tumor suppressor-related functions mediated by nuclear PTEN seem to be phosphatase or PI3K/AKT pathway independent. These include, among others, genome stability maintenance and DNA damage repair, regulation of cell cycle, apoptosis and differentiation, p53 stabilization, and gene expression control [16, 34, 17, 35, 36]. The specific functions of nuclear PTEN are relevant for the implementation, in tumors lacking PTEN, of alternative therapies to those based on directly targeting the PI3K/AKT pathway [37]. In this regard, PTEN loss sensitizes tumors for synthetic lethality triggered by PARP inhibitors [38].

The crystal structure of PTEN reveals an N-terminal protein tyrosine phosphatase (PTP) domain followed by a C2-membrane-binding domain, whose integrity determines proper subcellular localization and catalysis [39]. In addition, the N- and C-terminal tails of PTEN, as well as an internal loop in its C2 domain (C2 loop), are intrinsically disordered unstructured regions [40], which play important roles in the control of PTEN membrane binding and nuclear accumulation (Fig. 1a). Interaction of PTEN with the plasma membrane involves $PI(4,5)P_2$ binding through an N-terminal $PI(4,5)P_2$-binding motif (PBM), as well as electrostatic interactions between positively charged residues at PTEN PTP- and C2-domains and anionic lipids at the plasma membrane [41, 42, 39, 43–45]. Sumoylation of Lys266 at one of these PTEN regions (CBR3) facilitates plasma membrane association [46]. A variety of mechanisms have been proposed for PTEN nuclear accumulation,

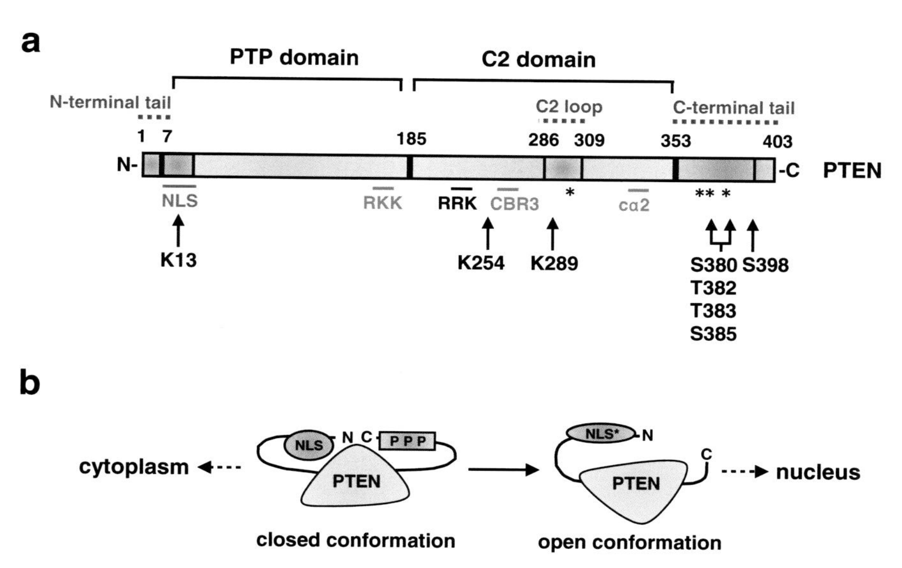

Fig. 1 (**a**) Molecular determinants involved in nuclear PTEN accumulation. PTEN protein (amino acids 1–403) is schematically depicted, with indication of its structured PTP and C2 domains. The PTEN N- and C-terminal tails (residues 1 to 7, and 353 to 403) and the internal unstructured C2 loop at the C2 domain (residues 286 to 309) are indicated in *blue*. Amino acids that suffer posttranslational modifications important for PTEN nuclear accumulation (Lys ubiquitylation [K13, K269] or Lys sumoylation [K254]; Ser/Thr phosphorylatable residues [S380, T382, T383, S385, S398]) are indicated. Note that multiple kinases are involved in the phosphorylation of the PTEN C-terminal tail [72]. Asterisks indicate caspase-3 cleavage sites [73]. The N-terminal nuclear localization sequence (NLS)-like (in *red*) (residues 8–32), the nuclear exclusion motifs (in *green*) RKK-PTP (RKK, residues 159–164), CBR3 (residues 263–269), and cα2 (residues 327–335), and the motif RRK-C2 (RRK, residues 233–237) are indicated [58]. The nuclear exclusion motifs RKK, CBR3, and cα2 are located on the PTEN structural elements TI-loop, CBR3-loop, and cα2-helix, respectively, and the RRK motif is located in the cβ3′-helix. Note that the CBR3-loop and the cα2-helix are also involved in binding to membranes [39]. In addition, the N-terminal NLS overlaps with the PI(4,5)P2-binding motif (residues 6–15) and with a cytoplasmic localization sequence (residues 19–25) [32, 74]. The N-terminal NLS is equivalent to the motif NLS1, and the RKK, RRK, and CBR3 motifs are equivalent to the motifs NLS2, NLS3, and NLS4 defined in [48]. See text for additional details. (**b**) Schematic of the PTEN nuclear accumulation model monitored with our experiments. In the PTEN closed conformation, the N-terminal NLS is non-competent to accumulate PTEN in the nucleus. Alterations on the PTEN C-terminal tail or on the PTEN nuclear exclusion motifs favor the PTEN open conformation, which display an N-terminal competent NLS (NLS*) to accumulate PTEN in the nucleus. See text for additional details

which are likely to operate in physiological conditions in a cell type- and cell stimulation-dependent manner. These include passive diffusion through the nuclear envelope [47]; nuclear entry facilitated by calcium-dependent interaction with the major vault protein (MVP) [48, 49]; nuclear accumulation by monoubiquitylation at residues Lys13 and Lys289 [50] or by sumoylation at residue Lys254 [51], which are under the control of different phosphorylation events [51–53]; nuclear or cytoplasmic retention by binding to nuclear or cytoplasmic proteins [54, 55]; CRM1-dependent nuclear export [56, 57]; cytoplasmic/nuclear shuttling regulated by a cytoplasmic localization signal [32]; and active shuttling by the Ran GTPase involving an N-terminal nuclear localization sequence (NLS) and various nuclear exclusion motifs [58]. Remarkably, several molecular determinants that govern PTEN nuclear accumulation are also directly involved in the targeting of PTEN to the plasma membrane. Such is the case of the overlapping N-terminal NLS and PBM, or the CBR3 and the cα2 regions in the C2 domain [16, 39, 59, 60] (Fig. 1a). This suggests the existence of a coordinated regulation of the targeting of both nuclear and plasma membrane PTEN. Some tumor-associated mutations at the *PTEN* gene prevent PTEN translocation to the plasma membrane and/or favor its nuclear accumulation [61, 62, 45]. However, other disease-associated mutations hamper PTEN nuclear accumulation [50] (Gil et al., unpublished), suggesting a complex relation between PTEN mutation in human disease and PTEN nuclear localization. Thus, the forced enhancement of plasma membrane binding of PTEN could be beneficial in cancer therapy, although the consequences on nuclear redistribution need to be considered.

The current model of PTEN regulation relies on the dynamic equilibrium between a closed cytoplasmic inactive conformation and an open plasma membrane/nuclear active conformation [16, 63]. PTEN conformational changes driving this equilibrium are thought to be mainly triggered by the integrity and phosphorylation status of the PTEN C-terminal tail, which affects its intramolecular interaction with the core of the protein [64–66] (Fig. 1b). A variety of mutations at the PTEN C-terminus, including C-terminal truncations or amino acid substitutions at the phosphorylatable- or caspase-3-targeted-residues, as well as mutations at the nuclear exclusion motifs, cause PTEN nuclear accumulation in a manner dependent on the PTEN N-terminal NLS [58, 16]. Here, using this model as a paradigm, we provide rapid subcellular fractionation and immunofluorescence methods to study PTEN nuclear accumulation on cells ectopically expressing PTEN. In addition, we provide an IHC method suitable to analyze the PTEN subcellular localization pattern found in tumor tissues.

2 Materials

All solutions are prepared in double-distilled, deionized MilliQ filtered water. In the case of the IHC protocol (*see* Subheading 2.3), commercial solutions are used for standardization. Cell culture and transfection procedures require sterile conditions.

2.1 PTEN Subcellular Fractionation

1. Tissue culture plates.
2. Simian kidney COS-7 cells, suitable for transfection and transient overexpression of recombinant proteins (*see* **Note 1**).

3. Complete medium: Dulbecco's modified Eagle medium (DMEM) containing high glucose supplemented with 5 % heat-inactivated FBS, 1 mM L-glutamine, 100 U/ml penicillin, and 0.1 mg/ml streptomycin.
4. Trypsin-EDTA solution.
5. cDNAs of PTEN wild type (wt) and mutations, cloned into suitable mammalian expression vectors (*see* **Note 2**).
6. Transfection reagents (*see* **Note 3**).
7. Ice-cold phosphate-buffered saline (PBS): 1.85 mM NaH_2PO_4, 8.4 mM $NaHPO_4$, and 150 mM NaCl.
8. Buffer A (hypotonic lysis buffer): 5 mM Na_2HPO_4 pH 7.4, 50 mM NaCl, 150 mM sucrose, 5 mM KCl, 2 mM DTT, 1 mM $MgCl_2$, 0.5 mM $CaCl_2$, 1 mM PMSF.
9. Buffer A containing 0.2 % Nonidet P-40: NP-40; IGEPAL CO-630.
10. Buffer B: 30 % w/v sucrose, 2.5 mM Tris–HCl pH 7.4, 10 mM NaCl.
11. Buffer C: 50 mM Tris–HCl pH 7.4, 300 mM NaCl, 0.5 % Triton X-100.
12. Sodium dodecyl sulfate-polyacrylamide gel electrophoresis (SDS-PAGE) loading buffer (2×; reducing): 100 mM Tris–HCl, pH 6.8, 4 % SDS, 4 % bromophenol blue, 20 % glycerol, 4 % 2-mercaptoethanol.
13. Polyvinylidene fluoride (PVDF) protein transfer membranes.
14. Transfer buffer: 48 mM Tris base, 39 mM glycine, 0.037 % SDS, 20 % methanol.
15. Prestained molecular weight standard protein markers.
16. Primary antibodies: The primary antibodies used in the experiments shown to illustrate this technique are anti-PTEN (Upstate) and anti-GAPDH (Sigma-Aldrich) rabbit polyclonal antibodies, and anti-PCNA mouse monoclonal antibody (mAb) (Santa Cruz Biotechnology) (*see* **Note 4**).
17. Horseradish peroxidase (HRP)-conjugated anti-rabbit and anti-mouse secondary antibodies (*see* **Note 5**).
18. Incubation buffer for immunoblot analysis (NET-gelatin buffer): 50 mM Tris–HCl, pH 7.5, 150 mM NaCl, 5 mM EDTA, 0.05 % Triton X-100, 0.25 % gelatin.
19. Stripping solution: 0.2 N NaOH, 1 % SDS.
20. Immunoblot chemiluminescence kit (*see* **Note 5**).

2.2 PTEN Immunofluorescence

1. See materials 1 to 7 in Subheading 2.1.
2. Poly-L-lysine, glass cover slips and slides (*see* **Note 6**).

3. −20 °C Methanol.
4. PBS-BSA solution: PBS containing 3 % bovine serum albumin (BSA).
5. Primary antibodies: The primary antibody used in the experiments shown to illustrate this technique is the anti-PTEN 425A mouse mAb, which recognizes an epitope within the PTEN C2 domain [67] (*see* **Note 4**).
6. Fluorescein (FITC)-conjugated anti-mouse secondary antibody.
7. Hoechst 33258 (Molecular Probes) (nuclear staining).
8. FluorSave™ Reagent (Calbiochem) (mounting fluid).
9. Fluorescence microscope (*see* **Note 7**).

2.3 PTEN Immunohistochemistry

1. Formalin-fixed, paraffin-embedded (FFPE) tissue sections.
2. Xylene.
3. Ethanol: 96° and absolute.
4. EnVision™ FLEX+ detection kit (Dako), and Dako PT Link and Dako Autostainer Plus instruments (*see* **Note 8**).
5. PBS (Biomerieux).
6. Primary antibodies: PTEN was detected using the anti-PTEN 425A and 17.A mAbs (culture supernatants). 425A mAb recognizes an epitope within the PTEN C2 domain; 17.A mAb recognizes an epitope at the PTEN C-terminal tail [67, 68] (*see* **Note 4**).

3 Methods

The methods presented are suitable to analyze PTEN subcellular localization from mammalian cells overexpressing recombinant PTEN or from tumor tissue sections. The *first method* (*see* Subheading 3.1) is modified from ref. [69], and permits the rapid isolation of nuclear and cytoplasmic cell fractions by hypotonic cell lysis followed by membranes solubilization and centrifugation steps. PTEN protein content in the fractions is analyzed by immunoblot using anti-PTEN antibodies. The *second method* (*see* Subheading 3.2) allows visualization by immunofluorescence of the subcellular distribution of ectopically expressed PTEN, using anti-PTEN antibodies. When performed in combination, these two methods provide reliable information on the nucleus/cytoplasm subcellular distribution of overexpressed recombinant PTEN. This is of utility to analyze the intrinsic mechanisms that regulate PTEN subcellular localization, such as those affecting PTEN conformation and involving specific targeting motifs at PTEN amino acid sequence. Examples are provided using a PTEN compound mutation targeting PTEN C-terminal phosphorylation sites (PTEN QMA; mutation S370A/S380A/T382A/T383A/S385A) and a PTEN C-terminal truncation (PTEN 1–375; mutation Δ376–403). Ectopically expressed PTEN QMA and PTEN 1–375 accumulate in the nucleus of COS-7 and other cells in a manner dependent on the PTEN N-terminal NLS [58]. The overexpression of PTEN in these conditions could mask the influence of other extrinsic regulatory mechanisms of PTEN subcellular localization, such as binding to targeting or regulatory proteins. In these cases, the analysis of PTEN expressed at physiologic levels is recommended. The *third method* (*see* Subheading 3.3) illustrates the visualization of subcellular localization of endogenous PTEN from tumor tissue sections, using an IHC protocol and two anti-PTEN mAb recognizing different PTEN epitopes. Information obtained using this technique has potential clinical relevance, and it is expected that standardized IHC analysis of PTEN expression and subcellular distribution from tumor tissue sections will be routinely performed in the clinics.

3.1 PTEN Subcellular Fractionation

1. Plate cells on 10 cm diameter tissue culture plates (2×10^6 cells/plate).
2. After 24 h of culture, transfect with 2 μg of plasmid containing PTEN wt or mutations, or with empty vector (*see* **Notes 2** and **3**).
3. After 48 h of culture, harvest the cells with trypsin-EDTA solution, centrifuge $500 \times g$, 6 min, 4 °C, and wash the pellet twice with ice-cold PBS.
4. Resuspend the cell pellet in buffer A (200 μl/10 cm plate), and keep in ice for 30 min (*see* **Note 9**).
5. Centrifuge cellular extract at $1000 \times g$, 10 min, 4 °C. The supernatant after centrifugation will be considered as the cytoplasmic fraction.
6. Resuspend the pellet with buffer A (200 μl/10 cm plate) containing 0.2 % NP-40 (*see* **Note 9**), keep in ice 5 min, and centrifuge again as above to eliminate cytoplasmic remainders and solubilized protein membranes.
7. Carefully discharge the supernatant.
8. The pellet resulting will be resuspended in buffer A (100 μl/10 cm plate) and load dropwise onto a tube containing solution B (1 ml/10 cm plate), without mixing.
9. Centrifuge at $1000 \times g$, 10 min, 4 °C.
10. Carefully discharge the supernatant.
11. Repeat **steps 8–10**.
12. Extract the nuclear pellet with buffer C (40 μl/10 cm plate) for 30 min in ice.

13. Centrifuge the nuclear extract at 10,000 × *g*, 10 min, 4 °C. The supernatant will be considered as the nuclear fraction.
14. Quantify cytoplasmic and nuclear fractions relative to a BSA standard using a Bradford (or other suitable) protein concentration assay (*see* **Note 10**).
15. Mix equal volumes of 2× SDS-PAGE reducing-loading buffer and fractions containing 10–20 μg of protein (*see* **Note 11**).
16. Boil for 3 min, spin, and load the mix in a 10 % SDS-PAGE gel. Include a lane with prestained molecular weight standard protein markers.
17. Run the gel and transfer to a PVDF membrane. Cut a piece of the membrane in the range of 20–75 kDa (*see* **Note 12**).
18. *Optional:* Stain the membrane with Ponceau solution to check the transfer and presence of the proteins on the membrane.
19. Perform standard immunoblot and chemiluminescence techniques, using anti-PTEN primary antibodies (*see* **Notes 4**, **5**, and **13**) and specific secondary antibodies.
20. To monitor the subcellular fractionation, run additional samples and probe with anti-PCNA (nuclear marker) and anti-GAPDH (cytoplasmic marker) antibodies (*see* **Note 12**) (or with other suitable markers). Alternatively, the membranes used for PTEN immunoblot can be stripped with stripping solution at 20 °C for 30 min, and reprobed with subcellular fractionation markers.

An example is provided of the subcellular distribution of PTEN wt and a PTEN compound mutation targeting PTEN C-terminal phosphorylation sites (PTEN QMA; mutation S370A/S378A/T382A/T383A/S385A), overexpressed in COS-7 cells (Fig. 2). PTEN wt is present in both the cytoplasmic and the nuclear fractions, whereas PTEN QMA is highly enriched in the nuclear fraction. Note that this technique does not give information on PTEN subcellular distribution at the single-cell level, but rather on the whole-cell population analyzed.

3.2 PTEN Immunofluorescence

1. Place a poly-L-Lys-coated glass cover slip in the bottom of each well of a 12-w tissue culture plate (*see* **Note 6**), and plate the cells (6×10^4 cells/well).
2. After 24 h of culture, transfect the cells with 1 μg of plasmid containing PTEN wt or mutations, or with empty vector (*see* **Notes 2** and **3**).
3. After 48 h of culture, wash the cells twice with ice-cold PBS.
4. Fix and permeabilize the cells in 1 ml of −20 °C methanol, 5 min.
5. Block with 1 ml of ice-cold PBS-BSA, 10 min.

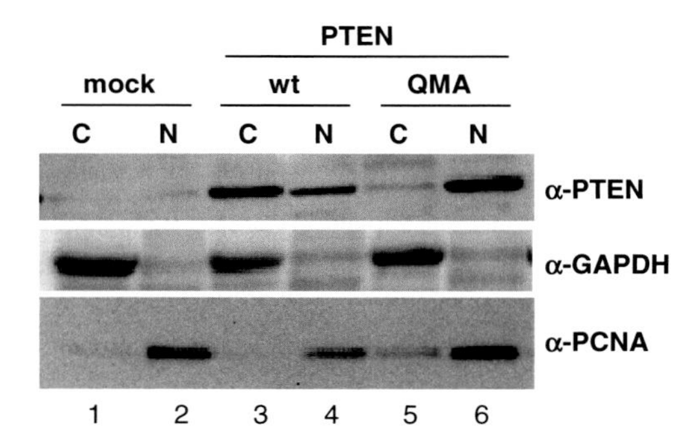

Fig. 2 Subcellular fractionation of COS-7 cells overexpressing PTEN. COS-7 cells were transfected with pRK5 alone (mock), pRK5 PTEN wild type (wt), or pRK5 PTEN QMA (mutation S370A/S380A/T382A/T383A/S385A), and cytoplasmic (C) and nuclear (N) fractions were isolated by subcellular fractionation. Samples were resolved by 10 % SDS-PAGE and subjected to immunoblot with anti-PTEN (rabbit polyclonal; Upstate; 1:1000), anti-GAPDH (cytoplasmic marker; rabbit polyclonal; Sigma-Aldrich; 1:5000), and anti-PCNA (nuclear marker; mouse monoclonal; Santa Cruz Biotechnology; 1:1000) antibodies. Antibodies were diluted in NET-gelatin buffer. PTEN wt was present in both the cytoplasmic and nuclear fractions, whereas PTEN QMA was enriched in the nuclear fraction

6. Remove the PBS-BSA and repeat **step 5** twice.
7. Transfer the plate to a wet chamber and lay on the cover slip 50 μl of the anti-PTEN antibody, diluted in PBS-BSA (*see* **Note 4**).
8. Incubate for 1 h, 37 °C.
9. Wash three times with PBS, 10 min, in a swing shaker.
10. Lay on the cover slip 50 μl of FITC-conjugated secondary antibody, diluted 1:200 in PBS-BSA (*see* **Note 14**).
11. Incubate in the wet chamber, 1 h, 20 °C, in darkness (*see* **Note 15**).
12. Wash three times with PBS, 10 min.
13. Add Hoechst 33258 (dilution 1:1000 in PBS-BSA) to stain the nucleus.
14. Using tweezers, carefully fish out the cover slips and mount, inverted, onto glass slides using FluorSave mounting-fluid. Leave it to solidify for 1 h, 20 °C.
15. Examine the cells using a fluorescence standard or confocal microscope (*see* **Note 7**).

Examples of visualization of the subcellular distribution of PTEN wt, PTEN 1–375 (mutation lacking the PTEN 376–403 C-terminal residues), and PTEN K13A (mutation targeting the PTEN N-terminal NLS) are provided, using COS-7 and U87MG cells (Fig. 3). At the single-cell level, PTEN wt distributes in the cytoplasm or in the cytoplasm and the nucleus, whereas PTEN

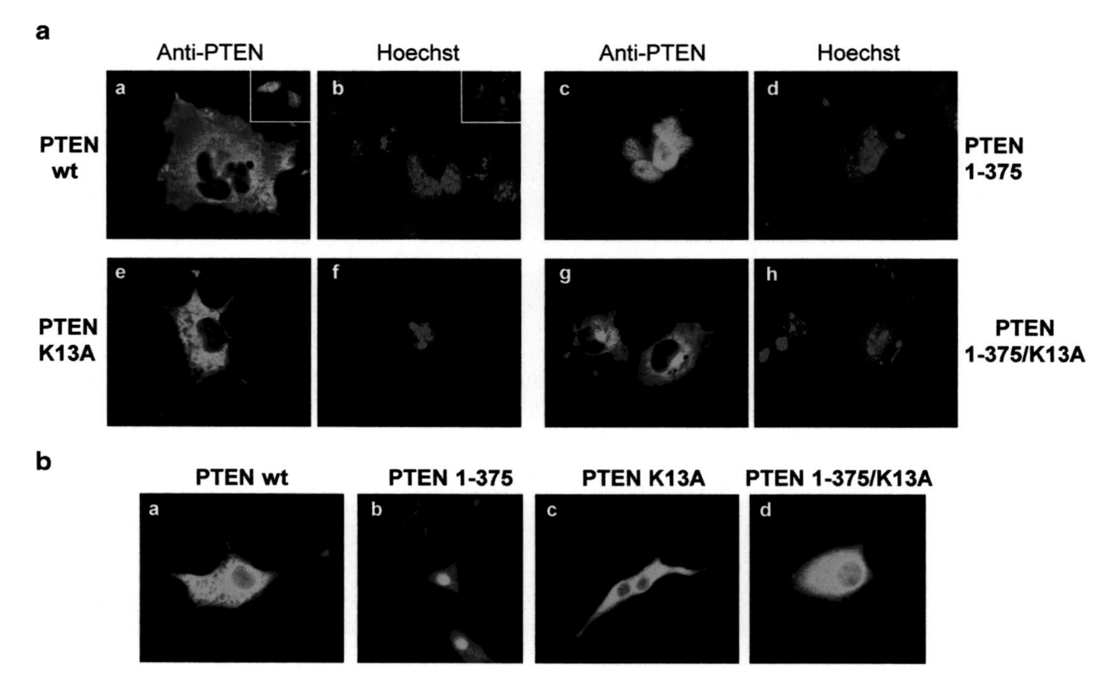

Fig. 3 Immunofluorescence of COS-7 and U87MG cells overexpressing PTEN. Cells were transfected with pRK5 PTEN wt or mutations, as indicated. PTEN 1–375 lacks the 376–403 C-terminal residues. The K13A mutation targets the PTEN N-terminal NLS. Cells were processed for immunofluorescence using anti-PTEN 425A (mouse monoclonal; ref. [67]; ascites fluid at dilution 1:200) and anti-mouse-FITC (*green*) antibodies. Antibodies were diluted in PBS-BSA. Nuclei were stained with Hoechst 33258. Representative images of the localization of each PTEN protein are shown. (**a**) Representative examples of localization of PTEN in COS-7 cells (*a, c, e, g*) are shown. Nuclei were stained with Hoechst (*b, d, f, h*). The inset in the PTEN wt panels (insets in *a* and *b*) illustrates cells expressing PTEN both in the cytoplasm and in the nucleus. (**b**) Representative examples of localization of PTEN wt or mutations in U87MG cells are shown

1–375 distributes mostly in the nucleus. Note that the K13A mutation impedes the nuclear accumulation of PTEN 1–375 ([58]; *see* Fig. 1b for a model). For quantitation of PTEN subcellular distribution, we score at least 100 PTEN-positive cells for each experiment, and we rate the cells as showing nuclear staining, cytoplasmic staining, or staining within both the nucleus and the cytoplasm [58].

3.3 PTEN Immunohistochemistry

1. Place the slides containing the FFPE tissue sections on a tray/rack and deparaffinize by three washes in xylene, 3 min, followed by two washes in absolute ethanol, 3 min.
2. Rinse with running cold tap water to hydrate.
3. Perform antigen retrieval, 20 min, 95 °C, in Dako PT Link instrument. For anti-PTEN 17.A mAb, citrate buffer, pH 6.1,

is used (EnVision™ FLEX Target Retrieval Solution, Low pH). For anti-PTEN 425A mAb, Tris/EDTA buffer, pH 9, is used (EnVision™ FLEX Target Retrieval Solution, High pH).

4. Soak the slides with PBS and transfer to Dako Autostainer Plus instrument.
5. Run the staining protocol, using the following steps and incubation times (*see* **Note 15**):
   - Endogenous peroxidase blocking, 10 min.
   - Antibody incubation, 30 min; anti-PTEN 17.A and 425A mAbs were used as undiluted culture supernatants (10–20 μg/ml antibody concentration).
   - Dextran-coupled horseradish peroxidase (HRP)-secondary antibody, 30 min.
   - Chromogen incubation, 10 min.
   - Hematoxylin counterstaining, 5 min.
6. Wash with PBS and cold tap water, twice.
7. Wash to dehydrate: twice with 96° ethanol, 2 min; twice with absolute ethanol, 2 min; twice with xylene, 2 min.
8. Mount with mounting fluid and examine under the microscope.

Tissue sections stained with anti-PTEN 17.A and 425A mAbs are shown in Fig. 4. Note the difference in staining intensities depending on the antibody used. In general, the anti-PTEN 425A mAb displayed weaker reactivity. The staining patterns are variable when different tumor specimens are compared. In Fig. 4a, a cytoplasmic staining pattern is observed in esophageal carcinoma (panels *b* and *c*) and gastric adenocarcinoma (panels *e* and *f*) specimens, whereas a cytoplasmic pattern with enrichment in plasma membrane staining is observed in clear cell renal cell carcinoma (panels *h* and *i*). In Fig. 4b, a esophageal carcinoma specimen displaying nuclear/cytoplasmic pattern in many of the cells is shown (panels *b* and *c*).

4 Notes

1. Other mammalian cell lines can be used for transient transfection, such as the commonly used human embryonic kidney HEK293T cell line. In these cell lines, the detection of endogenous PTEN is negligible in comparison with that of overexpressed recombinant PTEN. Cells lacking endogenous PTEN, such as the U87MG glioblastoma cell line [58], can also be used (*see* also Subheading 3.2).

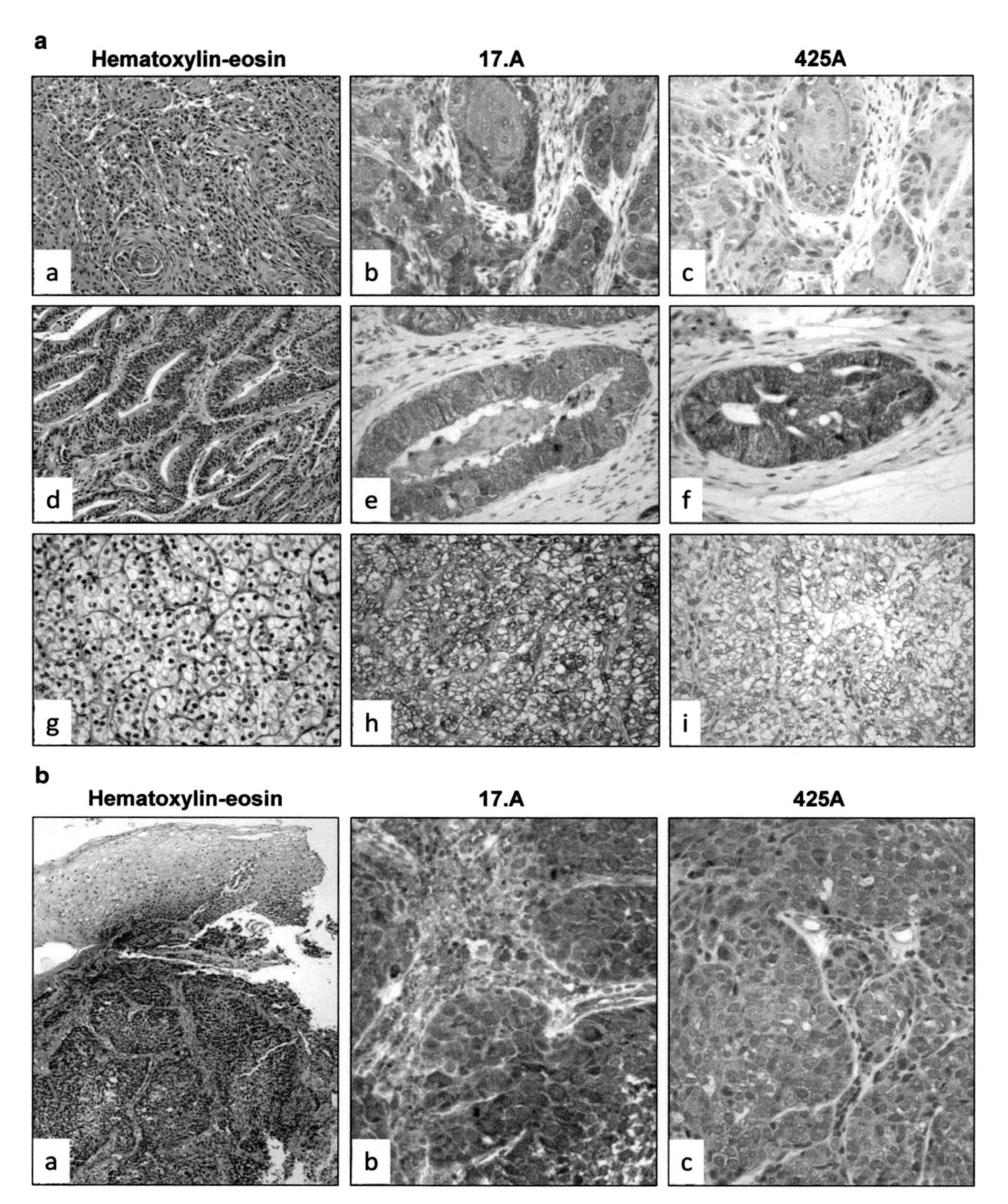

Fig. 4 Immunohistochemistry of tumor tissue sections using the 17.A and 425A anti-PTEN mAbs. (**a**) Patterns of cytoplasmic immunostaining of both anti-PTEN mAbs in well-differentiated squamous cell carcinoma of the esophagus (panels *b* and *c*), gastric adenocarcinoma (panels *e* and *f*), and clear cell renal cell carcinoma (panels *h* and *i*) (original magnification 400×). Panels *a*, *d*, and *g* show the hematoxylin-eosin staining of the same sections (original magnification 250×). (**b**) Nuclear and cytoplasmic immunostaining pattern of both anti-PTEN mAbs in a basaloid carcinoma of the esophagus (panels *b* and *c*) (original magnification 400×). Panel *a* shows the hematoxylin-eosin staining of the same section (original magnification 250×)

2. Our PTEN transient transfection experiments are made using the pRK5 mammalian expression vector, which contains the SV40 origin of replication and works efficiently in COS-7 or HEK293T cells expressing the SV40 large T-antigen. In the experiments shown here, we have used pRK5 PTEN (wt or mutations) plasmids, encoding untagged PTEN. Tagging of PTEN at its N-terminus can affect PTEN nuclear accumulation [58].
3. There are different suitable protocols for transfection of adherent mammalian cells that work well with some given cell lines. We routinely transfect COS-7 cells with good efficiency (20–40 % transfection efficiency) using the DEAE-dextran/DMSO method described in ref. [70], modified by adding chloroquine diphosphate (100 μl final concentration) to the DMSO solution [71]. Alternatively, commercial lipid transfection reagents also work well in COS-7 cells. We usually transfect U87MG cells with FuGENE 6.
4. Several commercial anti-PTEN antibodies are available. Dilution and washing conditions need to be individually tested for each antibody. See the figure legends for information on the usage of anti-PTEN antibodies in the experiments shown here.
5. If direct infrared fluorescence methods are used, suitable secondary infrared-emitting antibodies need to be used, and chemoluminiscence kit is not required.
6. Soak the cover slips in poly-L-Lys solution (10 μg/ml) for 4–16 h at 20 °C. Alternatively, commercial chamber plates can be used and handling of individual cover slips is not necessary.
7. In the experiments shown here, we have used a standard immunofluorescence microscope (Zeiss), which is well suited to visualize cytoplasmic versus nuclear localization. To visualize plasma membrane localization, a confocal microscope is better suited.
8. Buffers provided in the EnVision™ FLEX+ kit, and Dako PT Link and Dako Autostainer instruments, were used for standardization of staining conditions. *See* also **Note 16** for an alternate manual procedure.
9. Gentle pipetting up and down using a micropipette and yellow tip is advised. Cell integrity can be checked by diluting 5 μl of suspension in Trypan Blue solution.
10. Concentrations may range about 0.2–1 μg/μl.
11. Final loading volumes about 40–80 μl are convenient, depending on the size of your SDS-PAGE wells. A more concentrated SDS-PAGE sample buffer can be used to allow loading more volume of fraction sample.

12. Under reducing conditions, PTEN migrates as a band of about 55 kDa Mr. The PCNA nuclear marker migrates as a band of about 29 kDa. The GAPDH cytoplasmic marker migrates as a band of about 37 kDa.
13. The well transfected with empty vector gives information on the background of the primary antibody. Take into consideration the expression of endogenous PTEN, if any, depending on the cell line being used.
14. Other fluorescent conjugates, such as rhodamine-conjugates, can be used. To monitor the background of the secondary antibody, incubate an additional well only with it (no primary antibody in the assay). Dilution needs to be individually tested for each secondary antibody.
15. The secondary antibodies are light sensitive reagents. Incubations, washes, and mounting must be protected from light.
16. If the staining protocol is made manually, use the EnVision™ FLEX+ detection kit reagents and the incubation times indicated, washing with PBS between steps. Perform the incubations with the primary and secondary antibodies in a wet chamber.

Acknowledgements

The authors thank Mar González, Alicia Esteve, and Aida Larrañaga for their excellent immunohistochemical analysis. This work was supported in part by grants SAF2009-10226 from Ministerio Ciencia e Innovación (Spain and Fondo Europeo de Desarrollo Regional) and SAF2013-48812-R from Ministerio de Economía y Competitividad (Spain). A Gil has been the recipient of a research contract from Instituto de Salud Carlos III (CP04/00318) (Spain).

References

1. Song MS, Salmena L, Pandolfi PP (2012) The functions and regulation of the PTEN tumour suppressor. Nat Rev Mol Cell Biol 13(5):283–296. doi:10.1038/nrm3330
2. Worby CA, Dixon JE (2014) Pten. Annu Rev Biochem 83:641–669. doi:10.1146/annurev-biochem-082411-113907
3. Keniry M, Parsons R (2008) The role of PTEN signaling perturbations in cancer and in targeted therapy. Oncogene 27(41):5477–5485. doi:10.1038/onc.2008.248
4. Leslie NR, Foti M (2011) Non-genomic loss of PTEN function in cancer: not in my genes. Trends Pharmacol Sci 32(3):131–140. doi:10.1016/j.tips.2010.12.005
5. Blumenthal GM, Dennis PA (2008) PTEN hamartoma tumor syndromes. Eur J Hum Genet 16(11):1289–1300. doi:10.1038/ejhg.2008.162
6. Orloff MS, Eng C (2008) Genetic and phenotypic heterogeneity in the PTEN hamartoma tumour syndrome. Oncogene 27(41):5387–5397. doi:10.1038/onc.2008.237
7. Maehama T, Dixon JE (1999) PTEN: a tumour suppressor that functions as a phospholipid phosphatase. Trends Cell Biol 9(4):125–128

8. Carracedo A, Alimonti A, Pandolfi PP (2011) PTEN level in tumor suppression: how much is too little? Cancer Res 71(3):629–633. doi:10.1158/0008-5472.CAN-10-2488
9. Hollander MC, Blumenthal GM, Dennis PA (2011) PTEN loss in the continuum of common cancers, rare syndromes and mouse models. Nat Rev Cancer 11(4):289–301. doi:10.1038/nrc3037
10. Polivka J Jr, Janku F (2014) Molecular targets for cancer therapy in the PI3K/AKT/mTOR pathway. Pharmacol Ther 142(2):164–175. doi:10.1016/j.pharmthera.2013.12.004
11. Das S, Dixon JE, Cho W (2003) Membrane-binding and activation mechanism of PTEN. Proc Natl Acad Sci U S A 100(13):7491–7496. doi:10.1073/pnas.0932835100
12. Georgescu MM, Kirsch KH, Kaloudis P, Yang H, Pavletich NP, Hanafusa H (2000) Stabilization and productive positioning roles of the C2 domain of PTEN tumor suppressor. Cancer Res 60(24):7033–7038
13. Vazquez F, Matsuoka S, Sellers WR, Yanagida T, Ueda M, Devreotes PN (2006) Tumor suppressor PTEN acts through dynamic interaction with the plasma membrane. Proc Natl Acad Sci U S A 103(10):3633–3638. doi:10.1073/pnas.0510570103
14. Bononi A, Pinton P (2014) Study of PTEN subcellular localization. Methods. doi:10.1016/j.ymeth.2014.10.002
15. Kreis P, Leondaritis G, Lieberam I, Eickholt BJ (2014) Subcellular targeting and dynamic regulation of PTEN: implications for neuronal cells and neurological disorders. Front Mol Neurosci 7:23. doi:10.3389/fnmol.2014.00023
16. Gil A, Andrés-Pons A, Pulido R (2007) Nuclear PTEN: a tale of many tails. Cell Death Differ 14(3):395–399. doi:10.1038/sj.cdd.4402073
17. Planchon SM, Waite KA, Eng C (2008) The nuclear affairs of PTEN. J Cell Sci 121(Pt 3):249–253. doi:10.1242/jcs.022459
18. Li P, Wang D, Li H, Yu Z, Chen X, Fang J (2014) Identification of nucleolus-localized PTEN and its function in regulating ribosome biogenesis. Mol Biol Rep 41(10):6383–6390. doi:10.1007/s11033-014-3518-6
19. Carvalho KC, Maia BM, Omae SV, Rocha AA, Covizzi LP, Vassallo J, Rocha RM, Soares FA (2014) Best practice for PTEN gene and protein assessment in anatomic pathology. Acta Histochem 116(1):25–31. doi:10.1016/j.acthis.2013.04.013
20. Maiques O, Santacana M, Valls J, Pallares J, Mirantes C, Gatius S, Garcia Dios DA, Amant F, Pedersen HC, Dolcet X, Matias-Guiu X (2014) Optimal protocol for PTEN immunostaining; role of analytical and preanalytical variables in PTEN staining in normal and neoplastic endometrial, breast, and prostatic tissues. Hum Pathol 45(3):522–532. doi:10.1016/j.humpath.2013.10.018
21. Pallares J, Bussaglia E, Martinez-Guitarte JL, Dolcet X, Llobet D, Rue M, Sanchez-Verde L, Palacios J, Prat J, Matias-Guiu X (2005) Immunohistochemical analysis of PTEN in endometrial carcinoma: a tissue microarray study with a comparison of four commercial antibodies in correlation with molecular abnormalities. Mod Pathol 18(5):719–727. doi:10.1038/modpathol.3800347
22. Sangale Z, Prass C, Carlson A, Tikishvili E, Degrado J, Lanchbury J, Stone S (2011) A robust immunohistochemical assay for detecting PTEN expression in human tumors. Appl Immunohistochem Mol Morphol 19(2):173–183. doi:10.1097/PAI.0b013e3181f1da13
23. Gimm O, Perren A, Weng LP, Marsh DJ, Yeh JJ, Ziebold U, Gil E, Hinze R, Delbridge L, Lees JA, Mutter GL, Robinson BG, Komminoth P, Dralle H, Eng C (2000) Differential nuclear and cytoplasmic expression of PTEN in normal thyroid tissue, and benign and malignant epithelial thyroid tumors. Am J Pathol 156(5):1693–1700. doi:10.1016/S0002-9440(10)65040-7
24. Perren A, Komminoth P, Saremaslani P, Matter C, Feurer S, Lees JA, Heitz PU, Eng C (2000) Mutation and expression analyses reveal differential subcellular compartmentalization of PTEN in endocrine pancreatic tumors compared to normal islet cells. Am J Pathol 157(4):1097–1103. doi:10.1016/S0002-9440(10)64624-X
25. Whiteman DC, Zhou XP, Cummings MC, Pavey S, Hayward NK, Eng C (2002) Nuclear PTEN expression and clinicopathologic features in a population-based series of primary cutaneous melanoma. Int J Cancer 99(1):63–67
26. Tachibana M, Shibakita M, Ohno S, Kinugasa S, Yoshimura H, Ueda S, Fujii T, Rahman MA, Dhar DK, Nagasue N (2002) Expression and prognostic significance of PTEN product protein in patients with esophageal squamous cell carcinoma. Cancer 94(7):1955–1960
27. Bai Z, Ye Y, Chen D, Shen D, Xu F, Cui Z, Wang S (2007) Homeoprotein Cdx2 and nuclear PTEN expression profiles are related to gastric cancer prognosis. APMIS 115(12):1383–1390. doi:10.1111/j.1600-0463.2007.00654.x
28. McCall P, Witton CJ, Grimsley S, Nielsen KV, Edwards J (2008) Is PTEN loss associated with

clinical outcome measures in human prostate cancer? Br J Cancer 99(8):1296–1301. doi:10.1038/sj.bjc.6604680
29. Hsu CP, Kao TY, Chang WL, Nieh S, Wang HL, Chung YC (2011) Clinical significance of tumor suppressor PTEN in colorectal carcinoma. Eur J Surg Oncol 37(2):140–147. doi:10.1016/j.ejso.2010.12.003
30. Jang KS, Song YS, Jang SH, Min KW, Na W, Jang SM, Jun YJ, Lee KH, Choi D, Paik SS (2010) Clinicopathological significance of nuclear PTEN expression in colorectal adenocarcinoma. Histopathology 56(2):229–239. doi:10.1111/j.1365-2559.2009.03468.x
31. Bakarakos P, Theohari I, Nomikos A, Mylona E, Papadimitriou C, Dimopoulos AM, Nakopoulou L (2010) Immunohistochemical study of PTEN and phosphorylated mTOR proteins in familial and sporadic invasive breast carcinomas. Histopathology 56(7):876–882. doi:10.1111/j.1365-2559.2010.03570.x
32. Denning G, Jean-Joseph B, Prince C, Durden DL, Vogt PK (2007) A short N-terminal sequence of PTEN controls cytoplasmic localization and is required for suppression of cell growth. Oncogene 26(27):3930–3940. doi:10.1038/sj.onc.1210175
33. Lindsay Y, McCoull D, Davidson L, Leslie NR, Fairservice A, Gray A, Lucocq J, Downes CP (2006) Localization of agonist-sensitive PtdIns(3,4,5)P3 reveals a nuclear pool that is insensitive to PTEN expression. J Cell Sci 119(Pt 24):5160–5168. doi:10.1242/jcs.000133
34. Ming M, He YY (2012) PTEN in DNA damage repair. Cancer Lett 319(2):125–129. doi:10.1016/j.canlet.2012.01.003
35. Salmena L, Carracedo A, Pandolfi PP (2008) Tenets of PTEN tumor suppression. Cell 133(3):403–414. doi:10.1016/j.cell.2008.04.013
36. Yin Y, Shen WH (2008) PTEN: a new guardian of the genome. Oncogene 27(41):5443–5453. doi:10.1038/onc.2008.241
37. Dillon LM, Miller TW (2014) Therapeutic targeting of cancers with loss of PTEN function. Curr Drug Targets 15(1):65–79
38. Mendes-Pereira AM, Martin SA, Brough R, McCarthy A, Taylor JR, Kim JS, Waldman T, Lord CJ, Ashworth A (2009) Synthetic lethal targeting of PTEN mutant cells with PARP inhibitors. EMBO Mol Med 1(6-7):315–322. doi:10.1002/emmm.200900041
39. Lee JO, Yang H, Georgescu MM, Di Cristofano A, Maehama T, Shi Y, Dixon JE, Pandolfi P, Pavletich NP (1999) Crystal structure of the PTEN tumor suppressor: implications for its phosphoinositide phosphatase activity and membrane association. Cell 99(3):323–334
40. Malaney P, Pathak RR, Xue B, Uversky VN, Dave V (2013) Intrinsic disorder in PTEN and its interactome confers structural plasticity and functional versatility. Sci Rep 3:2035. doi:10.1038/srep02035
41. Campbell RB, Liu F, Ross AH (2003) Allosteric activation of PTEN phosphatase by phosphatidylinositol 4,5-bisphosphate. J Biol Chem 278(36):33617–33620. doi:10.1074/jbc.C300296200
42. Iijima M, Huang YE, Luo HR, Vazquez F, Devreotes PN (2004) Novel mechanism of PTEN regulation by its phosphatidylinositol 4,5-bisphosphate binding motif is critical for chemotaxis. J Biol Chem 279(16):16606–16613. doi:10.1074/jbc.M312098200
43. Lumb CN, Sansom MS (2013) Defining the membrane-associated state of the PTEN tumor suppressor protein. Biophys J 104(3):613–621. doi:10.1016/j.bpj.2012.12.002
44. Shenoy S, Shekhar P, Heinrich F, Daou MC, Gericke A, Ross AH, Losche M (2012) Membrane association of the PTEN tumor suppressor: molecular details of the protein-membrane complex from SPR binding studies and neutron reflection. PLoS One 7(4):e32591. doi:10.1371/journal.pone.0032591
45. Walker SM, Leslie NR, Perera NM, Batty IH, Downes CP (2004) The tumour-suppressor function of PTEN requires an N-terminal lipid-binding motif. Biochem J 379(Pt 2):301–307. doi:10.1042/BJ20031839
46. Huang J, Yan J, Zhang J, Zhu S, Wang Y, Shi T, Zhu C, Chen C, Liu X, Cheng J, Mustelin T, Feng GS, Chen G, Yu J (2012) SUMO1 modification of PTEN regulates tumorigenesis by controlling its association with the plasma membrane. Nat Commun 3:911. doi:10.1038/ncomms1919
47. Liu F, Wagner S, Campbell RB, Nickerson JA, Schiffer CA, Ross AH (2005) PTEN enters the nucleus by diffusion. J Cell Biochem 96(2):221–234. doi:10.1002/jcb.20525
48. Chung JH, Ginn-Pease ME, Eng C (2005) Phosphatase and tensin homologue deleted on chromosome 10 (PTEN) has nuclear localization signal-like sequences for nuclear import mediated by major vault protein. Cancer Res 65(10):4108–4116. doi:10.1158/0008-5472.CAN-05-0124
49. Minaguchi T, Waite KA, Eng C (2006) Nuclear localization of PTEN is regulated by Ca(2+) through a tyrosil phosphorylation-independent conformational modification in major vault protein. Cancer Res 66(24):11677–11682. doi:10.1158/0008-5472.CAN-06-2240
50. Trotman LC, Wang X, Alimonti A, Chen Z, Teruya-Feldstein J, Yang H, Pavletich NP, Carver BS, Cordon-Cardo C, Erdjument-

Bromage H, Tempst P, Chi SG, Kim HJ, Misteli T, Jiang X, Pandolfi PP (2007) Ubiquitination regulates PTEN nuclear import and tumor suppression. Cell 128(1):141–156. doi:10.1016/j.cell.2006.11.040

51. Bassi C, Ho J, Srikumar T, Dowling RJ, Gorrini C, Miller SJ, Mak TW, Neel BG, Raught B, Stambolic V (2013) Nuclear PTEN controls DNA repair and sensitivity to genotoxic stress. Science 341(6144):395–399. doi:10.1126/science.1236188
52. Li Z, Li J, Bi P, Lu Y, Burcham G, Elzey BD, Ratliff T, Konieczny SF, Ahmad N, Kuang S, Liu X (2014) Plk1 phosphorylation of PTEN causes a tumor-promoting metabolic state. Mol Cell Biol 34(19):3642–3661. doi:10.1128/MCB.00814-14
53. Wu Y, Zhou H, Wu K, Lee S, Li R, Liu X (2014) PTEN phosphorylation and nuclear export mediate free fatty acid-induced oxidative stress. Antioxid Redox Signal 20(9):1382–1395. doi:10.1089/ars.2013.5498
54. Andrés-Pons A, Gil A, Oliver MD, Sotelo NS, Pulido R (2012) Cytoplasmic p27Kip1 counteracts the pro-apoptotic function of the open conformation of PTEN by retention and destabilization of PTEN outside of the nucleus. Cell Signal 24(2):577–587. doi:10.1016/j.cellsig.2011.10.012
55. Kavela S, Shinde SR, Ratheesh R, Viswakalyan K, Bashyam MD, Gowrishankar S, Vamsy M, Pattnaik S, Rao S, Sastry RA, Srinivasulu M, Chen J, Maddika S (2013) PNUTS functions as a proto-oncogene by sequestering PTEN. Cancer Res 73(1):205–214. doi:10.1158/0008-5472.CAN-12-1394
56. Beckham TH, Cheng JC, Lu P, Marrison ST, Norris JS, Liu X (2013) Acid ceramidase promotes nuclear export of PTEN through sphingosine 1-phosphate mediated Akt signaling. PLoS One 8(10):e76593. doi:10.1371/journal.pone.0076593
57. Liu JL, Mao Z, LaFortune TA, Alonso MM, Gallick GE, Fueyo J, Yung WK (2007) Cell cycle-dependent nuclear export of phosphatase and tensin homologue tumor suppressor is regulated by the phosphoinositide-3-kinase signaling cascade. Cancer Res 67(22):11054–11063. doi:10.1158/0008-5472.CAN-07-1263
58. Gil A, Andrés-Pons A, Fernández E, Valiente M, Torres J, Cervera J, Pulido R (2006) Nuclear localization of PTEN by a Ran-dependent mechanism enhances apoptosis: Involvement of an N-terminal nuclear localization domain and multiple nuclear exclusion motifs. Mol Biol Cell 17(9):4002–4013. doi:10.1091/mbc.E06-05-0380
59. Nguyen HN, Afkari Y, Senoo H, Sesaki H, Devreotes PN, Iijima M (2013) Mechanism of human PTEN localization revealed by heterologous expression in Dictyostelium. Oncogene. doi:10.1038/onc.2013.507
60. Nguyen HN, Yang JM, Afkari Y, Park BH, Sesaki H, Devreotes PN, Iijima M (2014) Engineering ePTEN, an enhanced PTEN with increased tumor suppressor activities. Proc Natl Acad Sci U S A 111(26):E2684–E2693. doi:10.1073/pnas.1409433111
61. Lobo GP, Waite KA, Planchon SM, Romigh T, Nassif NT, Eng C (2009) Germline and somatic cancer-associated mutations in the ATP-binding motifs of PTEN influence its subcellular localization and tumor suppressive function. Hum Mol Genet 18(15):2851–2862. doi:10.1093/hmg/ddp220
62. Nguyen HN, Yang JM Jr, Rahdar M, Keniry M, Swaney KF, Parsons R, Park BH, Sesaki H, Devreotes PN, Iijima M (2014) A new class of cancer-associated PTEN mutations defined by membrane translocation defects. Oncogene. doi:10.1038/onc.2014.293
63. Vazquez F, Devreotes P (2006) Regulation of PTEN function as a PIP3 gatekeeper through membrane interaction. Cell Cycle 5(14):1523–1527
64. Bolduc D, Rahdar M, Tu-Sekine B, Sivakumaren SC, Raben D, Amzel LM, Devreotes P, Gabelli SB, Cole P (2013) Phosphorylation-mediated PTEN conformational closure and deactivation revealed with protein semisynthesis. eLife 2:e00691. doi:10.7554/eLife.00691
65. Odriozola L, Singh G, Hoang T, Chan AM (2007) Regulation of PTEN activity by its carboxyl-terminal autoinhibitory domain. J Biol Chem 282(32):23306–23315. doi:10.1074/jbc.M611240200
66. Rahdar M, Inoue T, Meyer T, Zhang J, Vazquez F, Devreotes PN (2009) A phosphorylation-dependent intramolecular interaction regulates the membrane association and activity of the tumor suppressor PTEN. Proc Natl Acad Sci U S A 106(2):480–485. doi:10.1073/pnas.0811212106
67. Andrés-Pons A, Valiente M, Torres J, Gil A, Roglá I, Ripoll F, Cervera J, Pulido R (2005) Functional definition of relevant epitopes on the tumor suppressor PTEN protein. Cancer Lett 223(2):303–312. doi:10.1016/j.canlet.2004.09.047
68. Torres J, Navarro S, Rogla I, Ripoll F, Lluch A, Garcia-Conde J, Llombart-Bosch A, Cervera J, Pulido R (2001) Heterogeneous lack of expression of the tumour suppressor PTEN protein in

human neoplastic tissues. Eur J Cancer 37(1): 114–121

69. Ishida N, Hara T, Kamura T, Yoshida M, Nakayama K, Nakayama KI (2002) Phosphorylation of p27Kip1 on serine 10 is required for its binding to CRM1 and nuclear export. J Biol Chem 277(17):14355–14358. doi:10.1074/jbc.C100762200
70. Lopata MA, Cleveland DW, Sollner-Webb B (1984) High level transient expression of a chloramphenicol acetyl transferase gene by DEAE-dextran mediated DNA transfection coupled with a dimethyl sulfoxide or glycerol shock treatment. Nucleic Acids Res 12(14): 5707–5717
71. Ausubel FM (2002) Short protocols in molecular biology: a compendium of methods from Current protocols in molecular biology, 5th edn. Wiley, New York
72. Fragoso R, Barata JT (2014) Kinases, tails and more: regulation of PTEN function by phosphorylation. Methods. doi:10.1016/j.ymeth.2014.10.015
73. Torres J, Rodriguez J, Myers MP, Valiente M, Graves JD, Tonks NK, Pulido R (2003) Phosphorylation-regulated cleavage of the tumor suppressor PTEN by caspase-3: implications for the control of protein stability and PTEN-protein interactions. J Biol Chem 278(33):30652–30660. doi:10.1074/jbc.M212610200
74. Maehama T, Taylor GS, Dixon JE (2001) PTEN and myotubularin: novel phosphoinositide phosphatases. Annu Rev Biochem 70:247–279. doi:10.1146/annurev.biochem.70.1.247

Chapter 13

Methods to Study PTEN in Mitochondria and Endoplasmic Reticulum

Sonia Missiroli*, Claudia Morganti*, Carlotta Giorgi, and Paolo Pinton

Abstract

Although PTEN has been widely described as a nuclear and cytosolic protein, in the last 2 years, alternative organelles, such as the endoplasmic reticulum (ER), pure mitochondria, and mitochondria-associated membranes (MAMs), have been recognized as pivotal targets of PTEN activity.

Here, we describe different methods that have been used to highlight PTEN subcellular localization.

First, a protocol to extract nuclear and cytosolic fractions has been described to assess the "canonical" PTEN localization. Moreover, we describe a protocol for mitochondria isolation with proteinase K (PK) to further discriminate whether PTEN associates with the outer mitochondrial membrane (OMM) or resides within the mitochondria. Finally, we focus our attention on a subcellular fractionation protocol of cells that permits the isolation of MAMs containing unique regions of ER membranes attached to the outer mitochondrial membrane (OMM) and mitochondria without contamination from other organelles. In addition to biochemical fractionations, immunostaining can be used to determine the subcellular localization of proteins; thus, a detailed protocol to obtain good immunofluorescence (IF) is described. The employment of these methodological approaches could facilitate the identification of different PTEN localizations in several physiopathological contexts.

Key words PTEN, Subcellular fractionation, Nuclear extraction, Immunofluorescence, Mitochondria, Endoplasmic reticulum, Mitochondria-associated membranes (MAMs)

1 Introduction

A phosphatase and tensin homolog deleted on chromosome 10 (PTEN) [1] is among the most commonly lost or mutated tumor suppressors in human cancers [2], and it is a key regulator of a wide range of biological functions in addition to tumor suppression. Recently, PTEN has been demonstrated to localize or associate with organelles and specialized subcellular compartments, such as the nucleus, nucleolus, mitochondria, ER, and MAMs [3–7].

*Both are contributed equally to this work

Leonardo Salmena and Vuk Stambolic (eds.), *PTEN: Methods and Protocols*, Methods in Molecular Biology, vol. 1388, DOI 10.1007/978-1-4939-3299-3_13, © Springer Science+Business Media New York 2016

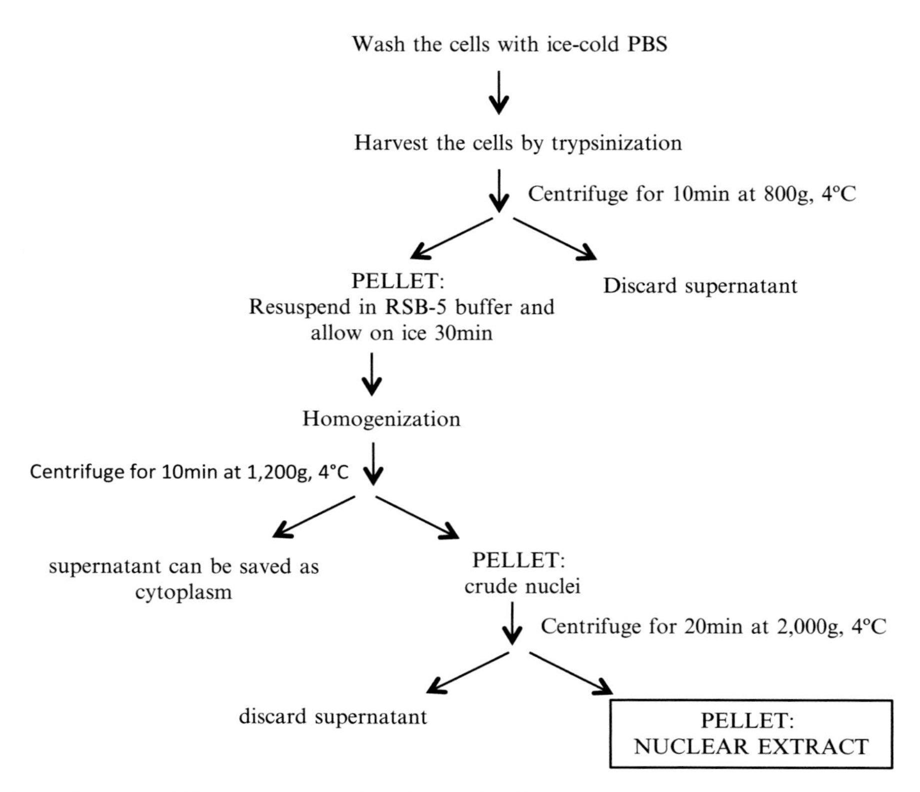

Fig. 1 Schematic steps of the nuclear extraction from cell culture

Here, we provide detailed protocols to investigate PTEN localization in the cell. A nuclear extraction protocol from cell culture can be used to determine the known localization of PTEN at the nucleus (*see* Fig. 1). Nevertheless, mitochondria isolation followed by PK digestion provides a detailed localization of a protein in the mitochondria (*see* Fig. 2). Moreover, the employment of PK enables discrimination between cytosolic proteins loosely bound to the OMM or intramitochondrial proteins integrated to outer or inner membranes. Using these different methods, it has been demonstrated that PTEN is not able to enter the mitochondria [8] and is predominately loosely bound to the OMM [3].

Using different approaches, including subcellular fractionation protocol [9] and the IF technique, Bononi et al. demonstrated that a fraction of PTEN localizes also to the ER and MAMs [3] (*see* Fig. 3, 4, 5, 6, and 7). These intracellular domains are involved in calcium (Ca^{2+}) transfer from the ER to mitochondria and apoptosis induction. Indeed, MAMs represent a specific molecular platform to receive Ca^{2+} signals from ER and transmit them to mitochondria, for the regulation of several processes including cell death [10–14].

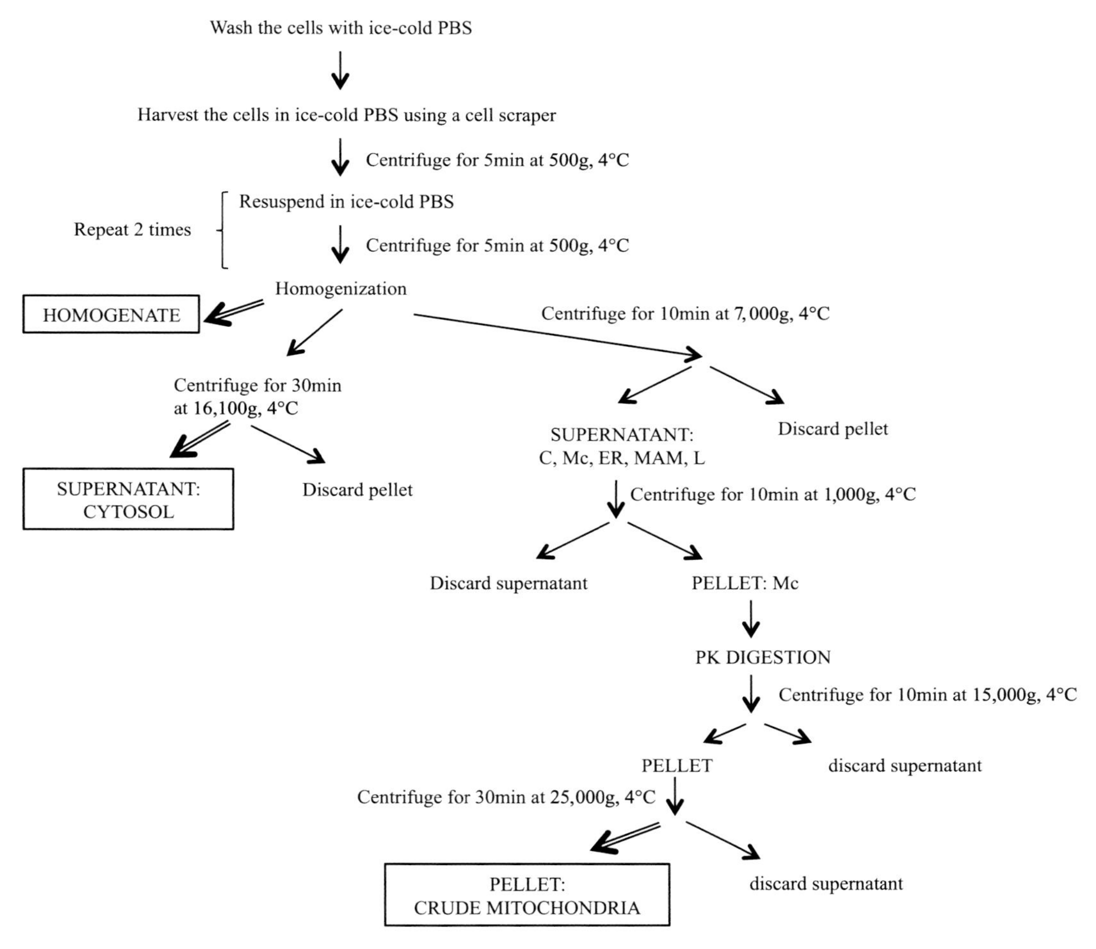

Fig. 2 Schematic representation of the mitochondrial isolation steps for the PK assay protocol. *C* cytosol; *H* homogenate; *ER* endoplasmic reticulum; *L* lysosomes; *MAMs* mitochondria-associated membranes; *Mc* crude mitochondria; *Mp* pure mitochondria; *PK* proteinase K

At the ER and MAMs, PTEN interacts with the inositol 1,4,5-trisphosphate receptors (IP3Rs) and regulates Ca^{2+} release from the ER in a protein phosphatase-dependent manner that counteracts Akt activation; thus, it can inhibit Akt-mediated phosphorylation of IP3R3 [15], which protects from Ca^{2+}-mediated apoptosis [3].

Intracellular fractionation techniques should be supported by other methods to further validate the localization of PTEN [16]. IF can be used to compare the localization of a protein of interest against known markers of intracellular structures. Here, we have provided a standard protocol of IF that can be used to simultaneously compare the localization of PTEN against known markers of the nucleus, mitochondria and ER (*see* Fig. 7). This protocol

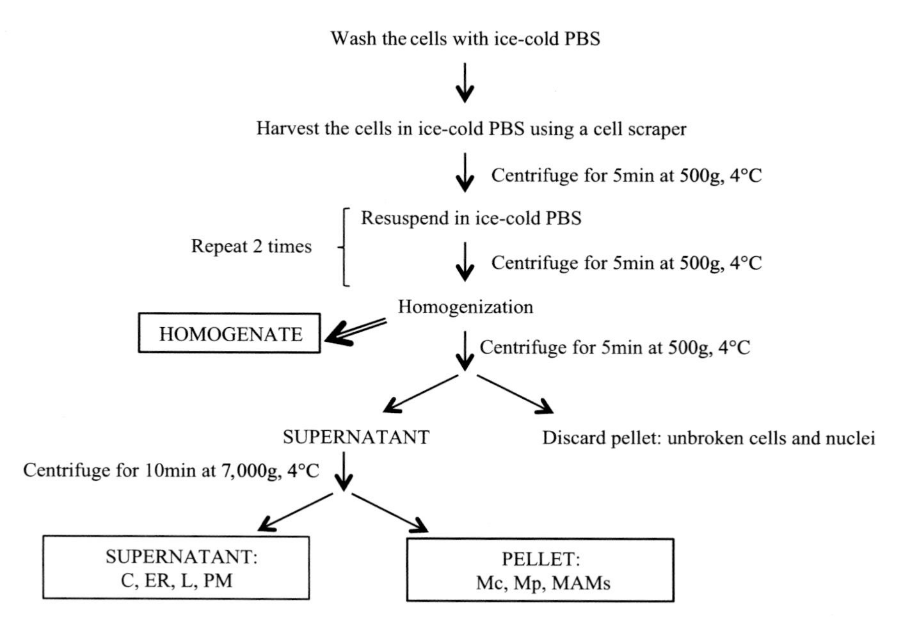

Fig. 3 Schematic steps of the subcellular fractionation protocol from cells. *C* cytosol; *ER* endoplasmic reticulum; *L* lysosomes; *MAMs* mitochondria-associated membranes; *Mc* crude mitochondria; *Mp* pure mitochondria; *PM* plasma membrane

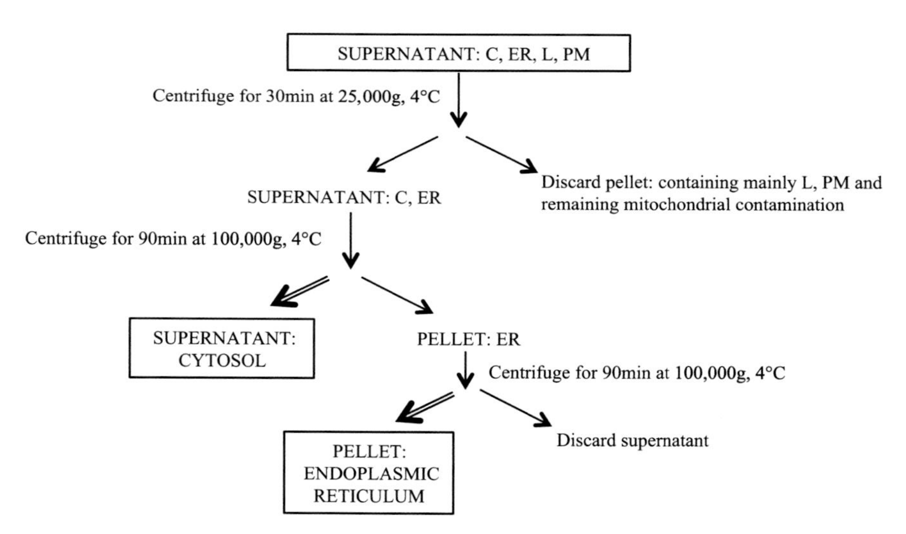

Fig. 4 Schematic steps of the subcellular fractionation protocol from cells for ER and C isolation. The final steps of the presented procedure result in the isolation of pure microsomes (endoplasmic reticulum (ER)) and cytosolic fraction. *C* cytosol; *ER* endoplasmic reticulum; *L* lysosomes; *PM* plasma membrane

can be used as a model to analyze unknown protein localization compared with markers of subcellular compartments, including MAMs. Tables 1, 2, and 3 indicate suitable alternative markers that can be used in IF localization analysis to ensure that fluorescence emission overlap is avoided.

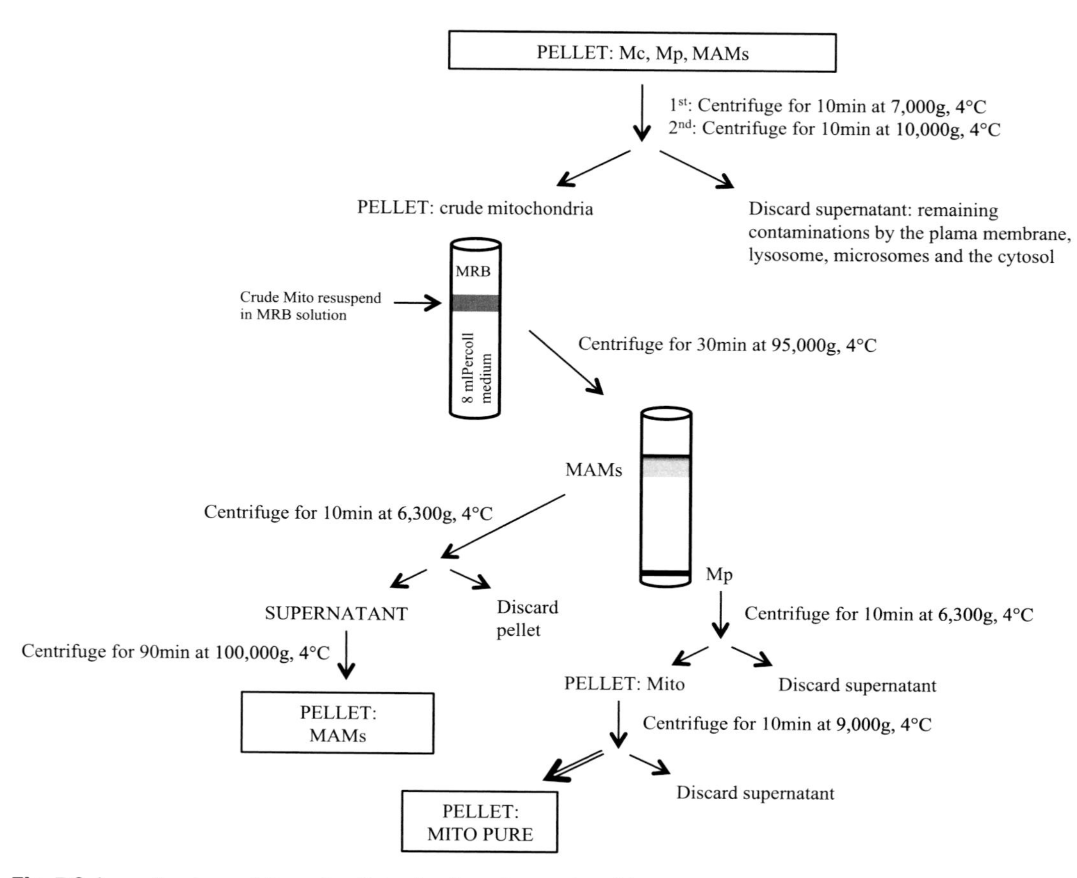

Fig. 5 Schematic steps of the subcellular fractionation protocol from cells for MAMs and Mp isolation. The final steps of the presented procedure result in the isolation of crude mitochondria, pure mitochondria, and mitochondria-associated membrane (MAM) fractions. MAMs, mitochondria-associated membranes; *Mc* crude mitochondria; *Mp* pure mitochondria

2 Materials

2.1 Nuclear Extraction from Cell Culture

2.1.1 Equipment

1. Cell culture dishes, 10 cmϕ.
2. Cell scrapers.
3. Pasteur pipette.
4. 1.5 ml Eppendorf microfuge test tubes.
5. Sigma rotor angular 6 × 30 ml.
6. Refrigerated Sigma low-speed centrifuge (Sigma (Braun), Model 2 K15, tabletop).
7. 7 ml Dounce tissue homogenizer.
8. Microcentrifuge.

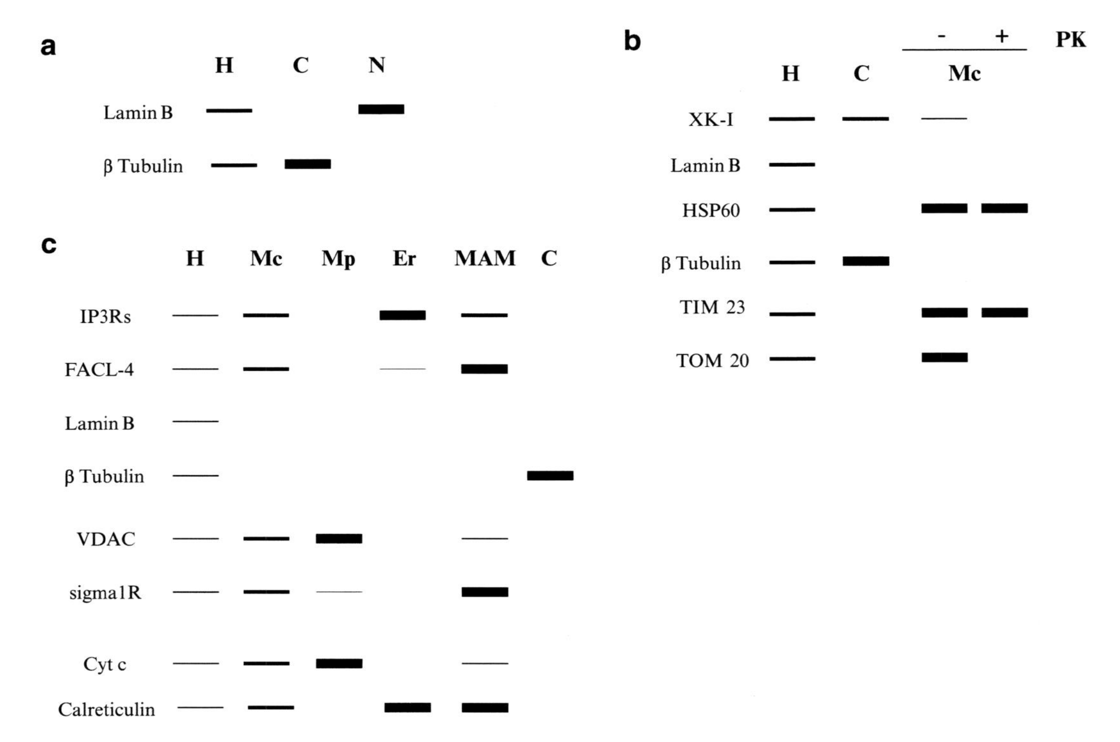

Fig. 6 Schematic representation of how markers for fractions should be distributed *H* homogenate; *Mc* crude mitochondria; *Mp* pure mitochondria fraction; *ER* endoplasmic reticulum; *MAMs* mitochondria-associated membranes; *C* cytosol; *N* nucleus. (**a**) Schematic model of markers for nuclear extraction. (**b**) schematic representation for mitochondrial isolation for PK. (**c**) Schematic representation for subcellular localization

2.1.2 Reagents

1. Dulbecco's phosphate-buffered saline (D-PBS), liquid, without Ca^{2+} and Mg^{2+}.
2. IGEPAL CA-630.
3. Sodium chloride (NaCl).
4. Mg acetate tetrahydrate (Mg acetate).
5. Magnesium chloride ($MgCl_2$).
6. Phenylmethanesulfonyl fluoride (PMSF).
7. Protease inhibitor cocktail (100×).
8. Sodium dodecyl sulfate (SDS).
9. Sodium fluoride (NaF).
10. Sodium orthovanadate (Na_3V0_4).
11. Sucrose.
12. Trizma-Base.
13. Trypsin, 0.25 % (1×) with EDTA $4Na^+$, liquid.

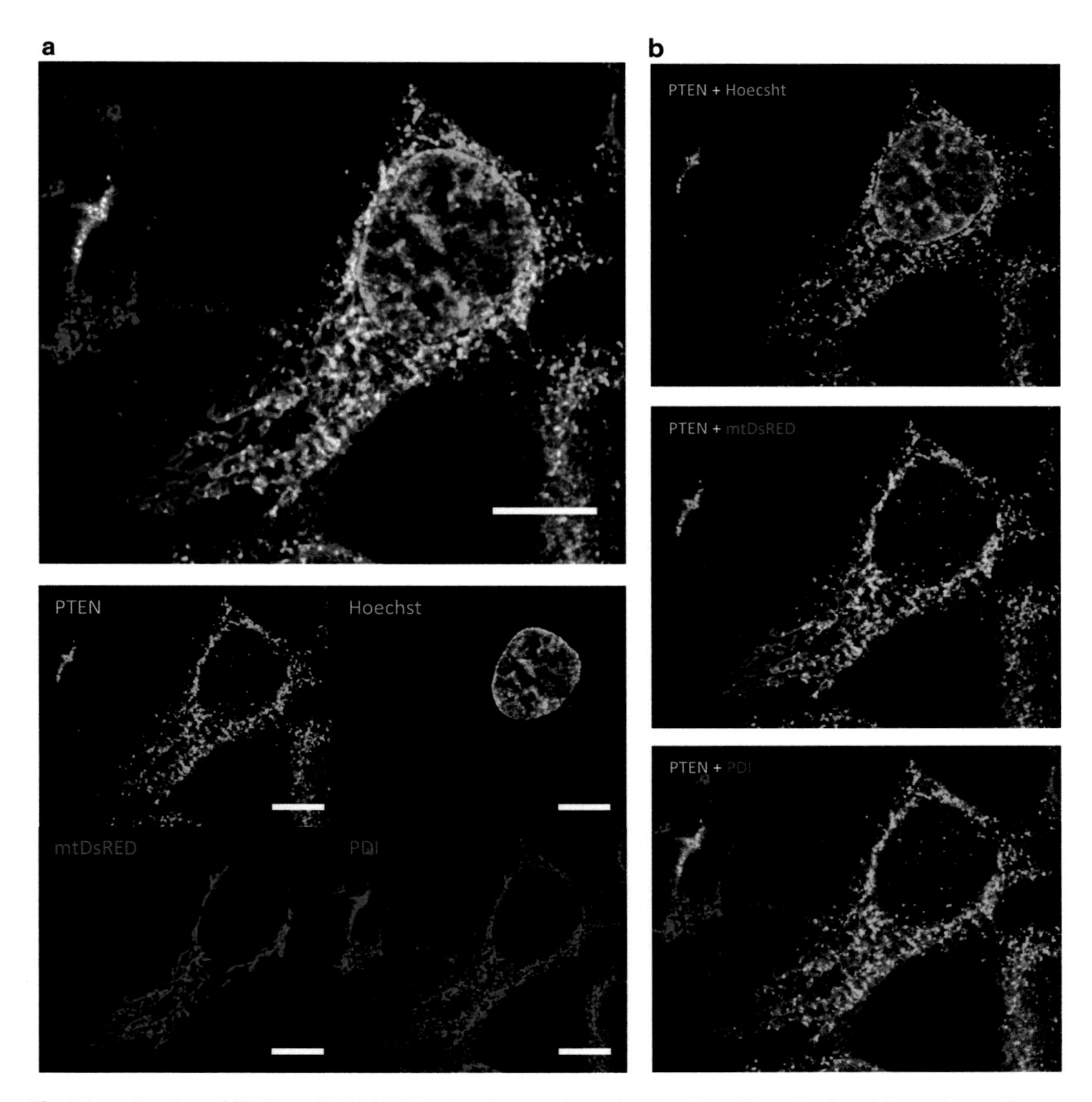

Fig. 7 Localization of PTEN by IF. (**a**) HEK-293 cells transfected with mtDsRED (mitochondria, *red*) were immunostained for PTEN (*green*), PDI (ER, *blue*), and loaded with Hoechst (nucleus, *cyan*). Representative merge images of the four channels and single images of each one are shown. (**b**) Merge images below show a succession of overlapping signals between PTEN and Hoechst, PTEN and mtDsRED, and PTEN and PDI. Scale bar, 10 mm. Figure modified from [3]

2.1.3 Reagent Preparation

1. *1 M Tris–HCl* (pH 7.4): Dissolve 121.14 g of Trizma-Base in 500 ml of bi-distilled water, adjust the pH to 7.4 with HCl, bring the solution to 1 l with bidistilled water, and store at 4 °C. The solution remains stable for a long time period.
2. *RSB-5 buffer*: 10 mM Tris–HCl, 10 mM NaCl, and 5 mM $MgCl_2$ (pH 7.4). Dissolve 0.0584 g of NaCl and 0.1017 g of $MgCl_2$ in 50 ml of bi-distilled water and add 1 ml of Tris–HCl.

Table 1
DNA constructs

	Acronym	Reference
Mitochondria	mtDsRED	
	mtGFP	[18]
Nucleus	H2BRFP	
ER	er-RFP	
	GFP-Sec61	[19]
PTEN	GFP-PTEN	
	FLAG-PTEN	[20]
	PTEN-YFP	[21]
	HA-tagget	
	PTEN	[22]
	cherry-PTEN	[23]

Table 2
Staining

Compartment	Staining
Mitochondria	Mito Tracker® Green, Red, Deep Red, Orange
Nucleus	Hoechst 33342
	Dapi
	DRAQ5™
ER	CellLight® ER-RFP

If necessary, adjust the pH to 7.4 and bring the solution to a final volume of 100 ml with bi-distilled water; store at 4 °C.

3. *0.88 M sucrose and 5 mM Mg acetate*: Dissolve 15.06 g of sucrose and 0.0536 g Mg acetate in 50 ml of bi-distilled water. Vortex carefully. If the powder is not completely dissolved in the solution, increase the temperature to put it at 37 °C. Store at −20 °C to avoid contamination.

4. *RIPA buffer*: 50 mM Tris–HCl (pH 8), 150 mM NaCl, 1 % IGEPAL CA-630, 0.5 % sodium deoxycholate (DOC), and 0.1 % SDS. Add 2.5 ml Tris–HCl 1 M (pH 8), 1.5 ml NaCl 5 M, 500 μl IGEPAL, 2.5 ml DOC stock 10 %, and 500 μl SDS 10 % and bring the solution to 50 ml with bi-distilled water.

Table 3
Antibodies

	Antibodies	Reference
Mitochondria	HSPD1 (1:100, Abcam)	[24]
ER	PDI (1 : 100, Abcam) Sec61 (1:100, Abcam) IP3R (1:50, BD Bioscience)	
MAMs	SIGMAR1 (1:100, Sigma)	
PTEN	PTEN (6H2.1) (1:200, Cascade Bioscience)	[25]
	PTEN (1:100, Cell Signaling Technology)	[7]

2.2 Mitochondria Isolation for PK Assay

2.2.1 Equipment

1. Cell culture dishes, 10 cmϕ.
2. Cell scrapers.
3. Pasteur pipettes.
4. 7 ml glass/Dounce tissue grinder.
5. 1 ml Dounce tissue grinder.
6. 1.5 ml Eppendorf microfuge test tubes.
7. Sigma rotor angular 6 × 30 ml.
8. Refrigerated Sigma low-speed centrifuge.
9. Olympus microscope LKX31.
10. Microcentrifuge.

2.2.2 Reagents

1. D-Mannitol.
2. D-PBS, liquid, without Ca^{2+} and Mg^{2+}.
3. Ethylene-bis(oxyethylenenitrilo) tetraacetic acid (EGTA).
4. HEPES.
5. Percoll.
6. PMSF.
7. Protease inhibitor cocktail (100×).
8. Sodium fluoride.
9. Sodium orthovanadate.
10. Sucrose.
11. Trizma-Base.

2.2.3 Reagent Preparation

Prepare all solutions using bi-distilled water. Extreme care should be taken to avoid contamination with ice and tap water in all preparations. Wash all glassware with bi-distilled water to avoid Ca^{2+} contamination, which can cause swelling of the mitochondria and rupture of the OMM.

1. *0.1 M EGTA/Tris* (pH 7.4): Dissolve 3.8 g of EGTA in 90 ml of bi-distilled water, adjust pH to 7.4 with Tris powder, bring the solution to 100 ml with bi-distilled water and store at 4 °C. The solution remains stable for a long time period.
2. *0.1 M Tris-MOPS* (pH 7.4): Dissolve 0.6057 g of Trizma-Base in 40 ml of bi-distilled water, adjust pH to 7.4 using MOPS powder, bring the solution to 50 ml with bi-distilled water and store at 4 °C.
3. *1 M sucrose*: In 1 l of bi-distilled water, dissolve 342.3 g of sucrose, mix well and prepare 25 ml aliquots. Store at −20 °C.
4. *Mitochondria buffer (MT)*: 200 mM sucrose, 10 mM Tris/MOPS and 0.5 mM EGTA-Tris (pH 7.4). To prepare 100 ml of MT solution, add 10 ml of 0.1 M Tris-MOPS and 100 μl of 0.1 M EGTA/Tris to 25 ml of 1 M sucrose. Bring the solution to 100 ml with bi-distilled water, adjust pH to 7.4 using Tris or MOPS powder and store at 4 °C (*see* **Note 1**).
5. *10× Proteinase K (PK)*: In 10 ml of MT buffer dissolve 10 mg of proteinase K. Store on ice (*see* **Note 2**).

2.3 Subcellular Fractionation Protocol from Cells

2.3.1 Equipment

1. Cell culture dishes, 10 or 15 cmϕ.
2. Cell scrapers.
3. Pasteur pipettes.
4. 40 ml glass/Dounce tissue grinder.
5. 1 ml Dounce tissue grinder. Loose pestle.
6. Ultra-Clear 14-ml polybrene tubes (Beckman, cat. no 344060) for SW40 rotor.
7. Polycarbonate tubes with cap assembly (Beckman, cat. no. 355618) for 70-Ti rotor.
8. 1.5 ml Eppendorf microfuge test tubes.
9. 70-Ti rotor (Fixed angle, 8 × 39 ml, 70,000 rpm, 504,000 × *g*) (Beckman, cat. no. 337922).
10. SW 40 rotor (Swinging bucket, 6 × 14 ml, 40,000 rpm, 285,000 × *g*) (Beckman, cat. no. 331302).
11. Sigma rotor angular 6 × 30 ml (Sigma, cat. no. 12139).
12. Beckman Coulter Optima L-100 XP Ultracentrifuge.
13. Refrigerated Sigma low-speed centrifuge (Sigma (Braun), Model 2 K15, tabletop).
14. Olympus microscope LKX31.
15. Microcentrifuge.

2.3.2 Reagents

1. D-Mannitol.
2. D-PBS, liquid, without Ca^{2+} and Mg^{2+}.
3. Ethylene-bis(oxyethylenenitrilo)tetraacetic acid (EGTA).

4. HEPES.
5. Percoll.
6. PMSF.
7. Protease inhibitor cocktail (100×).
8. Sodium fluoride.
9. Sodium orthovanadate.
10. Sucrose.
11. Trizma-Base.

2.3.3 Reagent Preparation

Prepare all solutions using bi-distilled water. Extreme care should be taken to avoid contamination with ice and tap water in all preparations. Wash all glassware with bi-distilled water to avoid Ca^{2+} contamination, which can cause swelling of the mitochondria and rupture of the OMM.

1. *0.5 M EGTA* (pH 7.4): Dissolve 19 g of EGTA in 70 ml of bi-distilled water, adjust the pH to 7.4 with KOH, bring the solution to 100 ml with bi-distilled water and store at 4 °C. The solution remains stable for a long time period.
2. *0.5 M HEPES* (pH 7.4): Dissolve 59.57 g of HEPES in 400 ml of bi-distilled water, adjust the pH to 7.4 with KOH, bring the solution to 500 ml with bi-distilled water and store at 4 °C. The solution remains stable for a long time period.
3. *1 M Tris–HCl* (pH 7.4): Dissolve 121.14 g of Trizma-Base in 500 ml of bi-distilled water, adjust the pH to 7.4 with HCl, bring the solution to 1 l with bi-distilled water and store at 4 °C. The solution remains stable for a long time period.
4. *Homogenization buffer*: 225 mM D-mannitol, 75 mM sucrose, and 30 mM Tris–HCl (pH 7.4). Dissolve 12.30 g of D-MANNITOL and 7.7 g of sucrose in 250 ml of bi-distilled water and add 9 ml Tris–HCl (1 M pH 7.4). Leave the buffer for approximately 30 min at 4 °C to cool down. If necessary, adjust the pH to 7.4 with KOH (if too low) or HCl (if too high); bring the solution to a final volume of 300 ml with bi-distilled water and store at 4 °C (*see* **Note 1**).
   Immediately prior to use, add 60 μl of 0.5 M EGTA pH 7.4 and 3 ml of PMSF. EGTA is recommended to remove traces of Ca^{2+}.
5. *Mitochondria resuspending buffer (MRB)*: 250 mM D-mannitol, 5 mM HEPES (pH 7.4), and 0.5 mM EGTA. To prepare 50 ml of MRB solution, dissolve 2.28 g D-mannitol in 45 ml of bi-distilled water and add 500 μl of 0.5 M HEPES pH 7.4. Leave the buffer for approximately 30 min at 4 °C to cool down. If necessary, adjust the pH to 7.4 with KOH (if too low) or HCl (if too high); bring the solution to a final volume

of 50 ml with bi-distilled water and store at 4 °C (*see* **Note 1**). Immediately prior to use, add 50 μl of 0.5 M EGTA (pH 7.4).

6. *Percoll medium*: 225 mM D-mannitol, 25 mM HEPES (pH 7.4), 1 mM EGTA and 30 % Percoll (vol/vol). Dissolve 2.052 g D-mannitol in 25 ml of bi-distilled water and add 2.5 ml of 0.5 M HEPES (pH 7.4). Leave the buffer for approximately 30 min at 4 °C to cool down. If necessary, adjust the pH to 7.4 with KOH (if too low) or HCl (if too high); bring the basal solution to a final volume of 35 ml with bi-distilled water and store at 4 °C (*see* **Note 1**).

   Immediately prior to use, add 100 μl of 0.5 M EGTA (pH 7.4) and 15 ml of Percoll.

2.4 Cell Imaging

2.4.1 Equipment

1. Aluminum foil.
2. Cell culture dishes, 10 or 15 cmϕ.
3. Filter paper.
4. Forceps.
5. 6 wells, multi-well plate.
6. Nail polish.
7. Needle.
8. Parafilm.
9. Round glass coverslip (*see* **Note 3**).

2.4.2 Reagents

1. Bovine serum albumin (BSA) (Sigma-Aldrich, cat. no.A6003).
2. D-PBS, liquid, without Ca^{2+} and Mg^{2+}.
3. Hoechst 33342 (Life Technologies, cat. no. H3570).
4. Paraformaldehyde (Merk, cat. no. 8.18715.1000).
5. Poly-L-lysine (Sigma-Aldrich, cat. no. P36934).
6. Prolong Gold antifade reagent (Life Technologies, cat. no. P36934).
7. Triton X-100.
8. Vector encoding mtDsRED.
9. Primary antibodies: (a) mouse monoclonal to PTEN (A2B1) (Santa Cruz, cat. no. sc-7974), 1:50, and (b) rabbit polyclonal to PDI-ER marker (Abcam, cat. no. ab3672), 1:100.
10. Secondary antibodies: (a) Alexa Fluor 488 goat anti-mouse (Life Technologies, cat. no. A-11001), 1:1000, and (b) Alexa Fluor 633 goat anti-rabbit (Life Technologies, cat. no. A-21070), 1:1000.

2.4.3 Reagent Preparation

1. *Fixing solution*: 4 % paraformaldehyde. Dissolve 4 g of paraformaldehyde in 100 ml of PBS. The solution should be freshly prepared. Store at −20 °C.

2. *Permeabilization solution*: 0.1 % Triton X-100. Add 100 μl of Triton X-100 in 100 ml of PBS. Undiluted Triton X-100 is a clear viscous liquid; thus, use a wide orifice pipette tip or cut off the tip of a regular pipette tip to enlarge the opening.
3. *Blocking solution*: 1 % BSA. Dissolve 100 mg of BSA in 10 ml of PBS.

3 Methods

3.1 Nuclear Extraction from Cell Culture

Starting material: 4 human embryonic kidney (HEK) 293 cell confluent plates (10 cmϕ) or 8 MEF confluent plates (10 cmϕ).

1. Remove the medium and wash the cells with ice-cold PBS (Ca^{2+} and Mg^{2+} free).
2. Harvest the cells by trypsinization.
3. Centrifuge at 800 × *g* for 10 min at 4 °C.
4. Resuspend in 10 ml of RSB-5 buffer per gram of cell pellet.
5. Allow cells to swell on ice for 30 min (*see* **Note 4**).
6. Add NP40 to a final concentration of 0.3 %.
7. Transfer the cell suspension to a pre-cooled 7 ml Dounce tissue homogenizer. Homogenize 20 strokes using a tight pestle while maintaining the homogenizer on ice (*see* **Note 5**). An aliquot is saved as homogenate.
8. Centrifuge the homogenate at 1,200 × *g* for 10 min to sediment crude nuclei (the supernatant fraction can be saved as cytoplasm).
9. Resuspend nuclei in 20 volumes (of the pellet) of 0.88 M sucrose and 5 mM Mg acetate.
10. Centrifuge at 2,000 × *g* for 20 min.
11. Discard the supernatant and resuspend the pellet (nuclear extract) in RIPA buffer (please note: the nuclear pellet is hard to resuspend).
12. Maintain on ice for 20 min and vortex every 5 min.
13. Centrifuge at 16,000 × *g* for 10 min. Save the supernatant as nuclear homogenate.

3.2 Mitochondria Isolation for PK Assay

Starting material: 10 HEK confluent plates (15 cmϕ).

1. Remove the medium and wash the cells with ice-cold PBS (Ca^{2+} and Mg^{2+} free, supplemented with 2 mM Na_3V0_4 and 2 mM NaF).
2. Harvest the cells in ice-cold PBS using a cell scraper.
3. Centrifuge at 500 × *g* for 5 min at 4 °C.

4. Resuspend in ice-cold PBS.
5. Centrifuge at 500 × *g*, 5 min at 4 °C (repeat two times and combine the cells in one tube).
6. Resuspend the cell pellet in MT buffer using a ratio of 5 ml of buffer per 1 ml of pellet (*see* **Note 6**).
7. Transfer the cell suspension in a 7 ml glass/Dounce tissue grinder.
8. Homogenize the cells with tight pestle. Every five strokes control the cell integrity under a microscope. Finish homogenization when 80–90 % of the cells are broken (*see* **Note 7**). Take ~50 μl of homogenate and immediately add protease inhibitor cocktail.
9. To obtain the cytosolic fraction, obtain a 200 μl aliquot of homogenate and centrifuge at 16,100 × *g* for 30 min at 4 °C in a microcentrifuge. The supernatant is the cytosolic fraction.
10. From **step 8**, centrifuge at 600 × *g* for 5 min at 4 °C.
11. Collect the supernatant and discard the pellet (which contains unbroken cells and nuclei); repeat 2–3 times (until pellet is no longer present).
12. Collect supernatant and centrifuge at 7,000 × *g* for 10 min at 4 °C.

The supernatant is a cytosolic fraction that contains lysosomes and microsomes, whereas the pellet contains mitochondria.

13. Wash the pellet carefully with MT buffer and detach (*see* **Note 8**).
14. Centrifuge at 10,000 × *g* for 10 min at 4 °C.
15. Discard the supernatant. Wash the pellet with MT buffer, detach 1 ml of buffer, and then carefully transfer in a 1 ml Dounce tissue grinder. Resuspend using the loose pestle (*see* **Note 9**).
16. Quantify the protein content.
17. Prepare equal aliquots of organelles in Eppendorf tubes: 100 μg in a final volume of 900 μl. Add 100 μl of MT buffer to one aliquot (*see* **Note 10**); add 100 μl of 10× Proteinase K to another aliquot (*see* Fig. 2).
18. Incubate on ice for 5 min.
19. Centrifuge at 15,000 × *g* for 10 min at 4 °C in a microcentrifuge.
20. Discard the supernatants and wash the pellets with MT buffer. Add 1 mM PMSF and detach in 1 ml.
21. Centrifuge at 15,000 × *g* for 10 min at 4 °C in a microcentrifuge.
22. Discard the supernatants and resuspend the pellets (mito crude fraction) in 50 μl of sample buffer.

3.3 Subcellular Fractionation Protocol for Cells

Starting material (10^9 cells): 40 HEK confluent plates (15 cmϕ) or 150 MEF confluent plates (10 cmϕ).

1. Remove the medium and wash the cells with ice-cold PBS (Ca^{2+} and Mg^{2+} free).
2. Harvest the cells in ice-cold PBS using a cell scraper.
3. Centrifuge at 500 × *g* for 5 min at 4 °C.
4. Resuspend in ice-cold PBS.
5. Centrifuge at 500 × *g* for 5 min at 4 °C (repeat 2 times and combine the cells in one tube).
6. Resuspend the cell pellet in Homogenization buffer using a ratio of 3 ml of buffer per 1 ml of pellet (*see* **Note 6**).
7. Transfer the cell suspension in a 40 ml glass/Dounce tissue grinder.
8. Homogenize the cells with tight pestle. Every 25 strokes, control the cell integrity under a microscope. Finish homogenization when 80–90 % of the cells are broken (*see* **Note 7**).
   Take ~100 μl of homogenate and immediately add 2 mM Na_3V0_4, 2 mM NaF and protease inhibitor cocktail.
9. Transfer the homogenate to a 30 ml polypropylene centrifugation tube and centrifuge at 600 × *g* for 5 min at 4 °C.
10. Collect the supernatant and discard the pellet (which contains unbroken cells and nuclei); repeat 2–3 times (until pellet is no longer present).
11. Collect the supernatant and centrifuge at 7,000 × *g* for 10 min at 4 °C.

    The supernatant is a cytosolic fraction that contains lysosomes and microsomes, whereas the pellet contains mitochondria (*see* Fig. 3).

    PAUSE POINT. Store the supernatant at 4 °C (on ice) if there is a plan to proceed with further separation of the cytosolic, lysosomal, and ER fractions (refers to point 25 and below).
12. Gently resuspend the pellet that contains mitochondria in 20 ml of ice-cold MRB (*see* **Note 8**); centrifuge the mitochondrial suspension at 7,000 × *g* for 10 min at 4 °C.
13. Discard the supernatant, resuspend the mitochondrial pellet as previously described in 20 ml of ice-cold MRB, and centrifuge the mitochondrial suspension at 10,000 × *g* for 10 min at 4 °C (*see* **Note 11**).
14. Discard the supernatant and gently resuspend the crude mitochondrial pellet in 1 ml of ice-cold MRB.
15. Carefully transfer in a 1 ml Dounce tissue grinder and resuspend using the loose pestle (*see* **Note 12**).

16. Add 8 ml of Percoll medium to the 14 ml thin-wall, polybrene ultracentrifuge tubes.
17. Layer the suspension of mitochondria collected on top of 8 ml Percoll medium in the ultracentrifuge tube; then, gently layer the MRB solution (approximately 3.5 ml) on top of the mitochondrial suspension to fill up the centrifuge tube (*see* **Note 13**).
18. Centrifuge at 95,000×*g* for 30 min at 4 °C in a Beckman Coulter Optima L-100 XP Ultracentrifuge (SW40 rotor, Beckman). A dense band that contains purified mitochondria is localized at approximately the bottom of the ultracentrifuge tube (*see* Fig. 4). MAM is visible as the diffused white band located above the mitochondria.
19. Collect the MAM fraction from the Percoll gradient with a Pasteur pipette and dilute with 15 ml of homogenization buffer. Collect the pure mitochondrial band with a Pasteur pipette and dilute with 15 ml of homogenization buffer.
20. Centrifuge the MAM and mitochondrial suspension at 6,300×*g* for 10 min at 4 °C (refrigerated Sigma low-speed centrifuge).
21. Discard the pellet obtained from the MAM and transfer the MAM supernatant to polycarbonate tubes with a cap assembly. Centrifuge at 100,000×*g* for 90 min (70-Ti rotor, Beckman) at 4 °C.
22. Discard the mitochondrial supernatant obtained in **step 20** (which contains MAM contamination), gently resuspend the pellet in a small volume of MRB (300–500 μl), and centrifuge at 9,000×*g* for 10 min at 4 °C in a microcentrifuge.
23. Resuspend the pellet (pure mitochondria) in 50–60 μl of homogenization buffer and immediately add 2 mM Na_3V0_4, 2 mM NaF and protease inhibitor cocktail.
24. Discard the supernatant (from **step 21**) and collect the pellet using a pipette; resuspend in a small volume of MRB (100–200 μl), and immediately add 2 mM Na_3V0_4, 2 mM NaF and protease inhibitor cocktail.
25. From **step 11**, to proceed with further separation of the cytosolic, lysosomal and ER fractions, centrifuge the supernatant at 25,000×*g* for 30 min at 4 °C.
26. Discard the pellet that contains lysosome, plasma membrane, and remaining mitochondrial contamination. Transfer the supernatant to polycarbonate tubes with a cap assembly and centrifuge the supernatants at 100,000×*g* for 90 min at 4 °C (*see* **Note 14**).
27. The supernatant consists of cytosolic fraction; collect 1 ml and immediately add 2 mM Na_3VO_4, 2 mM NaF and protease

inhibitor cocktail. The pellet consists of ER fraction; carefully wash with 1 ml of homogenization buffer and then detach and resuspend in 15 ml of homogenization buffer (do not pipette the pellet).

28. Centrifuge at 100,000 × *g* for 90 min at 4 °C (*see* **Note 14**).
29. Carefully discard the supernatant. Using a pipette with the cut-out end, collect the pellet (ER fraction) in a small volume of homogenization buffer (200–300 μl) (*see* Fig. 5).

    Transfer in a 1 ml Dounce tissue grinder and resuspend using the loose pestle. The result is the ER fraction. Immediately add 2 mM Na_3VO_4, 2 mM NaF and protease inhibitor cocktail.

The quality of protocol preparation can be checked by western blot analysis. Using different markers for the fractions obtained, it is possible to control the purity of each fraction and the presence of contaminations from other compartments. Figure 6 reports, as an example, the marker distribution that characterizes each fractionation from cells:

- Lamin B1 antibody is extremely useful as a nuclear loading control.
- Tubulin is used as a cytosolic marker.
- Hexokinase I (HXK-I) is used as a digestion control in mitochondrial isolation for the PK assay. It should be absent after PK treatment.
- HSP60, a constitutively expressed mitochondrial protein, localizes in the mitochondrial matrix; thus, after digestion with PK, it should be still present in the mito crude fraction.
- TIM 23, a translocase of the inner membrane, is an integral membrane protein of the mitochondrial protein translocation machinery; thus, after digestion with PK, it should be still present in the mito crude fraction.
- TOM 20, a translocase of the outer membrane, is a multi-subunit protein complex that facilitates the import of nucleus-encoded precursor proteins across the OMM. After digestion with PK, it should be absent in the mito crude fraction.
- IP3Rs, an ER channel responsible for agonist-dependent ER-Ca^{2+} release, is used as an ER marker.
- Calreticulin, an ER-Ca^{2+}-buffering protein that binds misfolded proteins and prevents them from exportation from the ER to the Golgi apparatus, is used as ER and MAM markers and should be present in both fractions at comparable levels.
- Voltage-dependent anion channel (VDAC), which is localized on the OMM and is responsible for the permeability of the membrane, is used as a mitochondrial marker. It should be

present in the crude mitochondria, but it must be enriched in the Mp fraction.

- Cytochrome c (Cyt c) is a component of the electron-transfer chain localized in the mitochondrial inter-membrane space; it is used as a mitochondrial marker and should be extremely enriched in the Mp fraction.
- The sigma-1 receptor (Sig1R) as a MAM marker should be enriched in this fraction.
- Long-chain fatty-acid CoA synthase (FACL)-4 is important for the synthesis of cellular lipids and β-oxidation degradation. As a MAM marker, it should be extremely enriched in the MAM fraction.

3.4 Cell Imaging

The cell preparation, transfection methods, staining, optimal dilutions, and incubation times for the primary and secondary antibodies will need to be empirically determined. IF requires careful interpretation. There may be autofluorescence, nonspecific binding of the secondary antibody, and the primary antibody may not only bind to the expected protein. Thus, it is necessary to include controls: (1) a non-processed sample to control for autofluorescence (*see* **Note 15**); (2) a sample incubated with secondary antibody only to determine the nonspecific staining of the secondary antibody; and (3) a non-immune serum or isotype Ig (which matches the primary antibody) to determine the non-specific staining of the primary antibody.

1. Prepare the coverslip (*see* **Note 16**): place glass coverslips (24 mm in diameter) in a 6-well plate and coat them with 1 % poly-L-Lysine in PBS for 1 h at RT; wash 3 times with sterile ultrapure H_2O and allow the coverslips to dry completely (*see* **Note 17**).
2. Seed the cells on the dry coverslips (*see* **Note 18**) and allow them to grow.
3. When the cells have reached 50 % confluence, transfect them with DNA (Table 1), e.g., 3 μg of vector that encodes mtDsRED per well (*see* **Note 19**).
4. After 36–48 h (*see* **Note 20**), remove the medium and wash 3 times with 2 ml PBS per well.
5. Stain (Table 2) the nuclei with 0.25 μl/ml Hoechst 33342 (Life Technologies) in PBS for 10 min at 37 °C in the dark (*see* **Note 21**).
6. Remove the staining solution and wash 3 times with 2 ml PBS per well.
7. To fix the cells, add 2 ml per well of fixing solution and incubate for 10 min at 37 °C (*see* **Note 22**).

8. Wash 3 times with PBS for 5 min with gentle shaking to remove the residual paraformaldehyde.
9. Add 2 ml per well of permeabilization solution (*see* **Note 23**) for 15 min at RT with gentle shaking; then wash 2 times with 2 ml PBS per well.
10. Incubate with 2 ml of blocking solution per well for 20 min at RT (*see* **Note 24**).
11. Prepare the primary antibodies (Table 3) by diluting them in blocking buffer: e.g., (a) mouse monoclonal to PTEN (A2B1) 1:50 (Santa Cruz); (b) rabbit polyclonal to PDI-ER marker 1:100 (Abcam).
12. Remove the blocking buffer by holding each coverslip by draining it onto a sheet of fiber-free paper (*see* **Note 25**); turn it cell-side down on a 50 μl drop (*see* **Note 26**) of the primary antibody solution placed on a layer of parafilm. Incubate overnight at 4 °C in the humidified chamber (*see* **Note 27**).
13. Carefully turn the coverslip cell-side up and wash 3 times for 10 min each with 2 ml of 1 % Triton X-100 with gently shaking.
14. Prepare the secondary antibodies (*see* **Note 28**) conjugated with two different fluorochromes by diluting them in blocking buffer, e.g., (a) Alexa Fluor 488 goat anti-mouse (Life Technologies) 1:1000 and (b) Alexa Fluor 633 goat anti-rabbit (Life Technologies) 1:1000.
15. Add at least 200 μl (*see* **Note 29**) of the secondary antibody mix on the cells and incubate for 1 h at RT in the dark.
16. Remove the secondary antibodies and wash 3 times for 10 min each with 2 ml of 1 % Triton X-100 with gently shaking; at the end, leave the coverslip in PBS.
17. Mounting (*see* **Note 30**): invert each coverslip on a slide that contains a drop of mounting medium (Prolong Gold antifade reagent) and allow it to dry.
18. The next day, seal the edges of each coverslip with transparent nail polish and store in the dark at 4 °C until image acquisition.
19. Acquire each field over 26 *z*-planes spaced 0.4 μm on an Axiovert 220 M microscope equipped with an ×100 oil immersion Plan-Neofluar objective (NA 1.3, from Carl Zeiss, Jena, Germany) and a CoolSnap HQ CCD camera. Image sampling was below the resolution limit and calculated according to the Nyquist calculator (*see* **Note 31**).
20. After acquisition, deconvolute the *z*-stacks with the "Parallele Iterative Deconvolution" plugin of the open source Fiji software (*see* **Note 32**).

21. Merger of the fluorescence images is the most prevalent visual method used to evaluate the colocalization of two probes (*see* **Note 33**).

A representative image of the result of this protocol is shown in Fig. 7.

3.4.1 Quantification of Co-localization

1. Open the images of the separated channel using Fiji software.
2. Background correction (*see* **Note 34**).
3. Colocalization should be measured for individual cells; thus, hand-draw a "region of interest" (ROI) over the image.
4. Estimate a global threshold to restrict the analysis to the pixel with intensity greater than the threshold value (*see* **Note 35**).
5. Measure the colocalization using Just another Colocalization Plugin (JACoP), which is a free Fiji plug-in. Select the images and the analysis to perform; there are two coefficients that can be used for the colocalization analysis: Pearson's coefficient and Manders' coefficient.
6. Include the threshold, the correct information regarding the microscope used to acquire the images and the correct wavelengths of the images; press "analyze."
7. Pearson's correlation coefficient (PCC) (*see* **Note 36**) values range from 1 for two images whose fluorescence intensities are perfectly and linearly related to −1 for two images whose fluorescence intensities are perfectly, but inversely, related to one another. Values near zero reflect the distributions of probes that are uncorrelated with one another. Manders' colocalization coefficients (MCC) measure the fraction of one protein that colocalizes with a second protein (*see* **Note 37**). The MCC values range from 0 and 1 and correspond to completely mutually exclusive and perfect colocalization, respectively. To select the most appropriate methods to evaluate the colocalization and an appreciation of the factors that must be considered for meaningful interpretation of colocalization studies, refer to a review of Dunn and colleagues [17].
8. Acquire and analyze sufficient images to perform a statistical analysis.

An alternative widespread approach that has been used to study PTEN localization is the generation of fusion proteins tagged with GFP or its derivatives. Targeting is a powerful and commonly used approach to investigate the functions and localization of proteins in eukaryotic cells. We used this strategy to target PTEN to different organelles and cell compartments, which thus generates novel tools for the investigation of the spatial complexity of PTEN functions. The available compartment-specific PTEN chimeras are reported in Table 4. In all cases, correct intracellular localization of

Table 4
Compartment-specific PTEN chimeras

Acronym	Intracellular localization
GFP-PTEN-NLS^{SV40}	Nucleus
GFP-PTEN-NES^{PKI}	Cytoplasm
snap25-PTEN	Plasma membrane (inner surface)
AKAP-PTEN	Outer mitochondrial membrane (cytoplasmatic surface)
ER-PTEN	Endoplasmic reticulum

the PTEN chimeras has been verified through IF, using antibodies against PTEN and specific markers for the various intracellular compartments.

4 Notes

1. This buffer must be prepared fresh for the experiment and must be free of protease and phosphatase inhibitor cocktails to avoid sample alterations.
2. This buffer must be prepared fresh immediately prior to use.
3. Use 1.5 mm thick coverslips because most microscope objectives are designed to work optimally with this thickness. Coverslips must be sterilized and subsequently placed in a multi-well plate: 18 mm coverslips in a 12-well plate, 12–13 mm coverslips in a 24-well plate, or 24 mm coverslips in a 6-well plate.
4. The cells should be swollen, but should not burst. It is imperative to maintain the cell suspension on ice during this step.
5. The number of required strokes depends on the cell type used. It is therefore necessary to check the homogenized cells under a microscope after every 10 strokes. Stop when >90 % of the cells have burst, thereby leaving intact nuclei, with various amounts of cytoplasmic material attached.
6. From this point, perform all procedures in a 4 °C room to minimize the activation of proteases and phospholipases.
7. Precool the glassware and homogenizer with pestle in an ice bath for 5 min prior to the initiation of the homogenization step. Pestle homogenization at a higher force can affect mitochondrial integrity; thus, the movement of the pestle should be slow and performed with "fluidity."
8. Do not pipette the pellet because it will result in mitochondria breakage.

9. It is very important to maintain the sample on ice.
10. The aliquot without PK will be considered a mito crude control sample.
11. If crude mitochondria are contaminated by other organelles, add one additional 10,000 ×*g* centrifugation step for 10 min at 4 °C.
12. For the resuspension process, two or three hand-made strokes are sufficient. Store a small amount (60 μl) of crude mitochondrial fraction in 1.5 ml Eppendorf microfuge for future investigations (western blot). Add 2 mM Na_3VO_4, 2 mM NaF and protease inhibitor cocktail. Freeze at −80 °C if the sample will not be used immediately.
13. The suspension should remain 4–5 mm below the top of the tubes. Filling up the tubes is important to protect them against damage during centrifugation.
14. Ultracentrifugation was performed using a Beckman L8-70 M ultracentrifuge, 70-ti rotor (tube: Beckman, cat. no. 355618).
15. Autofluorescence can be quenched using a 100 mM glycine solution for 30 min prior to the blocking step.
16. This procedure can be used for HEK-293 and other non-adherent cell lines.
17. This passage requires at least 1 h.
18. HEK-293 cells were grown in Dulbecco's modified Eagle's medium (DMEM) supplemented with 10 % fetal bovine serum (FBS).
19. The amount of DNA depends on the transfection method: 3 μg is suggested for a standard calcium phosphate procedure.
20. This time is required to enable the protein expression of the constructs.
21. From this point, minimize the light exposure: turn off the light in the room and cover the sample with aluminum foil.
22. Cell fixing can damage the endoplasmic reticulum; if this occurs, you could reduce the time of fixing.
23. Permeabilization helps get the antibodies into now-fixed cells. In addition to Triton X-100, a range of different detergents can be used, including NP-40, Tween-20 and Saponin. Fixing and permeabilization steps affect the cell morphology and the availability of the antigen of interest. You may obtain different results with different reagents, times and concentrations; thus, there is a need for protocol optimization. The distortion of cell morphology is something to consider during image interpretation.

24. This step is essential to block unspecific binding of the antibodies. BSA or nonfat dry milk can be used as a blocking buffer. The optimization of the percentage and time of blocking depends on the antibodies used.
25. Holding the coverslip on its edge with forceps is facilitated with a needle. It is recommended to hold the coverslip on the edge to minimize damage to the cell layer.
26. Antibody incubation requires a small volume; flipping the coverslip on the drop allows tight contact between the cell and antibody.
27. The humidified chamber can be easily constructed from a 15 cmϕ cell culture dish upholstered by aluminum foil in which a layer of parafilm is placed on PBS soaked filter paper. The humidified chamber is fundamental to avoid sample drying.
28. Centrifuge the secondary antibody solutions (e.g., 2 min at 10,000 × *g, at 4 °C*) and add only the supernatant to the blocking solution. This step eliminates protein aggregates that may have formed during storage, which thereby reduces the nonspecific background staining.

    When two secondary antibodies are used, ensure that they are produced in a third different species (e.g., for mouse-anti PTEN and rabbit-anti ER, use secondary antibody anti-mouse and anti-rabbit produced in goat to avoid cross reaction between the antibodies).
29. Use the smallest amount of solution but cover all samples.
30. Mounting medium helps preserve your sample and raises the refractive index to provide good performance with oil objectives. Mountants often have scavengers, which soak up free radicals and reduce photobleaching (these sometimes can reduce the initial brightness of the samples).
31. Available at http://www.svi.nl/NyquistCalculator.
32. Freely available at http://fiji.sc/.
33. For example, colocalization of Alexa Fluor 488 and mtDsRED can be apparent in structures that appear yellow because of the combined contributions of green and red fluorescence, respectively.
34. Run process/subtract background (BG) in which a rolling ball radius of 50 pixels is suggested or select a BG ROI and Run Plugins/ROI/BG subtract from the ROI.
35. It is fundamental to select the same threshold value for the analysis of all images. The Costes method for threshold estimation is a robust and reproducible method that can be easily automated. The method has been implemented in Fiji/ImageJ plug-ins (JACoP).

36. The formula for PCC is provided for a typical image that consists of red and green channels:

$$\mathrm{PCC} = \frac{\sum_i (R_i - R)x(G_i - G)}{\sqrt{\sum_i (R_i - R)^2 x \sum_i (G_i - G)^2}}$$ where R_i and G_i refer to the intensity values of the red and green channels, respectively, of pixel i, and R and G refer to the mean intensities of the red and green channels, respectively, across the entire image.

37. For two probes, denoted R and G, two different MCC values are derived. M_1 is the fraction of R in compartments that contain G, and M_2 is the fraction of G in compartments that contain R. These coefficients are calculated as follows: $M_1 = \frac{\sum_i R_i,\mathrm{colocal}}{\sum_i R_i}$, where R_i,colocal = R_i if $G_i > 0$ and R_i,colocal = 0 if $G_i = 0$ and $M_2 = \frac{\sum_i G_i,\mathrm{colocal}}{\sum_i G_i}$, where G_i,colocal = G_i if $R_i > 0$ and G_i,colocal = 0 if $R_i = 0$.

Acknowledgements

The financial support of Telethon GGP11139B to P.P., the Italian Association for Cancer Research (IG-14442 to P.P. and MFAG-13521 to C.G.), the Italian Cystic Fibrosis Research Foundation (grant #19/2014), and the Italian Ministry of Education, University and Research (COFIN, FIRB, and Futuro in Ricerca) to P.P. are gratefully acknowledged.

References

1. Pulido R, Baker SJ, Barata JT, Carracedo A, Cid VJ, Chin-Sang ID, Dave V, den Hertog J, Devreotes P, Eickholt BJ, Eng C, Furnari FB, Georgescu MM, Gericke A, Hopkins B, Jiang X, Lee SR, Losche M, Malaney P, Matias-Guiu X, Molina M, Pandolfi PP, Parsons R, Pinton P, Rivas C, Rocha RM, Rodriguez MS, Ross AH, Serrano M, Stambolic V, Stiles B, Suzuki A, Tan SS, Tonks NK, Trotman LC, Wolff N, Woscholski R, Wu H, Leslie NR (2014) A unified nomenclature and amino acid numbering for human PTEN. Sci Signal 7(332):pe15. doi:10.1126/scisignal.2005560
2. Salmena L, Carracedo A, Pandolfi PP (2008) Tenets of PTEN tumor suppression. Cell 133(3): 403–414. doi:10.1016/j.cell.2008.04.013
3. Bononi A, Bonora M, Marchi S, Missiroli S, Poletti F, Giorgi C, Pandolfi PP, Pinton P (2013) Identification of PTEN at the ER and MAMs and its regulation of Ca(2+) signaling and apoptosis in a protein phosphatase-dependent manner. Cell Death Differ 20(12):1631–1643. doi:10.1038/cdd.2013.77
4. Bassi C, Ho J, Srikumar T, Dowling RJ, Gorrini C, Miller SJ, Mak TW, Neel BG, Raught B, Stambolic V (2013) Nuclear PTEN controls DNA repair and sensitivity to genotoxic stress. Science 341(6144):395–399. doi:10.1126/science.1236188
5. Li P, Wang D, Li H, Yu Z, Chen X, Fang J (2014) Identification of nucleolus-localized PTEN and its function in regulating ribosome

biogenesis. Mol Biol Rep 41(10):6383–6390. doi:10.1007/s11033-014-3518-6

6. Trotman LC, Wang X, Alimonti A, Chen Z, Teruya-Feldstein J, Yang H, Pavletich NP, Carver BS, Cordon-Cardo C, Erdjument-Bromage H, Tempst P, Chi SG, Kim HJ, Misteli T, Jiang X, Pandolfi PP (2007) Ubiquitination regulates PTEN nuclear import and tumor suppression. Cell 128(1):141–156. doi:10.1016/j.cell.2006.11.040
7. Zhu Y, Hoell P, Ahlemeyer B, Krieglstein J (2006) PTEN: a crucial mediator of mitochondria-dependent apoptosis. Apoptosis 11(2):197–207. doi:10.1007/s10495-006-3714-5
8. Liang H, He S, Yang J, Jia X, Wang P, Chen X, Zhang Z, Zou X, McNutt MA, Shen WH, Yin Y (2014) PTENalpha, a PTEN isoform translated through alternative initiation, regulates mitochondrial function and energy metabolism. Cell Metab 19(5):836–848. doi:10.1016/j.cmet.2014.03.023
9. Wieckowski MR, Giorgi C, Lebiedzinska M, Duszynski J, Pinton P (2009) Isolation of mitochondria-associated membranes and mitochondria from animal tissues and cells. Nat Protoc 4(11):1582–1590. doi:10.1038/nprot.2009.151
10. Giorgi C, Baldassari F, Bononi A, Bonora M, De Marchi E, Marchi S, Missiroli S, Patergnani S, Rimessi A, Suski JM, Wieckowski MR, Pinton P (2012) Mitochondrial Ca(2+) and apoptosis. Cell Calcium 52(1):36–43. doi:10.1016/j.ceca.2012.02.008
11. Marchi S, Lupini L, Patergnani S, Rimessi A, Missiroli S, Bonora M, Bononi A, Corra F, Giorgi C, De Marchi E, Poletti F, Gafa R, Lanza G, Negrini M, Rizzuto R, Pinton P (2013) Downregulation of the mitochondrial calcium uniporter by cancer-related miR-25. Curr Biol 23(1):58–63. doi:10.1016/j.cub.2012.11.026
12. Marchi S, Patergnani S, Pinton P (2014) The endoplasmic reticulum-mitochondria connection: one touch, multiple functions. Biochim Biophys Acta 1837(4):461–469. doi:10.1016/j.bbabio.2013.10.015
13. Patergnani S, Suski JM, Agnoletto C, Bononi A, Bonora M, De Marchi E, Giorgi C, Marchi S, Missiroli S, Poletti F, Rimessi A, Duszynski J, Wieckowski MR, Pinton P (2011) Calcium signaling around Mitochondria Associated Membranes (MAMs). Cell Commun Signal 9:19. doi:10.1186/1478-811X-9-19
14. Giorgi C, Missiroli S, Patergnani S, Duszynski J, Wieckowski M, Pinton P (2015) Mitochondria-associated membranes (MAMs): composition molecular mechanisms and physiopathological implications. Antioxid Redox Signal. doi:10.1089/ars.2014.6223
15. Marchi S, Marinello M, Bononi A, Bonora M, Giorgi C, Rimessi A, Pinton P (2012) Selective modulation of subtype III IP(3)R by Akt regulates ER Ca(2)(+) release and apoptosis. Cell Death Dis 3:e304. doi:10.1038/cddis.2012.45
16. Bononi A, Pinton P (2014) Study of PTEN subcellular localization. Methods. doi:10.1016/j.ymeth.2014.10.002
17. Dunn KW, Kamocka MM, McDonald JH (2011) A practical guide to evaluating colocalization in biological microscopy. Am J Physiol Cell Physiol 300(4):C723–C742. doi:10.1152/ajpcell.00462.2010
18. Bonora M, De Marchi E, Patergnani S, Suski JM, Celsi F, Bononi A, Giorgi C, Marchi S, Rimessi A, Duszynski J, Pozzan T, Wieckowski MR, Pinton P (2014) Tumor necrosis factor-alpha impairs oligodendroglial differentiation through a mitochondria-dependent process. Cell Death Differ 21(8):1198–1208. doi:10.1038/cdd.2014.35
19. Friedman JR, Webster BM, Mastronarde DN, Verhey KJ, Voeltz GK (2010) ER sliding dynamics and ER-mitochondrial contacts occur on acetylated microtubules. J Cell Biol 190(3):363–375. doi:10.1083/jcb.200911024
20. Wu X, Hepner K, Castelino-Prabhu S, Do D, Kaye MB, Yuan XJ, Wood J, Ross C, Sawyers CL, Whang YE (2000) Evidence for regulation of the PTEN tumor suppressor by a membrane-localized multi-PDZ domain containing scaffold protein MAGI-2. Proc Natl Acad Sci U S A 97(8):4233–4238
21. Vazquez F, Matsuoka S, Sellers WR, Yanagida T, Ueda M, Devreotes PN (2006) Tumor suppressor PTEN acts through dynamic interaction with the plasma membrane. Proc Natl Acad Sci U S A 103(10):3633–3638. doi:10.1073/pnas.0510570103
22. Lachyankar MB, Sultana N, Schonhoff CM, Mitra P, Poluha W, Lambert S, Quesenberry PJ, Litofsky NS, Recht LD, Nabi R, Miller SJ, Ohta S, Neel BG, Ross AH (2000) A role for nuclear PTEN in neuronal differentiation. J Neurosci 20(4):1404–1413
23. Howitt J, Lackovic J, Low LH, Naguib A, Macintyre A, Goh CP, Callaway JK, Hammond V, Thomas T, Dixon M, Putz U, Silke J, Bartlett P, Yang B, Kumar S, Trotman LC, Tan SS (2012) Ndfip1 regulates nuclear Pten import in vivo to promote neuronal survival

following cerebral ischemia. J Cell Biol 196(1):29–36. doi:10.1083/jcb.201105009

24. Bonora M, Bononi A, De Marchi E, Giorgi C, Lebiedzinska M, Marchi S, Patergnani S, Rimessi A, Suski JM, Wojtala A, Wieckowski MR, Kroemer G, Galluzzi L, Pinton P (2013) Role of the c subunit of the FO ATP synthase in mitochondrial permeability transition. Cell Cycle 12(4):674–683. doi:10.4161/cc.23599
25. Song MS, Salmena L, Carracedo A, Egia A, Lo-Coco F, Teruya-Feldstein J, Pandolfi PP (2008) The deubiquitinylation and localization of PTEN are regulated by a HAUSP-PML network. Nature 455(7214):813–817. doi:10.1038/nature07290

Part V

Methods to Study *PTEN* Structure

Chapter 14

Methods in the Study of PTEN Structure: X-Ray Crystallography and Hydrogen Deuterium Exchange Mass Spectrometry

Glenn R. Masson, John E. Burke, and Roger L. Williams

Abstract

Despite its small size and deceptively simple domain organization, PTEN remains a challenging structural target due to its N- and C-terminal intrinsically disordered segments, and the conformational heterogeneity caused by phosphorylation of its C terminus. Using hydrogen/deuterium exchange mass spectrometry (HDX-MS), it is possible to probe the conformational dynamics of the disordered termini, and also to determine how PTEN binds to lipid membranes. Here, we describe how to purify recombinant, homogenously dephosphorylated PTEN from a eukaryotic system for subsequent investigation with HDX-MS or crystallography.

Key words HDX-MS, Mass spectrometry, Insect cell expression, Protein purification, Liposome preparation

1 Introduction

The high-resolution structure of PTEN's dual-specificity phosphatase domain and C2 domain was determined by X-ray crystallography some time ago [1]. This has proved to be an excellent resource, providing insight into the mechanism of substrate catalysis [2, 3], the individual roles of specific residues in the active site [4], and also the possible mechanisms of oncogenic mutations in PTEN [1, 5]. However, in order to facilitate the crystallization of PTEN, a construct was used that removed intrinsically disordered segments of the protein, such as the N-terminal PIP_2 binding domain, and a 50 amino acid C-terminal region. Both of these segments are critical to the regulation and function of PTEN, dictating PIP_2 binding [6, 7] and membrane localization [8]. Furthermore, the recent discovery of PTEN-Long [9–11], a variant of PTEN that is expressed with a 176 amino acid intrinsically disordered [12] N-terminal extension, has highlighted the necessity for techniques

Leonardo Salmena and Vuk Stambolic (eds.), *PTEN: Methods and Protocols*, Methods in Molecular Biology, vol. 1388, DOI 10.1007/978-1-4939-3299-3_14, © Springer Science+Business Media New York 2016

that are capable of shedding light on these areas of intrinsic disorder.

Hydrogen deuterium exchange mass spectrometry (HDX-MS) is a technique that is capable of determining structural and conformational information on proteins that harbor areas of intrinsic disorder [13]. We have used HDX-MS extensively to determine the mechanism of membrane binding for phosphatidylinositol 3-kinases (PI3Ks), and how oncogenic mutations in these enzymes induce conformational shifts that lead to an increase in kinase activity [14]. HDX-MS can also be used to determine the sites of protein:protein interactions, such as the interaction of p110β and the Gβγ subunit [15], and the site of Rab11 binding to PI4KIIIβ [16], and also has been used to identify areas of intrinsic disorder within proteins that can be subsequently removed to facilitate their crystallization [16].

Here we describe how to purify recombinant PTEN expressed in a eukaryotic system, and then use phosphatases and kinases to produce homogenousely dephosphorylated or phosphorylated material. We also detail how to use HDX-MS on this material to determine the membrane-binding footprint of PTEN.

2 Materials

For either X-ray crystallography or HDX-MS, or indeed any structural biology method, it is necessary to obtain purified proteins. Although expression of recombinant PTEN has been carried out in bacteria, our experience indicates that the material obtained from baculovirus expression in Sf9 cells produces correctly folded, soluble protein of a reasonable yield (typically 0.5 mg per liter of cells). All buffers were prepared with ultrapure water and analytic grade reagents. All buffers used in the purification of PTEN were chilled to 4 °C prior to use.

2.1 PTEN Purification

1. 1 M Tris–HCl pH 8.0: Dissolve 121.14 g of Tris–HCl in 100 ml of water in a glass beaker. Adjust the pH to 8.0 (room temperature) using HCl. Bring to 1 l with water. Filter through a 0.22 μm Corning filter.
2. 5 M NaCl: Dissolve 292.2 g of NaCl in 1 l of water. Filter through a 0.22 μm Corning filter.
3. 1 M Imidazole: Dissolve 68.07 g of imidazole in 200 ml of water. Adjust the pH to 8.0 using HCl. Add water to a volume of 1 l. Filter through a Corning 0.22 μm filter.
4. 1 M HEPES pH 7.4: Dissolve 238.3 g of HEPES in 200 ml of water. Adjust the pH to 7.4 (RT) using NaOH. Add water to a volume of 1 l. Filter through a 0.22 Corning μm filter.

5. Lysis buffer: 20 mM Tris–HCl pH 8.0 (RT), 300 mM NaCl, 10 mM imidazole pH 8.0, 5 % glycerol, 2 mM β-mercaptoethanol, 0.5 % Triton X-100 (from a 10 % stock solution), with one complete EDTA-free protease inhibitor (Roche) tablet added per 50 ml of lysis buffer.
6. HisTrap buffer A: 20 mM Tris–HCl pH 8.0, 300 mM NaCl, 10 mM imidazole pH 8.0, 5 % glycerol, 2 mM β-mercaptoethanol.
7. HisTrap buffer B: 20 mM Tris–HCl pH 8.0, 100 mM NaCl, 200 mM imidazole pH 8.0, 5 % glycerol, 2 mM β-mercaptoethanol.
8. Dialysis buffer: 20 mM Tris–HCl pH 8.0, 200 mM NaCl, 5 % glycerol, 2 mM TCEP.
9. Dilution buffer: 20 mM Tris–HCl pH 8.0, 10 % glycerol, 1 mM DTT.
10. Ion exchange buffer A: 20 mM Tris–HCl pH 8.0, 50 mM NaCl, 10 % glycerol, 1 mM DTT.
11. Ion exchange buffer B: 20 mM Tris–HCl pH 8.0, 1 M NaCl, 10 % glycerol, 1 mM DTT.
12. Gel filtration buffer: 20 mM HEPES pH 7.4, 200 mM NaCl, 2 mM TCEP.
13. TEV Protease (Life Technologies).
14. 5 ml HisTrap HP (GE Healthcare Life Sciences).
15. 5 ml HiTrap Q HP (GE Healthcare Life Sciences).
16. HiLoad 16/60 Superdex 75 Gel Filtration Column (GE Healthcare Life Sciences).
17. SnakeSkin™ Dialysis Tubing, 3.5 K MWCO (Thermo Scientific).
18. ÄKTA Purification System (GE Healthcare Life Sciences).
19. Amicon® Ultra 15 ml 10 K MWCO Concentrators (Millipore).
20. Lambda Protein Phosphatase (New England Biolabs).
21. CK2 Protein Kinase (New England Biolabs).
22. GSK3β Protein Kinase (New England Biolabs).
23. 0.1 M ATP stock solution (pH 8.0).
24. Probe sonicator (such as a Sonics Vibra-Cell VCX 500).

2.2 Liposome Production

1. 10 mg/ml Porcine Brain L-α-phosphatidylethanolamine (Avanti Polar Lipids, Inc) in chloroform (*see* **Note 1**).
2. 10 mg/ml Porcine Brain L-α-phosphatidylserine (sodium salt) (Avanti Polar Lipids, Inc).
3. 10 mg/ml Porcine Brain L-α-phosphatidylcholine (sodium salt) (Avanti Polar Lipids, Inc).

4. 1 mg/ml Porcine Brain L-α-phosphatidylinositol-4,5-bisphosphate (Avanti Polar Lipids, Inc).
5. 10 mg/ml Cholesterol (Avanti Polar Lipids, Inc).
6. 10 mg/ml Porcine Brain Sphingomyelin (Avanti Polar Lipids, Inc).
7. Liquid Nitrogen.
8. A 43 °C water bath.
9. A vacuum desiccator.
10. A bath sonicator.
11. A vortex mixer.
12. Parafilm®.
13. Hamilton® Gastight Syringe, Model 1002 (Volume 2.5 ml).
14. Whatman® Anontop ten Inorganic Membrane Filters (0.1 μm pore size).
15. 0.2 M Stock solution of EGTA, pH 8.0: Dissolve 3.8 g of EGTA in 10 ml of ultrapure water. Adjust the pH of the solution to pH 8.0 using a concentrated NaOH solution. Bring to 50 ml with ultrapure water.
16. 1 M KCl stock solution: Dissolve 74.55 g of KCl in 1 l of ultrapure water.
17. Lipid Buffer: 20 mM HEPES pH 7.5, 100 mM KCl, 1 mM EGTA (0.22 μm filtered) (*see* **Note 2**).

2.3 Membrane-Binding Hydrogen Deuterium Exchange Sample Components

1. Corning® Costar® Spin-X® centrifuge tube filters (0.22 μm pore size).
2. Deuterium Oxide, 99.8 % atom D (Arcos Organics, New Jersey, USA).
3. D_2O buffer solution: 10 mM HEPES pH 7.5, 50 mM NaCl, 2 mM TCEP (with the stock solutions of HEPES, NaCl and TCEP diluted in D_2O rather than ultrapure water).
4. H_2O buffer solution: 10 mM HEPES pH 7.5, 50 mM NaCl, 2 mM TCEP (with the stock solutions of HEPES, NaCl, and TCEP diluted with ultrapure water).
5. Quench buffer: 2 M guanidinium chloride, 2.4 % formic acid. This should be made fresh, and kept on ice (dilute 1.91 g of guanidinium chloride, add 240 μl of formic acid, and make up to 10 ml with ultrapure water).
6. Purified dephosphorylated PTEN.
7. Liposomes (5 % brain phosphatidylinositol [4, 5] bisphosphate, 20 % brain phosphatidylserine, 45 % brain phosphatidylethanolamine, 15 % phosphatidylcholine, 10 % cholesterol, and 5 % sphingomyelin (all percentages given as weight/volume)).

8. Lipid buffer (20 mM HEPES pH 7.5, 100 mM KCl, 1 mM EGTA (0.22 μm filtered)).
9. Gel filtration buffer (20 mM HEPES pH 7.4, 200 mM NaCl, 2 mM TCEP).
10. Liquid nitrogen.
11. A timer.
12. 500 μl Eppendorf tubes.

2.4 HDX-MS Equipment and Solvents

1. Climate-controlled UPLC system: We have built a fluidics system and placed this into an ice box. The commercially available HDX Manager, available with the SYNAPT® G2-Si HDMS system by Waters is another possibility. Both a dual gradient analytical pump and an isocratic pump are required.
2. Poroszyme® Immobilized Pepsin Cartridge (Applied Biosystems (Life)).
3. Acquity 1.7 μm particle, 100 mm × 1 mm C18 UPLC Column (Waters).
4. Mass Spectrometer equipped with an electrospray ionization source (we have used a Waters Xevo or a Thermo Scientific Orbitrap).
5. HDX-MS Software (we have used HDExaminer (Sierra Analytics), DynamX (Waters) is another possibility).
6. Access to database search software in order to identify peptide sequences (we use Mascot Database Search (Matrix Science)).
7. 0.1 % Formic acid solution (produced with HPLC grade water).
8. 100 % Acetonitrile (HPLC Grade).

3 Methods

This protocol is for the purification of a Sf9 cell expressed, TEV-cleavable N-terminally His-tagged PTEN construct using an ÄKTA Purification System. Wherever possible the chromatography should be carried out at 4 °C, using chilled buffers, and protein-containing fractions should be kept on ice. For X-ray crystallography, the same purification method should be employed, but the construct purified should be the multiply truncated version as used by Lee et al. [1]. Subsequent information on suitable crystallisation conditions etc. can also be found in the Lee et al. manuscript materials and methods.

3.1 PTEN Purification

1. Remove the 1-8 l Sf9 cell pellet from storage at −80 °C and place within a glass beaker on ice. Add lysis buffer to the cell pellet, typically 50 ml of lysis buffer for 1 l of Sf9 cells being purified. Throughout the purification, care should be taken

that any reducing agents are added freshly to buffers, as PTEN has the propensity to become inactivated via oxidation [17]. Sonicate the cell pellet on ice with a probe sonicator, until the pellet has completely broken apart and the cells are lysed, typically for 6 min, with 10 s bursts of sonication followed by 10 s of rest to prevent the cell lysate from overheating.

2. Place the cell lysate in an ultracentrifugation vial and centrifuge at 140,000 × *g* for 45 min at 4 °C.
3. Equilibrate a 5 ml HisTrap FF column in HisTrap Buffer A by passing 20 ml of HisTrap Buffer A over the column at 4 ml/min.
4. Remove the supernatant (the soluble fraction) from the ultracentrifugation vial, and filter through a 0.45 μm filter. Load the filtered supernatant onto the HisTrap Column at 4 ml/min.
5. Run a 4 ml/min method on the ÄKTA system where after an initial 40 ml HisTrap Buffer A wash, a 40 ml 5 % HisTrap Buffer B wash is conducted. Finally, elute PTEN from the column using 40 ml of HisTrap Buffer B. It is advisable to then run an SDS-PAGE of the collected fractions to determine the location and purity of PTEN at this stage.
6. At this stage it is often preferable to remove the His-tag from PTEN. However, it may be desirable to retain the His-tag for certain experiments, in which case this cleavage step, and the subsequent second HisTrap purification can be skipped. Pool the PTEN-containing fractions, estimate the protein concentration by OD_{280}, and add TEV protease in a ratio of 1:20 TEV to PTEN (w/w).
7. Soak the SnakeSkin™ dialysis membrane in 4 l of dialysis buffer for 5 min. Pour the protein solution into the dialysis membrane, seal adequately with binder clips, and leave the protein solution overnight at 4 °C to digest and dialyze (use a stir bar to keep the solution moving). The purpose of this dialysis step is to remove the imidazole present in the PTEN solution, so that the solution can be passed back over a HisTrap column to remove the His-tagged TEV protease.
8. Repeat **step 5**, using the Dialysis Buffer in place of HisTrap Buffer A. Note well that the cleaved PTEN will now no longer bind to the HisTrap Column, and as such, it is imperative that the whilst loading the column the flow-through is collected and kept on ice. After running the HisTrap Column it is again advisable able to run an SDS-PAGE to determine that the protein is in the flow-through and that it has been correctly cleaved.

9. If necessary, an ion-exchange chromatography step can now be run to further purify PTEN prior to gel filtration. Equilibrate a HiTrap Q HP column with 20 ml Ion Exchange Buffer A run at 4 ml/min. Dilute the PTEN solution with an equal volume of Dilution Buffer before passing over the HiTrap Q HP column at 4 ml/min. Run a 4 ml/min method on the ÄKTA system where after an initial wash of 20 ml of Ion Exchange Buffer A, an increasing concentration of Ion Exchange Buffer B is run over 30 column volumes. It is strongly advised to run an SDS-PAGE to determine where PTEN has eluted from the HiTrap Q column. PTEN typically elutes at 200 mM NaCl (*see* **Note 3**).
10. Pool all PTEN containing fractions, and concentrate using a 10,000 MWCO Amicon Ultracentrifugation concentrator, run at 4 °C until a volume of approximately 1 ml is achieved. If required, PTEN can be dephosphorylated at this stage using lambda protein phosphatase (*see* **Note 4**) (*see* Subheading 3.2).
11. Inject the concentrated PTEN solution onto a Superdex 75 16/60 gel filtration column, equilibrated in gel filtration buffer, and run at 1 ml/min at 4 °C.
12. Pool and concentrate the PTEN containing fractions to at least 1 mg/ml for subsequent HDX-MS analysis. Aliquot, flash freeze in liquid nitrogen, and store at −80 °C.

3.2 PTEN Dephosphorylation and Re-phosphorylation

When purified from Sf9 cells, the C-terminus of PTEN is found primarily in a partially phosphorylated state (*see* **Note 4**). In order to investigate the structure of PTEN, it is necessary to first remove these modifications to produce a homogenous sample, and then re-phosphorylate with recombinant kinases in vitro if the phosphorylated state of the enzyme is desired.

1. Prior to the size exclusion chromatography step in the purification of PTEN, PTEN can be dephosphorylated. Incubate 1 ml of PTEN with 10 μl of Lambda protein phosphatase in a buffer (supplied by NEB) consisting of 50 mM HEPES, 100 mM NaCl, 2 mM DTT, 0.01 % Brij 35, 1 mM $MnCl_2$, pH 7.5, for 90 min at 30 °C (*see* **Note 5**). The subsequent gel filtration step will then remove the lambda protein phosphatase and the required manganese.
2. In order to produce homogenously phosphorylated material, PTEN is subsequently re-phosphorylated in vitro using recombinant CK2 and GSK3β (both available from New England Biolabs). Following the gel filtration step (*after PTEN has been dephosphorylated using lambda protein phosphatase),* concentrate PTEN again, as previously described, to 500 μl total volume. Add 20 μl of 0.1 M ATP solution, 100 μl of the

GSK3β and CK2 reaction buffers supplied by New England Biolabs (*see* **Note 6**), and 5 μl each of the kinases GSK3β and CK2 solutions. Bring the final volume to 1 ml with the addition of 270 μl gel filtration buffer. The solution is then incubated at 25 °C for 24 h, and then the PTEN is purified using gel filtration as previously described.

3.3 Liposome Production

1. In a sterile, glass, screw-top vial mix the lipid stock solutions together in the desired ratio (*see* **Notes 7–9**). A plasma membrane mimicking composition [18] is the following: 5 % brain phosphatidylinositol 4,5-bisphosphate, 20 % brain phosphatidylserine, 45 % brain phosphatidylethanolamine, 15 % phosphatidylcholine, 10 % cholesterol, and 5 % sphingomyelin (all percentages given as weight/volume). A typical liposome stock preparation may be 1 ml of 5 mg/ml lipids.
2. Evaporate the organic solvents under a stream of nitrogen or argon gas, whilst turning the vial in your hands. Care should be taken that the stream of gas is not too powerful, as this may cause the lipids to be blown out of the vial. A smear of lipids can be observed on the side of glass vial, resembling spider's silk.
3. Place the vial in a vacuum desiccator (with the lid of the vial removed), and allow the organic solvents to be further evaporated at room temperature under vacuum for 1 h.
4. Remove the vial from the desiccator, and add 1 ml lipid buffer.
5. Place the lid back on to the vial, and further seal the lid with Parafilm®.
6. Vortex the vial for 3 min to resuspend the lipids.
7. Sonicate the vial in a bath sonicator for 2 min to remove any residual lipids still stuck to the glass.
8. Remove the lipids from glass vial and place in a 1.5 ml Eppendorf® tube.
9. Freeze the vial in liquid nitrogen until the lipid solution is opaque and frozen solid. Once frozen, rapidly remove the vial from the liquid nitrogen and place the vial in a 43 °C water bath until completely thawed. This is a single "freeze-thaw." Repeat this process a total of ten times (*see* **Note 10**).
10. Wash the Hamilton® Syringe sequentially in 70 % ethanol, water, and finally lipid buffer.
11. Aspirate the lipid solution into the syringe. Screw the membrane filter onto the end of the syringe until finger tight, and then pass the lipid solution through the filter, collecting the extruded lipid solution a suitable sterile container. Remove the filter from the syringe, and aspirate the solution once again.

Repeat this process until the lipids have been extruded a total of 11 times (*see* **Note 11**).

12. Aliquot the liposomes in a desired volume, freeze in liquid nitrogen and store at −80 °C. When using lipids in subsequent experiments, it is preferable to thaw them from −80 °C at room temperature.

3.4 HDX-MS Sample Preparation

In this experiment, a binding footprint for PTEN on a membrane bilayer is determined. The protein will be incubated with a D_2O containing buffer, either in the presence or absence of liposomes, at four different time points: 3, 30, 300, and 3000 s (*see* **Note 12**). This process can be automated using the LEAP Technologies Precise Adaptable Liquid-Handling system.

1. The composition of a single deuterated sample should be the following: 10 μl of 5 μM PTEN (dephosphorylated), 40 μl H_2O or D_2O solution, 20 μl quench solution (*see* **Note 13**). Each time point (of which there is four) is carried out in triplicate, in two conditions (with and without lipids), producing a total of 24 samples. In addition to this three non-deuterated samples are required for identification of PTEN's peptic peptides, bringing the total number of samples to 27.
2. Each sample will occupy a single 500 μl Eppendorf tube. It is necessary to label each tube with its contents (PTEN with/without liposomes), and time point, and also to clearly label the three non-deuterated samples.
3. Thaw the dephosphorylated PTEN on ice, and then remove any possible precipitated protein by using a spin filter (centrifuge at 13,000 × *g* for 10 min in a 4 °C). Measure the concentration of the flow-through. In this example, the concentration may be 40 μM.
4. Thaw the 5 mg/ml liposomes at room temperature (165 μl required).
5. Produce two 200 μl aliquots of two different protein solutions. Assuming a PTEN concentration of 40 μM, and a liposome stock concentration of 5 mg/ml:
   (a) PTEN Stock A (without liposomes): 25 μl of PTEN (final concentration 5 μM), 40 μl lipid buffer, 135 μl gel filtration buffer.
   (b) PTEN Stock B (with liposomes): 25 μl of PTEN, 40 μl of liposomes, 135 μl gel filtration buffer.
6. Aliquot 10 μl of the protein stocks into the 0.5 ml tubes and allow to equilibrate to room temperature for 10 min.
7. Produce 1 ml of the D_2O buffer solution. To 0.5 ml of this solution, add 125 μl of either 5 mg/ml liposomes or lipid buffer.

8. To each PTEN aliquot, add the appropriate D_2O Buffer solution, and mix thoroughly. Incubate PTEN with the D_2O with the appropriate timing, using a timer. It is critical that the timing is precise to about 1 s in order that replicates have similar rates of exchange. At the end of each time point, rapidly add 20 μl of the ice-cold quench solution, mix (using the pipette), close the Eppendorf tube, and freeze immediately in liquid nitrogen (*see* **Note 14**). For longer time points (300/3000 s) close the Eppendorf tube between adding the D_2O and quenching the reaction.
9. Prepare non-deuterated samples by diluting PTEN Stock A with the H_2O buffer and quenching (as no deuterium is being incorporated at this point the time of incubation is irrelevant).
10. Store samples at −80 °C until analysis.

3.5 Non-deuterated Samples: Identification of Peptides

In order to identify the peptides that are to be subsequently analyzed for perturbations in their rate of solvent accessibility, it is necessary to digest a non-deuterated PTEN sample, and run an MS/MS experiment (from which the sequences of the peptides can be determined). The mass spectrometer should be in positive ion mode and resolution mode, and calibrated (using, e.g., Glu-1-fibrinopeptide B) (*see* **Note 15**). Refer to Fig. 1 for a schematic of the fluidic system used, and how it is manipulated during the process.

1. Using the isocratic pump (with 0.1 % formic acid solution in Line A) equilibrate the immobilized pepsin cartridge with 0.1 % formic acid solution at a 0.3 ml/min flow rate for 10 min.
2. Using the dual-gradient analytical pump (with 0.1 % formic acid solution in Line A, and 100 % acetonitrile in Line B), equilibrate the C18 UPLC Column in 3 % acetonitrile (97 % 0.1 % formic acid) by flowing at 0.04 ml/min for 10 min. Ensure that the pressure on the system has stabilized, typically at 40 MPa bar for the dual-gradient pump, and 5 MPa for the isocratic pump.
3. Inject the PTEN sample onto the immobilized pepsin cartridge, and allow PTEN to be digested for 180 s.
4. Reverse the flow over the peptide trap, eluting the peptides from the trap and loading them onto the C18 Column.
5. Concurrently with **step 4**, start an increasing acetonitrile gradient, which over 24 min reaches 90 % acetonitrile.
6. Concurrently with **step 5**, begin collecting MS/MS data on the mass spectrometer.
7. Repeat the process after the data acquisition (**steps 1–6**), but collecting in MS mode.

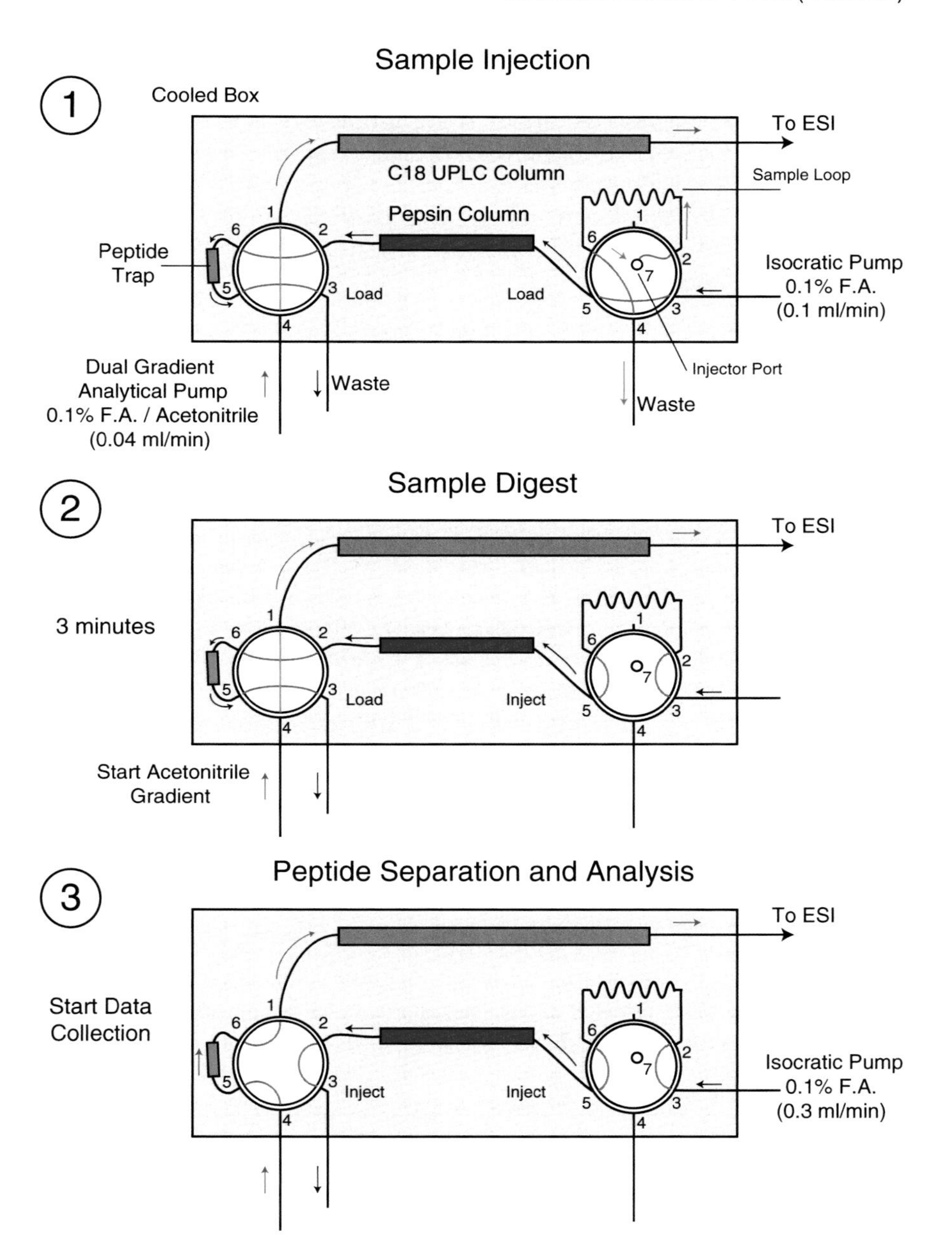

Fig. 1 Schematic of the UPLC system used in HDX-MS. (*1*) Both valves are in the "load" position. The 0.1 % formic acid solution from the isocratic pump flows over both the pepsin column and the peptide trap, eventually going to waste. The dual-gradient column pump flows onto the C18 UPLC column. The peptide sample is loaded at the injection point (labeled 7) onto the loop. (*2*) The right hand valve is switched to inject, emptying the contents onto the peptide trap. As the protein is digested, the peptides are immobilized onto the peptide trap. Digest the protein for 3 min. (*3*) The left hand valve is now switched to inject. The peptide trap is now in line with the acetonitrile gradient, but with the flow reversed. The peptide trap is also now in line with the C18 column. With the reversed flow, the peptides are eluted onto the column, and the acetonitrile gradient elutes the peptides from the C18 column to be subsequently analyzed using the mass spectrometer

3.6 Deuterated Samples: Identification of Peptides

The peptic digest and subsequent analysis of peptides of deuterated samples should be carried out as above, but in MS mode, rather than MS/MS mode. Care should be taken to ensure than the timing of the digestion and separation of peptides is carried out in a reproducible manner, as the deuterated sample is exposed to an aqueous environment, and back exchange will occur. For this reason, it is also imperative that the fluidics system maintains a low temperature (<4 °C).

3.7 Data Analysis

Currently we use HDExaminer (Sierra Analytics) for all data analysis, and further analyze the data produced with Microsoft Excel. Other software, such as DynamX (Waters) is available. The process of using your chosen software is beyond the remit of this chapter but can be briefly summated as follows:

1. Import the peptide list identified from the two MS/MS runs into the software (*see* **Note 16**).
2. Identify the non-deuterated peptides from the MS file. Manually check the peptides for correct selection and that each selected peptide is of suitable quality. This includes the peptide's retention time, charge state, and the presence of overlapping peptides that may influence the centroid mass calculation (*see* **Note 17**).
3. Add the files containing the data obtained from deuterated PTEN.
4. Ensure the deuterated peptides are of sufficient quality to produce reproducible data—repeating the quality assurance steps as mentioned for **Note 17** (*see* **Note 18**).
5. Compare the levels of deuterium incorporation for identical peptides in the two datasets to determine differences in HDX-rate (*see* **Note 19**).

4 Notes

1. Lipids can be ordered pre-dissolved in either a chloroform solvent or as a powdered stock, in which cause it is advisable to dissolve the powdered lipid in chloroform upon receipt and store at −80 °C. These materials are required to produce liposomes of the following composition: 5 % brain PIP_2, 20 % brain PS, 45 % brain PE, 15 % brain PC, 10 % cholesterol, 5 % sphingomyelin (all percentages given as weight/volume).
2. It is crucial to include EGTA in the lipid buffer to buffer Ca^{2+} in a systematic manner. This ion is vital for membrane fusion [19, 20], and in order to maintain reproducibility Ca^{2+} concentration must be controlled.

3. Two PTEN-containing peaks can often be observed when eluting from the HiTrap Q HP, one at 200 mM NaCl, and a second concentration of 400 mM NaCl or higher. This second peak is often misfolded protein, and will be present in the void of the gel filtration column.
4. If expressing PTEN with an intact C-terminus in a Eukaryotic expression system (i.e., containing residues 353–403) it is highly likely that the protein will be multiply and not homogeneously phosphorylated on the C-terminus [21, 22].
5. Mass spectrometry analysis found that there was no phosphorylated material after 90 min at 30 °C. The subsequent gel filtration step will then remove the lambda protein phosphatase (no phosphatase activity was determined in PTEN C124S that had been dephosphorylated using this method).
6. The NEB buffers contain the following (and are shipped as a 10× stock solution). CK2 reaction buffer: 200 mM Tris–HCl pH 7.5, 500 mM KCl, 100 mM $MgCl_2$. GSK3β reaction buffer: 200 mM Tris–HCl pH 7.5, 100 mM $MgCl_2$, 5 mM DTT. Therefore it is not necessary to add additional $MgCl_2$.
7. Pipetting organic solvents accurately can be challenging due to their low viscosity. Using a 200 μl pipette is preferable to a 1000 μl pipette as the larger bore size on the tip can lead to the solutions dribbling out.
8. Keep all lipid stocks dissolved in organic solvents sealed and on ice as much as possible. Owing to their volatility, the solvent will evaporate rapidly at room temperature, altering the concentrations of the stock solutions.
9. On addition of the PIP_2 to this lipid composition, the solution may turn cloudy, which is undesirable. Addition of 5 μl of methanol, and allowing the solution warm to room temperature will quickly reverse this.
10. Take note of the time required to freeze and thaw the lipids, and attempt to be as reproducible as possible for each freeze/thaw cycle. Do not, for instance, leave the lipids to thaw at 43 °C for longer than is absolutely necessary.
11. The unscrewing and rescrewing of the filter is difficult, and passing the lipid solution through the filter requires some effort to be applied to the syringe. Take care not to introduce air into the syringe, as bubbles can be formed in the liposome solution.
12. The “3 s” time point is difficult to conduct. Start by aspirating both the quench and the D_2O solution in the pipettes. Then add the D_2O Buffer, rapidly mix by aspirating up and down, and then add the quench, and flash freeze. Furthermore, by chilling the samples on ice, and also chilling the D_2O buffer

(and also preferably chilling the pipette tips and conducting the sample preparation step in a 4 °C room), a further time point of 3 s at 0 °C can be conducted. This is occasionally referred to as 0.3 s in the literature as there is an approximate tenfold decreased rate of solvent exchange between room temperature and 0 °C.

13. It is crucial to expose the two different set of samples (with and without lipids) to the exact same concentrations of D_2O, and also as far as possible, the exact same buffer, salt, and temperature conditions. Differing levels of D_2O between the samples will mask the perturbations of solvent exchange rate observed on membrane binding, and will prevent accurate data analysis.
14. Be sure to maintain sufficient distance between the liquid nitrogen container and the samples as to not chill them below room temperature during D_2O incubation.
15. If possible, an internal lock-spray calibration should also be run concurrently with the samples.
16. When exporting the identified peptides from the database, only export peptides that are well identified. This requires a ppm error that is commensurate with what would be expected from the instrument calibration, sufficient product ions to allow for confidence in the identification in the precursor ion, and an e-value that provides confidence that the identification is significant. The "ion score," an amalgam of these properties, can be used to this purpose. We use an ion score threshold of 15.
17. Peptides must meet certain criteria in order to ensure correct identification, and an accurate calculation of the centroid mass of the peptide. Ensure the following: that the charge state of the peptide (as seen within the software) matches the expected charge state; that there are no overlapping peptides—peptides that are also within the isotopic envelope of peptide of interest; that the intensity of the peptide is above background noise (typically about 5000 counts is the lower limit), and that the m/z shift is less than ±0.005.
18. The error between experiments, especially for the longer time points, should be certainly <3 %, and often within 1 %. The time point with the highest error associated is typically the 3 s time point, as this is carried out manually and has the greatest room for error. Peptides that are of poor quality, or incorrectly identified, will almost certainly have a higher error associated between the repeats. Hallmarks of incorrectly identified deuterated peptides are %D that are >100 %, large m/z shifts, and larger than 1-min deviations from the non-deuterated experiments, and negative values of deuteration.

19. We have two criteria for determining whether shifts in deuterium uptake are significant: the percentage difference (%D) and the absolute mass shift (Da). The %D is the percentage increase in mass that can be expected from a peptide (based on both the number of available imide groups, and the amount of deuterium in the buffer). A significant change in %D can be calculated with an unpaired *t*-test, and is typically on the order of 6 %. Obviously smaller peptides (e.g., 5 residues) that experience small mass differences will have exaggerated %D when compared to larger (10+ residue) peptides. To ensure confidence in the shifts seen in smaller peptides, we also require a mass difference of at least ±0.5 Da between the mean centroid mass value for both datasets.

Acknowledgement

This work was supported by the Medical Research Council (file reference U10518430).

References

1. Lee J.O. et al. (1999) Crystal structure of the PTEN tumor suppressor: implications for its phosphoinositide phosphatase activity and membrane association. *Cell* **99**, 323–334
2. Kolmodin K., Åqvist, J. (2001) The catalytic mechanism of protein tyrosine phosphatases revisited. *FEBS Lett* **498**, 208–213
3. Xiao Y. et al. (2007) PTEN catalysis of phospholipid dephosphorylation reaction follows a two-step mechanism in which the conserved aspartate-92 does not function as the general acid — Mechanistic analysis of a familial Cowden disease-associated PTEN mutation. *Cell Signal* **19**, 1434–1445
4. Rodriguez-Escudero I. et al. (2011) A comprehensive functional analysis of PTEN mutations: implications in tumor- and autism-related syndromes. *Hum Mol Genet* **20**, 4132–4142
5. Zhang X.C., Piccini A., Myers M.P., Van Aelst L., Tonks N.K. (2012) Functional analysis of the protein phosphatase activity of PTEN. *Biochem J* **444**, 457–464
6. Lumb C.N., Sansom M.S.P. (2013) Defining the membrane-associated state of the PTEN tumor suppressor protein. *Biophys J* **104**, 613–621
7. Shenoy S. et al. (2012) Membrane association of the PTEN tumor suppressor: molecular details of the protein-membrane complex from SPR binding studies and neutron reflection. *PLoS One* 7, e32591
8. Rahdar M. et al. (2009) A phosphorylation-dependent intramolecular interaction regulates the membrane association and activity of the tumor suppressor PTEN. *Proc Natl Acad Sci USA* **106**, 480–485
9. Hopkins B.D. et al. (2013) A secreted PTEN phosphatase that enters cells to alter signaling and survival. *Science* **341**, 399–402
10. Pulido R. et al. (2014) A unified nomenclature and amino acid numbering for human PTEN. *Sci Signal* 7, pe15
11. Liang H. et al. (2014) PTENα, a PTEN isoform translated through alternative initiation, regulates mitochondrial function and energy metabolism. *Cell Metab* **19**, 836–848
12. Malaney P., Uversky V.N., Davé V. (2013) The PTEN Long N-tail is intrinsically disordered: increased viability for PTEN therapy. *Mol Biosyst* **9**, 2877–2888
13. Konermann L., Pan J., Liu Y-H. (2011) Hydrogen exchange mass spectrometry for studying protein structure and dynamics. *Chem Soc Rev* **40**, 1224
14. Burke J.E., Perisic O., Masson G.R., Vadas O., Williams R.L. (2012) Oncogenic mutations mimic and enhance dynamic events in the natural activation of phosphoinositide 3-kinase

p110α (PIK3CA). *Proc Natl Acad Sci U S A* **109**, 15259–15264

15. Vadas O. et al. (2013) Molecular determinants of PI3Kγ-mediated activation downstream of G-protein-coupled receptors (GPCRs). *Proc Natl Acad Sci U S A* **110**, 18862–18867
16. Burke J.E. et al. (2014) Structures of PI4KIIIβ complexes show simultaneous recruitment of Rab11 and its effectors. *Science* **344**, 1035–1038
17. Lee S.R. et al. (2002) Reversible inactivation of the tumor suppressor PTEN by H2O2. *J Biol Chem* **277**, 20336–20342
18. Vance J., Steenbergen R. (2005) Metabolism and functions of phosphatidylserine. *Prog Lipid Res* **44**, 207–234
19. McNew J.A. et al. (2000) Compartmental specificity of cellular membrane fusion encoded in SNARE proteins. *Nature* **407**, 153–159
20. Nickel W. et al. (1999) Content mixing and membrane integrity during membrane fusion driven by pairing of isolated v-SNAREs and t-SNAREs. *Proc Natl Acad Sci U S A* **96**, 12571–12576
21. Miller S.J., Lou D.Y., Seldin D.C., Lane W.S., Neel B.G. (2002) Direct identification of PTEN phosphorylation sites. *FEBS Lett* **528**, 145–153
22. Torres J., Pulido R. (2000) The tumor suppressor PTEN is phosphorylated by the protein kinase CK2 at its C terminus -implications for PTEN stability to proteasome-mediated degradation. *J Biol Chem* **276**, 993–998

Part VI

Methods to Study *PTEN* and Stem Cell Function

Chapter 15

Methods for PTEN in Stem Cells and Cancer Stem Cells

Suzanne Schubbert, Jing Jiao, Marcus Ruscetti, Jonathan Nakashima, Shumin Wu, Hong Lei, Qinzhi Xu, Wenkai Yi, Haichuan Zhu, and Hong Wu

Abstract

PTEN (phosphatase and tensin homologue) is the first tumor suppressor identified to have phosphatase activity and its gene is the second most frequently deleted or mutated tumor-suppressor gene associated with human cancers. Germline *PTEN* mutations are the cause of three inherited autosomal dominant disorders. Phosphatidylinositol 3,4,5,-triphosphate (PIP3), the product of the PI3 kinase, is one of the key intracellular targets of PTEN's phosphatase activity, although PTEN's phosphatase-independent activities have also been identified. PTEN is critical for stem cell maintenance, which contributes to its controlled tumorigenesis. PTEN loss leads the development of cancer stem cells (CSCs) that share properties with somatic stem cells, including the capacity for self-renewal and multi-lineage differentiation. Methods to isolate and functionally test stem cells and CSCs are important for understanding PTEN functions and the development of therapeutic approaches to target CSCs without having adverse effects on normal stem cells. Here, we describe protocols for the isolation and functional analysis of PTEN deficient embryonic stem cells, hematopoietic stem cells and leukemia-initiating cells (LICs), neural stem cells, and prostate stem cells and CSCs.

Key words PTEN, Stem cell, Cancer stem cells, ES cell, HSC, Leukemia-initiating cell, NSC, Prostate cancer stem cell

1 Introduction

1.1 PTEN and Mouse Embryonic Stem Cells

Embryonic stem (ES) cells are derived from the inner cell mass of the blastocyst-stage embryo. The pluripotent ES cells are able to generate all cell types in the body, which distinguishes them from adult stem cells.

PTEN inactivation in the murine ES cells leads to an increased growth rate, proliferation, and survival even in the absence of serum [1]. ES cells lacking PTEN function also display advanced entry into S phase and elevated levels of PIP3. Consequently, PTEN deficiency leads to dosage-dependent increases in phosphorylation and activation of AKT/PKB, also a well-characterized target of the PI3 kinase signaling pathway [1], which is essential for

Leonardo Salmena and Vuk Stambolic (eds.), *PTEN: Methods and Protocols*, Methods in Molecular Biology, vol. 1388, DOI 10.1007/978-1-4939-3299-3_15, © Springer Science+Business Media New York 2016

ES cell growth [2]. PTEN also controls p53 level and transcription activity with phosphatase-dependent and -independent mechanisms [3]. *Pten−/−* mouse ES cells display altered differentiation into endodermal, ectodermal, and mesodermal derivatives [4]. During in vitro differentiation, *Pten* null ES cells still maintain relatively higher Nanog expression compared to wild-type (WT) ES cells, which may contribute to the emergence of a small number of aggressive, teratoma-initiating embryonic carcinoma cells in vitro and higher tumorigenesis in vivo [5].

1.2 PTEN and Hematopoietic Stem Cells and Leukemia-Initiating Cells

Hematopoietic stem cells (HSCs) are multipotent cells capable of self-renewal, differentiation and generating all of the cell types in the blood system [6]. HSCs typically reside in the bone marrow, but can be mobilized by various stimuli such as radiation or blood loss.

PTEN has been conditionally deleted in fetal and adult HSCs with a *Pten* floxed allele and various promoter driven Cre lines, including the interferon responsive *Mx1-Cre*, *vascular endothelial cadherin (VE-Cadherin)-Cre*, and the tamoxifen-inducible *Scl-CreERT* allele (*Scl-Cre*) [7–11]. Loss of PTEN in HSCs leads to transient expansion of HSCs, hematopoietic abnormalities and leukemia development [7–11]. PTEN deficient HSCs and leukemia are sensitive to mTOR inhibition [12, 13]. Leukemia-initiating cells (LICs), which are considered to be the subset of leukemia cells that maintain leukemia growth and are responsible for the initiation and propagation potential of the disease, have been identified and functionally studied in the *VE-Cad-Cre+;Ptenloxp/loxp* model and are responsible for transplantable acute T cell lymphoblastic leukemia (T-ALL). This PTEN T-ALL model mimics a subset of human T-ALL with PTEN loss, activated β-catenin and TCR α-c-Myc translocation, and provides a useful in vivo model for studying LIC-mediated therapeutic resistance and identifying potent LIC targeted reagents [9, 14, 15].

1.3 PTEN and Neural Stem Cells

Neural stem cells (NSCs) are derived from the neuroectoderm and are capable of differentiating into neurons and glia in embryonic and adult central nervous systems. In the developing brain, radial glia residing in the neuroepithelium divide asymmetrically to generate either neural or glial intermediate progenitor cells, which eventually give rise to neurons, astrocytes and oligodendrocytes, respectively [16]. In the adult brain, neurogenesis occurs the subgranular, subventricular zone, and the subcallosal zones [17]. There a number of assays developed to study NSCs, in vivo and in vitro.

PTEN plays an important role in regulating neural stem/progenitor cells in vivo and in vitro. *Pten* deletion in embryonic or adult neural stem cells exhibits enlarged, histoarchitecturally abnormal brains, which resulted from increased cell proliferation, decreased cell death, and enlarged cell size [18–21]. Neurosphere cultures revealed a greater proliferation capacity for tripotent Pten−/− CNS stem/progenitor cells. These increases in neural stem cell self-renewal and proliferation capacities are due, at least in part, to enhanced G0–G1 cell cycle transition, hypersensitivity to growth factor stimulation, and a shortened cell cycle. However, cell fate commitments of the progenitors are largely undisturbed, which fits well the concept of "cancer stem cells."

1.4 PTEN and Prostate Stem Cells and Cancer Stem Cells

Stem cells are important not only in the development of the prostate, but also in the maintenance of adult prostatic glandular structure. The growth of the prostate is dependent upon levels of the steroid hormone androgen. Although the prostate undergoes involution upon androgen withdrawal, it can become completely regenerated upon androgen restoration, suggesting the existence of long-lived prostate stem cell populations in the prostate gland [22–24]. Similarly, in prostate cancer patients, there exists a unique population of prostate tumor cells with the ability to survive therapy, such as androgen deprivation therapy (ADT), and eventually reconstitute the tumor in its entirety, giving rise to what is termed castration-resistant prostate cancer (CRPC). These cells, which harbor stem cell-like characteristics such as indefinite self-renewal, are thus referred to as prostate cancer stem cells (CSCs) [25, 26].

PTEN is one of the most commonly deleted or mutated tumor suppressor genes in human prostate cancer [27]. In the *PbCre4;PtenLox/Lox* murine model, it was identified that PTEN loss can lead to increased stem/progenitor cell proliferation, which in turn may be associated with prostate tumor initiation and progression [28]. Moreover, in *Pten*-null prostate mutants, the Lin(Ter119, CD45, CD31)$^{-}$ Sca-1^{+}CD49f^{hi} (LSC^{hi}) population, which marks the basal population, increases during tumor progression and harbors stem/progenitor function, as measured by its high sphere-forming activity and its capacity to be both necessary and sufficient for tumor initiation [29]. Not surprisingly, upregulation of the PI3K/AKT pathway, along with increased AR expression in human basal Lin^{-}Trop2^{+}CD49f^{hi} cells, the human equivalent of murine LSC^{hi} cells, is also necessary and sufficient for the initiation of human prostate cancer [30]. LSC^{hi} cell populations also have the capacity to generate differentiated luminal cells upon transplantation, an important feature for modeling prostate cancer as the human disease is marked clinically by the loss of the basal cell marker p63 [29–32]. The LSC^{hi} stem/progenitor cell population is also increased after castration in *Pten*-null animals, indicating that this population harbors an androgen-independent and castration-resistant phenotype that allows it to persist after hormone withdrawal and possibly contribute to CRPC [29]. More recent studies, however, have demonstrated that both prostate luminal and basal cells are capable of serving as the origin of cancer stem cells [33, 34]. Collectively, these data indicate that PTEN negatively regulates prostate stem/progenitor cells, and thus PTEN loss mediates the expansion of prostate cancer stem cells that may function as potential initiators of CRPC.

2 Materials

2.1 PTEN and Mouse Embryonic Stem Cells

2.1.1 In Vitro Culture of mESCs (Also Applies to Pten-Deficient mESCs)

1. mESC medium (*see* **Note 1**): Knockout Dulbecco's modified Eagle's medium (DMEM), 15 % v/v FBS, 100 units/ml penicillin, 100 μg/ml streptomycin, 1 % v/v L-glutamine, 1 % v/v nonessential amino acids, 100 μM 2-mercaptoethanol, 500–1000 U/ml units leukemia inhibitory factor (LIF).
2. Dulbecco's phosphate-buffered saline (DPBS).
3. Trypsin-EDTA.
4. 0.2 % Gelatin (0.2 g gelatin dissolved in 100 ml ddH_2O, autoclaved).
5. Mitotically inactivated primary mouse embryonic fibroblasts (MEFs).
6. 100 mm tissue culture dish.

2.1.2 Growth Competition Assay

1. mESC medium.
2. DPBS.
3. Trypsin-EDTA.
4. 0.2 % Gelatin (0.2 g gelatin dissolved in 100 ml ddH_2O, autoclaved).
5. 6-Well tissue culture dish.
6. Genomic DNA extraction kit.
7. Southern Blot detection system or real-time PCR detection system.

2.1.3 Cell Synchronization

1. mESC medium.
2. DPBS.
3. Trypsin-EDTA.
4. Culture flasks of desired size.
5. Colcemid.
6. 100 % Ethanol.
7. Propidium iodine (PI) solution: 50 μg/ml propidium iodine in PBS containing 0.2 % Triton X-100 and 20 μg/ml RNase A.
8. Flow cytometer.

2.1.4 Cell Differentiation

1. DPBS.
2. Trypsin-EDTA.

3. Differentiation medium: mESC medium without LIF.
4. Ultralow attachment 6-well plates.
5. Retinoic acid.

2.1.5 Teratoma Formation

1. mESC medium.
2. DPBS.
3. Trypsin-EDTA.
4. Mitotically inactivated primary MEF.
5. BD Matrigel matrix.
6. NSG (NOD.Cg-Prkdc^{scid} Il2rg^{tm1Wjl}/SzJ) mice.
7. 27-Gauge insulin syringe.

2.1.6 Enrichment for Highly Tumorigenic Pten−/− mESC Cell Subpopulation

1. mESC medium.
2. DPBS.
3. Trypsin-EDTA.
4. DPBS/bovine serum albumin (BSA): Dissolve 1 g BSA in 100 ml DPBS buffer.
5. Collagenase.
6. High-glucose DMEM.
7. SSEA1 antibody (DSHB; MC-480); c-kit antibody.
8. Cy5-conjugated goat anti-mouse IgM.
9. PE-conjugated goat anti-rat IgG.

2.2 PTEN and Hematopoietic Stem Cells and Leukemia-Initiating Cells

2.2.1 HSC Identification and Isolation by Immunophenotype

1. Antibodies (Clone): B220 (RA3-6B2), CD3 (145-2C11), CD4 (GK1.5), CD5 (53-7.3), CD8 (53-6.7), Gr-1 (RB6-8C5), CD11b (M1/70), Ter119 (TER-119), Sca-1 (D7), c-kit (2B8), CD34 (RAM34), Flk-2 (A2F10), CD150 (TC15-12F12.2), CD244.2 (2B4), CD48 (HM48-1) (*see* **Note 2**).
2. Viability dye: 7-AAD (7-aminoactinomycin D) or 4′,6-diamidino-2-phenylindole (DAPI).
3. HBSS/2 %FBS staining buffer: Hanks' buffered salt solution (without Ca^{2+} and Mg^{2+}), supplemented with 2 % fetal bovine serum, 10 mM HEPES, 100 units/ml penicillin and 100 μg/ml streptomycin.
4. 5 or 10 ml sterile syringe.
5. 25 G 7/8 needle.
6. 40 μm cell strainer.
7. Red blood cell (RBC) lysis buffer: 155 mM NH_4Cl, 12 mM $NaHCO_3$, 0.1 mM EDTA.
8. Hemocytometer.
9. FACS analyzer and sorter (*see* **Note 3**).

2.2.2 Colony-Forming Cell (CFC) Assays

1. Biosafety cabinet for sterile technique.
2. 37 °C incubator with 5 % CO_2.
3. Methylcellulose medium M3434 (Stem Cell Technologies; methylcellulose medium with recombinant cytokines and EPO for mouse cells) (*see* **Note 4**).
4. Mouse long-term culture medium with hydrocortisone: MyeloCult™ M5300 (Stem Cell Technologies) with hydrocortisone at 10^{-6} M (hydrocortisone 21-hemisuccinate sodium salt).
5. Iscove's Modified Dulbecco's Medium (IMDM) with 2 % FBS, 100 units/ml penicillin and 100 μg/ml streptomycin (IMDM/2 % FBS).
6. 35 mm petri dishes that are not tissue culture treated (for CFU-C assays with M3434).
7. 100 mm or 150 mm petri dishes (secondary dishes for 35 mm petri dishes).
8. 5 or 10 ml sterile syringe.
9. Sterile blunt end 16G needle.
10. RBC lysis buffer.
11. Feeder layer cells for LT-CTC: harvested mouse bone marrow cells, M2-10B4 bone marrow stromal cell line (ATCC), or AFT024 fetal liver stromal cell line (ATCC).
12. 0.25 % Trypsin-EDTA solution.
13. HBSS: Hanks' buffered saline solution.
14. Bouin's buffer.
15. Irradiator (gamma or X-ray).
16. Recipient mice for CFU-S assay.
17. Insulin syringe.
18. 96-Well tissue culture-treated plate (for LT-CTC assays with M5300).
19. Microscope.

2.2.3 Proliferation, Cell Cycle, and Quiescence of HSCs

1. 10 mg/ml BrdU solution in PBS.
2. BrdU Flow Kit (BD Pharmingen).
3. Antibodies: Ki67, BrdU.
4. DAPI.
5. Insulin syringes with 1 ml capacity and 28 ½ G beveled needle.
6. PBS/5 % FBS staining buffer: PBS, 5 % FBS, 100 units/ml penicillin and 100 μg/ml streptomycin.
7. Pyronin Y.
8. Hoechst 33342.

9. 70 % Ethanol.
10. PBS.
11. FACS analyzer.

2.2.4 Engraftment Assay for HSC Lodging and Homing

1. Carboxyfluorescein diacetate, succinimidyl ester (CFDA-SE).
2. DMSO.
3. PBS/5 % FBS staining buffer.
4. DMEM/10 % FBS media.
5. Irradiator (gamma or X-ray).
6. Insulin syringes with 1 ml capacity and 28 ½ G beveled needle.
7. Recipient and donor mice.
8. FACS analyzer.

2.2.5 Reconstitution Assays

1. HBSS/2 % FBS staining buffer.
2. Irradiator (gamma or X-ray).
3. Insulin syringes with 1 ml capacity and 28 ½ G beveled needle.
4. Recipient mice and donor mice: C57BL/6 J mice that express CD45.1, pan leukocyte marker also known as Ly5.1 and C57BL/6 J mice that express CD45.2 or Ly5.2.
5. RBC lysis buffer.
6. Hemocytometer.
7. Antibodies to distinguish host and donor and for analysis of engraftment: CD45.1 (A20), CD45.2 (104), B220 (RA3-6B2), CD11b (M1/70), CD3 (145-2C11), and Gr-1 (RB6-8C5).
8. FACS analyzer and sorter.

2.2.6 Leukemia-Initiating Cell Assays

1. Antibodies: CD45 and others as desired to identify PTEN deficient LICs (CD3, CD4, CD8, CD11b, Ter119, CD5, c-kit, Sca-1, Flk2, CD48).
2. HBSS/2 % FBS staining buffer.
3. Irradiator (gamma or X-ray).
4. Insulin syringes with 1 ml capacity and 28 ½ G beveled needle.
5. FACS analyzer and sorter.

2.3 PTEN and Neural Stem Cells

2.3.1 BrdU Pulse Labeling

1. Bromodeoxyuridine.
2. 0.9 % Sterile NaCl.
3. 1 N NaOH.
4. 27 1/4th-gauge needle and 1 ml syringe.

2.3.2 Immunohistochemsitry/ Immunofluorescence/ Immunocytochemistry (IHC/IF/ICC)

1. 0.1 M Phosphate-buffered saline (PBS).
2. Paraformaldehyde.
3. Tissue-Tek® O.C.T. Compound.
4. 0.1 M Tris-buffered saline.
5. Sucrose.
6. 1 N HCl or 0.01 M Sodium citrate.
7. Normal serum.
8. Triton X-100.
9. Primary antibodies.
10. Secondary antibodies (fluorescent or biotinylated for DAB).
11. ABC kit (Vector Laboratories).
12. Diaminobenzene kit (Biogenex).
13. Charged microscope slides.
14. Coverglass (round cover slips for ICC).
15. Antifade mounting medium (for fluorescence).
16. Permount mounting medium (for DAB).
17. Ethanol and xylene (for DAB).

2.3.3 Isolation of NSCs from Mouse Brain

1. Zivic Instruments Mouse Brain Slicer.
2. Single-edge razor blades No. 9.
3. Scalpel, size 11.
4. Fine forceps, size 5.
5. Dissecting scope.
6. Hanks' balanced salt solution.
7. Accumax.
8. 40–70 μm Cell strainers.

2.3.4 Flow Cytometry

1. Accumax (Sigma).
2. FACS buffer (PBS, 0.5–1 % BSA or 5–10 % FBS, 0.1 % sodium azide (NaN_3)).
3. FACS antibodies (http://www.ncbi.nlm.nih.gov/pubmed/21407814).
4. Flow cytometer.

2.3.5 Side Population Analysis

1. NSC media: DMEM F/12 supplemented with B27, penicillin (100 units/ml), streptomycin (100 μg/ml), heparin sulfate (5 μg/ml), bFGF (20 ng/ml), and EGF (50 ng/ml). See "In Vitro Neurosphere Assay."
2. HBSS+:Hanks' balanced salt solution supplemented with 10 mM HEPES and 2 % FBS.

3. Hoechst 33342: Hoechst 33342 powder is dissolved in distilled water and filter sterilized at 1 mg/ml concentration.
4. Red blood cell lysis buffer.
5. Verapamil.
6. 40–70 μm cell strainers.
7. 250 ml polypropylene tubes (Corning).
8. Circulating water bath at exactly 37 °C.
9. Flow cytometer.

2.3.6 CFSE Labeling

1. PBS /0.1 % (w/v) BSA.
2. 5 mM 5-(and -6)-carboxyfluorescein diacetate succinimidyl ester (CFDA-SE or CFSE).
3. RPMI 1640/10 % (v/v) FBS.
4. Culture medium.
5. Antibodies for immunophenotyping (optional).
6. Flow cytometer.

2.3.7 In Vitro Sphere Formation, Passaging, and Differentiation

1. Accumax.
2. Hanks' balanced salt solution.
3. 0.1 M PBS.
4. Neurosphere medium: DMEM F12, 1× B27 supplement, 100 units/ml penicillin, 100 μg/ml streptomycin, 5 μg/ml heparin, 20 ng/ml basic fibroblast growth factor (bFGF), 50 ng/ml epidermal growth factor (EGF).
5. Differentiation media: DMEM F12, 1× B27 supplement, 100 U/ml penicillin–streptomycin.
6. 40–70 μm Cell strainers.
7. T25 culture flasks.
8. Fire-polished glass pasture pipettes.
9. Rubber bulb.
10. 24-Well plates.
11. Poly-L-lysine.
12. Mouse laminin.
13. Round glass cover slips.

2.3.8 Immunocytochemistry of Differentiated Neurospheres

1. 0.1 M PBS.
2. 0.1 M Tris-buffered saline.
3. 1 N HCl or 0.01 M sodium citrate.
4. Normal serum.
5. Triton X-100.

6. Primary antibodies.
7. Secondary antibodies (fluorescent).
8. Charged microscope slides.
9. Coverglass (round cover slips for ICC).
10. Antifade mounting medium (for fluorescence).
11. Fine forceps to collect round glass cover slips.

2.4 PTEN and Prostate Cancer Stem Cells

2.4.1 Isolation of Prostate Cells

1. *Pb-Cre+/–*; *PtenL/L* mice.
2. *Pten*-null prostate cell lines (PC3, CaP2, etc.).
3. Dissecting scissors.
4. Dissecting forceps.
5. 10 cm petri dishes.
6. 1.5 ml Eppendorf tubes.
7. PBS.
8. Razor blades.
9. DMEM.
10. RPMI-1640.
11. Penicillin–streptomycin.
12. Fetal bovine serum (FBS).
13. HEPES buffer.
14. L-Glutamine.
15. Insulin.
16. Bovine pituitary extract (BPE).
17. Rocki (Y27632).
18. Prostate media: DMEM media, 100 units/ml penicillin, 100 μg/ml streptomycin, 10 % FBS, 4 mM L-glutamine, 10 mM HEPES, 5 μg/ml insulin, 25 μg/ml bovine pituitary extract, 10 μM ROCKi.
19. Collagenase I.
20. Collagenase media: 9 ml of prostate media, 1 ml of collagenase I (10 mg/ml in PBS).
21. Dispase.
22. Collagenase/dispase media: 8 ml of prostate media, 1 ml of collagenase I (10 mg/ml in PBS), 1 ml of dispase (10 mg/ml in PBS).
23. 15 and 50 ml Falcon tubes.
24. Parafilm.
25. DNase I.
26. Trypsin/0.05 % EDTA.
27. 18- and 21-gauge needles.

28. 10 cc syringe.
29. 40 and 100 μm nylon mesh filters.
30. Trypan blue.

2.4.2 Enrichment of Prostate Stem Cells

1. 1.5 ml Eppendorf tubes.
2. 5 ml Round-bottom FACS tubes with cell-strainer cap (12 × 75 mm).
3. DMEM.
4. RPMI-1640.
5. Penicillin–streptomycin.
6. FBS.
7. HEPES buffer.
8. L-Glutamine.
9. Insulin.
10. BPE.
11. Rocki (Y27632).
12. Prostate media.
13. ALDEFLUOR Kit (StemCell Technologies).
14. FACS Aria II.
15. Antibodies for FACS staining: CD45-PE, CD45-FITC, CD49f-PE, CD45-APC-eFluor780, CD24-FITC, CD133-PE, CD44-APC-Alexa fluor 750, CD166-PE, CD31-PE, CD31-FITC, Ter119-PE, Ter119-FITC, Sca1-PE-Cy7, CD49f-Alexa647, Epcam-APC-Cy7, Sca1-FITC, Sca-1-PE-Cy7, CD117-APC, CD44-PE, FITC-conjugated goat anti-mouse IgG, APC-conjugated goat anti-mouse IgG, Trop2-APC, CD133-APC, α2/β1.
16. FACS collection media: DMEM media, 50 % FBS, 100 units/ml penicillin, 100 μg/ml streptomycin, 1 % HEPES, 10 μM Rocki.
17. PrEGM media.
18. Matrigel.
19. 12-Well dish.
20. 15 ml Falcon tubes.
21. Trypsin/0.05 % EDTA.
22. Dispase.
23. 27-Gauge needle.
24. 1 cc syringe.
25. 40 μm nylon mesh filter.
26. The Lineage Cell Depletion Kit (Miltenyi Biotec).
27. Anti-Sca-1 Microbead Kit (FITC) (Miltenyi Biotec).

28. CD117 Microbead Kit (Miltenyi Biotec).
29. CD44 Microbead Kit (Miltenyi Biotec).
30. Type I collagen.
31. 5′-Bromo-2′-deoxyuridine (BrdU).
32. Background Sniper (Biocare Medical).
33. Goat anti-mouse IgG conjugated to Alexafluor 594.
34. BrdU mouse monoclonal antibody.
35. Prolong Gold Anti-Fade reagent with DAPI.
36. Triton X-100.
37. Verapamil.
38. Hoechst 3342.
39. 7-AAD.
40. Propidium iodide (PI).
41. Cytoslides.
42. Cytospin.
43. Tween 20.
44. Paraformaldehyde.
45. HCl.
46. Sodium borate.

2.4.3 Functional Assays for Assessment of Prostate Stem Cell Properties

1. 12-Well dish.
2. 6-Well dish.
3. 60 mm dish.
4. 1.5 ml Eppendorf tubes.
5. Type I collagen.
6. PBS.
7. PrEGM media.
8. Acetone.
9. Giemsa-modified solution.
10. Matrigel.
11. Neutral red solution.
12. Crystal violet.
13. Trypan blue.
14. NIH 3T3 cells.
15. DMEM.
16. Penicillin–streptomycin.
17. FBS.
18. L-Glutamine.
19. Cell scraper.

20. Agarose.
21. Nitrobluetetrazolium (NBT).
22. Neurobasal media.
23. Glutamax.
24. N-2 (chemically defined, serum-free supplement recommended for growth of neural cell cultures).
25. B27 (supplement to support growth of neural cell cultures).
26. Human FGF-basic.
27. Human recombinant EGF.
28. Sphere media: 200 ml Neurobasal media, 2 ml Glutamax (100×), 2 ml N-2 (100×), 4 ml B27 (50×), 40 μl FGF (100 μg/ml), 40 μl EGF (50 μg/ml).
29. Ultralow attachment plate (6-well).
30. Annexin-V PE Apoptosis Detection Kit.
31. Nikon Biostation Timelapse System.
32. Numb-GFP fusion retroviral reporter.
33. *NOD;SCID;IL2rγ-null* (*NSG*) male mice.
34. Insulin syringe.
35. 10 % Buffered formalin.
36. Ketamine.
37. Xylazine.
38. Caliper.
39. Collagen mixture: 500 μl Type I collagen, 11.5 μl 1 N NaOH, 56.8 μl 10× PBS.
40. NuSerum IV.
41. Dihydrotestosterone (DHT).
42. UGSM media: DMEM media (500 ml), 100 units/ml penicillin, 100 μg/ml streptomycin, 5 % FBS, 5 % NuSerum IV, 4 mM L-glutamine, 5 μg/ml insulin, 0.01 μM DHT.
43. Betadine.
44. Alcohol wipes.
45. 9 mm metal autoclips.
46. Wound clip applier.
47. Wound clip remover.
48. 5-0 Coated vicryl sutures.
49. Hamilton syringe.
50. 10 cc Syringe.
51. 18-Gauge needle.
52. Antibodies for immunohistochemical (IHC) staining: PTEN, P-AKT, P-S6, AR, Pan-Cytokeratin, Ki67.

3 Methods

3.1 PTEN and Mouse Embryonic Stem Cells

3.1.1 In Vitro Culture of mESCs (Also Applies to Pten-Deficient mESCs)

1. Pre-coat plate with 0.2 % gelatin and incubate at room temperature for 5 minutes. Then seed a feeder layer of mitotically inactivated primary mouse embryonic fibroblasts (MEFs) at least 4 h before plating mESCs in mESC culture medium.
2. Replace medium every day.
3. Routinely passage mESCs every 2–3 days (*see* **Note 5**).

   Aspirate medium off, wash once with DPBS, and add 1 ml trypsin/EDTA to each 100 mm cell culture dish. Incubate at 37 °C until colonies float off when flicking the plate. Add 3 ml mESC medium to quench trypsin activity and dissociate colonies into single cells by repetitive pipetting.
4. Pellet cells by low-speed centrifugation (table-top centrifuge) at 200 × *g* for 5 min. Split cells at ratio of 1:6 routinely (do not exceed 1:10 ratio).

3.1.2 Growth Competition Assay

1. 4×10^5 *Pten*–/– mESC cells are co-cultivated with equal numbers of *Wt* or *Pten*+/– mESC cells in a 6-well plate.
2. Passage cells every 3 days at the original density of 8×10^5 cells per well in 6-well plate.
3. Continuously culture for a total of three passages. Cells are collected at each passage and genomic DNAs are extracted.
4. For southern blot analysis, follow standard protocol. Briefly, gDNA is digested with *Eco*RV before applying to southern blot. 23 kb and 8.5 kb bands, corresponding to the WT allele or the *Pten*-deleted allele, respectively, can be detected using external probe (*see* **Note 6**).

3.1.3 Cell Synchronization

1. mESC cells are maintained in culture flasks, and are passaged twice without feeder cells to remove *Pten*+/+ feeder cells (*see* **Notes 7** and **8**).
2. Sub-confluent mESC cultures are shaken twice at a 1-h intervals to remove the loosely attached cells.
3. The medium is then replaced with pre-warmed medium containing 0.06 μg/ml colcemid and incubated for 4 h. The flasks are shaken again and the mitotic cells are collected.
4. The collected cells are washed twice in ice-cold medium without colcemid.
5. An equal number of mitotic cells are seeded in colcemid-free medium on gelatinized plates and incubated to allow synchronized cell cycle re-entry (this procedure yields typically 90–95 % mitotic cells, ~1 % of total number of cells).
6. Cells are collected at desired time points for PI staining and functional analysis.

7. For PI staining, at least 5×10^5 cells are resuspended thoroughly in 0.2 ml PBS, and adding 800 ml of 95 % (or 100 %) ethanol while vortexing at low speed. The permeabilized ES cells can be stored at 4 °C for weeks until final PI staining. The ES cells are then washed with PBS, stained with PI solution for 30 min at 37 °C, and analyzed by FACS.

3.1.4 Cell Differentiation

1. mESC cells are passaged twice without feeder cells to remove *Pten+/+* MEF feeder cells.
2. For the generation of embryonic bodies (EBs), mESC cells are trypsinized and plated at a density of 1×10^5 cells per well in an ultralow attachment 6-well plate in differentiation medium.
3. EBs are fed with fresh differentiation medium every other day.
4. Cells are collected at different days over the course of 2 weeks for differentiation progression analysis, such as loss of pluripotency markers (Oct4, Sox2, and Nanog) and gain of lineage-specific markers by immunofluorescence or Western blot analysis.
5. For detailed three lineage (endoderm, mesoderm, and ectoderm) specific differentiation protocol, refer to [35].
6. Retinoic acid can induce mESC differentiation. mESC cells can be treated with 10 μM retinoic acid in mESC differentiation medium (minus LIF) for 4 days without changing medium.

3.1.5 Teratoma Formation

1. Maintain mESCs in good growth condition with MEF feeder cells.
2. Freshly prepare 5×10^4 mESCs in a volume of 100 μl 0.5× Matrigel solution. Keep cell mixture on ice until use.
3. Inject subcutaneously onto the backs of NOD scid gamma (NOD.Cg-Prkdc^{scid} Il2rg^{tm1Wjl}/SzJ) mice ($n \geq 3$) using 27-gauge insulin syringe.
4. Tumor formation is observed at different time points (usually 3–6 weeks) followed by various analyses including tumor volume, histology, flow cytometry, or immunohistochemistry.

3.1.6 Enrichment of Tumorigenic Subpopulation from Differentiated Pten−/− mESCs (See Note 9)

The SSEA1$^+$/c-kit$^+$ subpopulation of differentiated *Pten−/−* mESCs has been shown to have high tumorigenic potential [5]. The enrichment of SSEA1$^+$/c-kit$^+$ subpopulation can be achieved by applying either in vitro-differentiated *cells* (*see* Subheading 3.1.4) or in vivo generated teratoma tissues (*see* Subheading 3.1.5).

1. For cultured cells—make single-cell suspension from in vitro retinoic acid-induced mESC cells.

   For teratoma—dissect the teratoma into 1 mm^2 pieces followed by incubation in 1 mg/ml collagenase in high glucose DMEM for 2 h at 37 °C in 5 % CO_2.

2. 1–2 × 10[6] cells are incubated with SSEA1 (1:100) and c-kit (1:200) antibodies in 1 ml DPBS/BSA for 20 min at 4 °C.
3. Cells are washed with DPBS/BSA, and incubated in DPBS/BSA for 5 min, and washed again.
4. Cells are incubated with Cy5-conjugated goat anti-mouse IgG and IgM (1:500) and PE-conjugated goat anti-rat IgG (1:1000) for 20 min at 4 °C (prevent from light).
5. Cells are washed and resuspended in 1 ml DPBS/BSA for sorting on a flow cytometer.

3.2 PTEN and Hematopoietic Stem Cells

3.2.1 HSC Identification and Isolation by Immunophenotype

Studies of PTEN loss in HSCs require methods for identification and isolation of HSCs and progenitor populations. Fluorophore-conjugated antibodies are used for flow cytometric analysis and cell sorting of HSCs based on cell surface antigens or immunophenotype. Sca-1 and c-kit combined with lineage markers can be used for HSC identification and isolation [36, 37]. The "LSK" cells ($Lin^-Sca\text{-}1^+c\text{-}Kit^+$) are a heterogeneous group containing LT-HSC and ST-HSC that can be discriminated by the marker Flk-2 [38]. SLAM family members (CD150, CD244, CD48) can be used to further purify hematopoietic stem cell and progenitors cells [39, 40].

1. Bone marrow (BM) is flushed from the long bones (tibias and femurs) in HBSS/2 % FBS using 5 or 10 ml syringe and 25 G 7/8 needle and filtered through 40 μm cell strainer to obtain a single-cell suspension. Alternatively, bones may be crushed into HBSS/2 % FBS and filtered through cell strainer.
2. Perform red blood cell lysis. Resuspend cells in red blood cell lysis buffer using about 3 ml volume of buffer per mouse BM. Incubate at room temperature for 5 min and promptly wash with HBSS/2 % FBS buffer. Resuspend cells in 1–5 ml HBSS/2 % FBS.
3. Count nucleated cells using hemocytomer (*see* **Note 10**).
4. Adjust concentration of cells in HBSS/2 % FBS to 1 to 10 × 10[7] cells per ml.
5. Incubate BM cells with antibodies against surface antigens in dark on ice using up to 1 × 10[7] cells per test in 100 μl HBSS/2 % FBS (*see* **Note 11**). Incubation times with surface staining antibodies are typically 30–60 min depending on the manufacturer's recommendation. Magnetic separation may be used for enrichment or depletion of cells prior to staining and/or sorting for LT-HSC (*see* **Note 12**).
6. Wash cells in HBSS/2 % FBS. Resuspend in HBSS/2 % FBS at concentration desired for flow cytometry or sorting, typically 1–10 × 10[6] cells/ml.
7. A viability dye can be added to samples for FACS analysis to identify dead cells. Add 5 μl of 50 μg/ml 7-AAD per 500 μl of sample for final concentration of 0.5 μg/ml 7-AAD.

Alternatively add 5 μl of 10 μg/ml DAPI per 500 μl of sample for 100 ng/ml final concentration. Vortex to mix. Incubate for 5–15 min and acquire immediately on FACS analyzer as viability dyes are toxic to cells.

8. Long-term repopulating HSCs (LT-HSC) are negative for lineage markers, B220, CD3, CD4, CD5, CD8, Gr-1, CD11b, and Ter119, and positive for c-kit and Sca-1. This population is referred to as LSK. LT-HSCs are LSK that are negative for Flk2, negative for CD34, positive for SLAM family member CD150, and negative for SLAM family members CD48 and CD224. Short-term repopulating HSCs (ST-HSC) are LSK that are positive for Flk2 and negative for CD34. Multipotent progenitor cells (MPP) are LSK that are positive for Flk2 and CD34, positive for CD224, and negative for CD150 and CD48. Highly Restricted Progenitors are LSK that are negative for CD150, and positive for CD224 and CD48. Note that CD34 is used as additional marker to discriminate MPP, but not routinely used for LT-HSC.

   LT-HSC: LSK that are $Flk2^{-}CD34^{-}CD150^{+}CD48^{-}CD224^{-}$.

   ST-HSC: LSK that are $Flk2^{+}CD34^{-}$.

   MPP: LSK that are $Flk2^{+}CD34^{+}CD150^{-}CD48^{-}CD224^{+}$.

   Highly restricted progenitors: LSK that are $CD150^{-}CD48^{+}CD224^{+}$.

9. HSC and progenitor populations isolated by immunostaining and cell sorting can be enumerated and/or tested functionally.

3.2.2 Colony-Forming Cell Assays

Colony-forming cell assays (CFCs) or colony-forming unit cell (CFU-C) assays are in vitro functional assays that can be used to evaluate the effects of PTEN loss in HSCs and progenitors. These assays are used to determine hematopoietic progenitor cell frequencies and quantitate colony-forming unit progenitors. A CFC assay with M3434 detects a mixture of multiple myeloid colonies including CFU-E (erythroid), CFU-GM (granulocyte, macrophage) and CFU-GEMM (granulocyte, erythrocyte, monocyte, and macrophage). Cobble stone area forming cell (CAFC) and long term culture-initiating cell (LTC-IC) assays are used to enumerate primitive mouse hematopoietic stem cells (HSCs) and relies on the two fundamental functions of HSCs: ability to self-renew and differentiation capacity. The CFU-spleen (CFU-S) assay is an in vivo functional assay used to enumerate HSC [41].

CFCs

1. Assays can be performed to enumerate hematopoietic progenitors and HSC in various hematopoietic tissues including BM, spleen, and thymus. BM is most commonly used.
2. Methocult M3434 should be pre-aliquoted into 14 ml sterile culture tubes and stored at −20 °C to avoid repeated freezing

and thawing of the bottle. Use a syringe and 16G blunt ended needle to dispense desired volume into 14 ml sterile culture tubes. M3434 is formulated to allow the addition of cells at a 1:10 (v/v) ratio. Aliquot 3 ml per tube to yield duplicate cultures of 1.1 ml each. Aliquot 4 ml per tube to yield triplicate cultures of 1.1 ml each.

3. For M3434, seed 2×10^4–2.5×10^4 cells per 35 mm dish. Prepare to seed cells in duplicate or triplicate.
4. Prepare single suspensions of BM (or other hematopoietic cells) in IMDM/2 % FBS. After counting, adjust cell concentration to be ten times the final plating concentration.
5. Add cells to M3434 aliquots (3 or 4 ml) at 1:10 (v/v) ratio and vortex. Let tube stand to allow bubbles to dissipate.
6. Dispense cell/M3434 methocult mixture into 35 mm petri dishes (1 ml per dish) using syringe and 16G blunt-end needle.
7. Place multiple 35 mm plates into larger 100 mm or 150 mm dishes. Add sterile water to larger dish to prevent drying out of methylcellulose.
8. Incubate mouse cells for 7–14 days (depending on desired colony) in humidified incubator at 37 °C, 5 % CO_2.
9. Count the colonies after appropriate duration using microscope and photograph.

LTC-IC

1. For long-term culture-initiating cell (LTC-IC) assays, first establish a stromal feeder layer using harvested mouse bone marrow cells, M2-10B4 bone marrow stromal cell line, or AFT024 fetal liver stromal cell line. Seed into 96-well tissue culture treated plate and incubate at 37 °C. Allow cells to reach confluence and then irradiate at 20 Gy using X-ray or gamma irradiation. The radiation prevents overgrowth of feeder layer while allowing the feeder cells to provide support for progenitor cells.
2. Prepare cells in M5300/1 μM hydrocortisone at different dilutions. An initial concentration for BM is 1×10^6 per ml, but it is best to do several limiting dilutions.
3. Carefully add cells in a 0.15 ml volume to each well and place the 96-well plate in a covered container (such as large petri dishes) with two to three uncovered 35 mm dishes containing sterile water. Incubate cultures at 37 °C.
4. Maintain cultures for 5 weeks. Perform half media change every week by removing half of the media (75 μl) and replacing with fresh M5300/1 μM hydrocortisone media (75 μl). Be careful to remove media from top of cultures and not disturb adherent cells. Use aseptic technique.

5. CAFC may be counted at 5 weeks. Cobblestone areas can be identified by phase-contrast microscopy and appear as dark centered circles under the feeder layer.
6. LTC-IC must be removed from the plates and tested in CFC assay to be enumerated. First, remove the non-adherent cells from each well of the 96-well plate and place into separate tubes (one tube per well). Then remove the adherent cells from each well using 0.25 % trypsin-EDTA. Wash adherent layer with HBSS to remove M5300 prior to trypsinization. After trypsinization, neutralize trypsin with 10 μl FBS. Combine the collected adherent cell portion from each well with the non-adherent cell portion. Wash the collected cells in HBSS/2% FBS, and leave in 100 μl of HBSS/2% FBS. Then set up the CFC assay with the collected cells using M3434 as described above. Seed in 35 mm dishes and incubate 7–14 days (depending on desired colony) in humidified incubator at 37 °C, 5 % CO_2. Enumerate colonies. If limiting dilutions of cells were seeded, use limiting dilution analysis to determine frequency of the LT-CTC based on number of positive wells per dilution.

CFU-S

1. For CFU-spleen (CFU-S) to enumerate HSC, transplant 1×10^5 BM (or other hematopoietic cells) into lethally irradiated mice using tail vein injection. Radiation dose depends on strain but typically is ~10 Gy.
2. After 16 days, sacrifice recipients and excise spleen.
3. Fix spleen in Bouin's buffer.
4. Count CFU-S using microscope and photograph.

3.2.3 Proliferation, Cell Cycle, and Quiescence of HSCs

PTEN negatively regulates the PI3K pathway, which controls proliferation, growth, differentiation, and apoptosis [42]. Proliferation, cell cycle, and quiescence assays are used to measure these effects. BrdU is a thymidine analogue that can be incorporated into DNA to label cells undergoing DNA synthesis. Ki67 is a nuclear protein expressed in proliferating cells that is often used as a marker for cell proliferation in solid tumors and hematological malignancies. In PTEN models, BrdU and Ki67 can be used to monitor cell cycle and proliferation in vivo and in vitro [7, 8, 15, 10] Hoechst combined with Pyronin Y staining of DNA/RNA can be used to assay quiescence of HSCs. The Hoechst low and Pyronin Y low fraction marks the quiescent population [43].

3.2.4 BrdU Incorporation and Ki67 for Cell Cycle Analysis

1. In vivo labeling of cells with BrdU can be performed by intraperitoneal injection of mice with 100–200 μl of 10 mg/ml BrdU solution (1–2 mg per mouse). Incorporation of BrdU can be detected in thymus and bone marrow as soon as 1 h post-injection (*see* **Note 13**).

2. After BrdU labeling, bone marrow and other hematopoietic cells of interest are harvested from mice in PBS/5 % FBS staining buffer.
3. Single-cell suspensions are made and stained with desired surface markers, such as antibodies used for HSC and progenitor immunophenotyping. Similar to immunophenotyping FACS, use 100 μl staining volume and $1–10 \times 10^6$ cells per test.
4. Fix and permeabilize cells using Cytofix™ and Cytoperm™ buffers in commercially available BrdU Flow Kit (BD Pharmingen). These reagents allow for staining of intracellular antigens. Follow manufacturer guidelines for incubation times and washes in PBS/5 % FBS staining buffer.
5. Treat cells with DNase (30 μg of DNase/10^6 cells) for 1 h at 37 °C to facilitate opening of DNA to expose BrdU label. Wash in PBS/5 % FBS staining buffer.
6. Incubate each test with 1 μl anti-BrdU FITC-conjugated antibody (BD Pharmingen) in 50 μl volume for 20–30 min at RT. Wash in PBS/5 % FBS staining buffer.
7. Use DAPI to measure DNA content with incorporation of BrdU. Add 5 μl of DAPI stain (100 μg/ml stock) to each test in 500 μl volume for 1 μg/ml final concentration per sample. Vortex to mix. Analyze samples by flow cytometry on FACS analyzer.
8. Proliferation can also be measured using Ki67. Stain cells with FITC conjugated anti-Ki67 antibody after surface staining and Cytofix™ and Cytoperm™ fixation and permeabilization steps.
9. Analyze samples by flow cytometry on FACS analyzer.

3.2.5 Pyronin Y/Hoechst Staining for Analysis of Quiescence

10. HSCs are sorted from bone marrow, spleen, or thymus using surface marker staining as described in immunophenotyping section.
11. Fix cells with 70 % ethanol for 3 h.
12. Wash samples two times with PBS.
13. Stain with PBS containing 0.5 μg/ml Pyronin Y and 2 μg/ml Hoechst 33342 for 10 min.
14. Analyze samples by flow cytometry on FACS analyzer. Quiescent cells (cells in G_0) reside in the population negative for Pyronin Y and Hoechst 33342 staining.

3.2.6 Engraftment Assay for HSC Lodging and Homing

Under steady-state or normal conditions, HSCs reside in the bone marrow niche. Upon transplantation, HSCs normally home and lodge to the BM. The role of PTEN in homing and lodging can be evaluated by transplantation of CFDA-SE-labeled cells into lethally irradiated and non-irradiated recipient mice followed by the collection of cells from tissues at different time points and flow cytometry to identify labeled cells [8]. CFDA-SE (carboxyfluorescein diacetate, succinimidyl ester), often called CFSE, is a cell-tracing dye that can be used to track hematopoietic cells in vivo. The dye is incorporated into labeled cells and inherited by daughter cells after cell division, which results in sequential halving of the CFDA SE dye intensity. CFDA-SE fluorescence is measured by flow cytometry with excitation at 492 nm and emission at 517 nm, similar to fluorescence of FITC. Analyze using a flow cytometer with 488 nm excitation and emission filters appropriate for FITC.

1. Follow manufacturer's protocol for CFDA-SE labeling. The CFDA-SE labeling protocol described below is adapted from Life Technologies. This protocol can be applied to study lodging and homing of bone marrow cells from control (WT) and PTEN-deficient mice.
2. Prepare a 5 mM stock solution of CFDA-SE in DMSO.
3. For transplantation and cell tracing in vivo, label $1–5 \times 10^7$ bone marrow cells with CFDA-SE. Pellet cells and resuspend in pre-warmed PBS/5 % FBS staining buffer at a final concentration of 1×10^6 cells/ml. Add 2 μl of 5 mM stock CFSE solution per ml of cells for a final working concentration of 10 μM.
4. Incubate cells with dye at 37 °C for 10 min.
5. Quench the staining by the addition of 5 volumes of ice-cold DMEM/10 % FBS media to the cells. Incubate for 5 min on ice.
6. Pellet cells by centrifugation. Wash the cells by resuspending the pellet in fresh DMEM/10 % FBS. Pellet and resuspend the cells in fresh media two more times for a total of three washes.
7. For the homing assay, transplant 2×10^6 CFDA-SE-labeled bone marrow cells into lethally irradiated recipient mice. For the lodging assay, transplant 10×10^6 CFDA-SE labeled cells into non-irradiated mice.
8. Euthanize recipient mice from the homing assay at 8 h post-transplantation. Euthanize recipient mice from the lodging assay at two time points: 6 and 18 h post-transplantation.
9. In both assays and at all time points, harvest BM and splenocytes from recipient mice. Prepare single cell suspensions of BM and splenocytes in PBS/5 % FBS staining buffer to evaluate levels of CFDA-SE by flow cytometry. The BM and splenocytes may be stained for other markers if desired. Analyze cells using a flow cytometer with 488 nm excitation and emission filters appropriate for FITC. Cells that underwent cell division during the lodging and homing assays will show a fluorescence histogram with multiple peaks as a result of sequential halving of CFDA-SE fluorescence. The peaks represent successive generations from cell divisions.

3.2.7 Reconstitution Assays

In vivo reconstitution and serial transplantation assays are gold standards for evaluating HSC function and maintenance [44, 45]. Competitive repopulation assays can be used to test the short- and long-term reconstitution ability of PTEN-deficient HSCs compared to normal (WT) HSCs [7, 8, 10].

1. CD45.1 and CD45.2 mouse backgrounds are used to perform a competitive repopulation assay. Recipient mice are lethally irradiated (single dose or split dose). Radiation dosage depends on mouse strain background. For C57/B6, performing two doses of 550 RAD each spaced 3 h apart is suggested.
2. Bone marrow cells are the most common source of cells for competitive repopulation assays. In this assay, if donor cells are CD45.2, the competitor cells and the recipient mice should be CD45.1. To evaluate the reconstitution capacity of PTEN-deficient HSCs, a competitive repopulation assay can compare the contribution of CD45.2 donor cells (PTEN-deficient or WT control) to that of CD45.1 recipient cells (WT) following transplantation into lethally irradiated CD45.1 mice.
3. Donor and competitor cells should be resuspended in HBSS/2 % FBS staining buffer at an appropriate ratio. Donor–competitor ratios of 1:1 or 1:2 are commonly used. However, PTEN-deficient bone marrow has been observed to show reduced repopulation capacity [7, 8]. Therefore, in order to evaluate long-term reconstitution greater ratios of PTEN-deficient cells to competitor cells may be desired, such as 5×10^5 PTEN-deficient CD45.2 donor cells to 1×10^5 WT CD45.1 competitor cells. Total injected volume should be between 100 and 200 μl per transplant. Be sure to have sufficient number of WT cells for recovery from lethal radiation (1×10^5).
4. Transplantation of mixed cells can be performed by tail vein injection or retro-orbital injection into the venous sinus. If injecting by retro-orbital sinus an anesthetic or sedative must be used to avoid permanent damage to the eye. An insulin syringe with 28G ½ in needle can be used for injections.
5. After transplantation, peripheral blood is collected from the recipient mice to investigate the percentage of donor-derived cells using flow cytometric analysis of CD45.2 and CD45.1. Analysis at 4–12 weeks is used to assess short-term engraftment and reconstitution, and analysis >16 weeks is used for long-term engraftment. Multi-lineage reconstitution can also be assessed at this time using lineage antibodies: CD3ε for T-cell lineage cells, B220 for B-cell lineage, and CD11b and Gr-1 for myeloid lineage.
6. Secondary transplantation should be performed to assess long-term engraftment capacity of HSCs. For this analysis, bone marrow is harvested from primary recipient mice and

transplanted into lethally irradiated secondary mice without a competitor population. Multi-lineage reconstitution is measured by lineage-specific antibodies, and used to evaluate long-term function and reconstitution capacity of HSCs.

3.3 PTEN and Leukemia-Initiating Cells

PTEN loss in HSCs leads to the formation of LICs and development of acute leukemia [7–10]. LICs are a subset of cells present in leukemia that are responsible for the growth and propagation potential of the disease. LICs may arise from hematopoietic stem cells or myeloid and lymphoid progenitors. LICs are defined functionally by the ability to cause malignancy upon transplantation into syngeneic or immunodeficient (if donor is mixed background) mice. These cells are often proposed to be responsible for therapeutic resistance.

The four assays below are useful to study PTEN-deficient LICs and leukemia.

(a) CD45-side scatter (CD45-SSC) analysis to identify leukemic blasts.
(b) Transplantation/reconstitution assays to test ability to cause malignancy upon transplantation into recipient mice.
(c) Limiting dilution analysis to determine LIC frequency.
(d) Serial transplantation to examine self-renewal capacity.

3.3.1 CD45-Side Scatter (CD45-SSC) Analysis to Identify Leukemic Blasts

CD45-side scatter (CD45-SSC) analysis is a flow cytometric approach to detect blasts in hematopoietic populations (BM, SP, Th, and PB) [9]. This method can be used to detect leukemia development in mice and assess disease burden.

1. Perform RBC lysis on PB samples. Wash and resuspend in HBSS/2 % FBS. Filter through 40 μm strainer.
2. Obtain single-cell suspensions of BM, SP, or Th in HBSS/2 % FBS. Adjust to concentration of up to 1×10^7 cells/ml.
3. Incubate with fluorophore conjugated anti-CD45 antibody in the dark, on ice for 30 min. Wash in HBSS+/2 % FBS and resuspend in HBSS+/2 % FBS. Refer to immunophenotyping section for details.
4. Add 7-AAD or DAPI to exclude dead cells.
5. Acquire on FACS analyzer and plot CD45 (*x*-axis) by SSC (*y*-axis). Identify left-shifted blast population.

3.3.2 Transplantation/Reconstitution Assays to Test Ability to Cause Malignancy upon Transplantation into Recipient Mice

1. Harvest BM, thymocytes (Th), or splenocytes (SP) from primary PTEN deficient mice with acute leukemia. Flush bone marrow from tibias and femurs using HBSS/2 % FBS, 5 or 10 ml syringe, and 25G 7/8 in needle. Collect into 50 ml conical through 40 μm cell strainer for single cell suspension of BM. Harvest splenocytes in biosafety cabinet. Place spleen in 10 cm plate with 1–4 ml RBC lysis buffer at RT. For large

spleen (leukemic mice) use 4 ml, but begin mashing in less volume (~1 ml) and add RBC lysis buffer to final volume of 4 ml. For smaller spleen use 1–4 ml depending on size (1 ml for WT). Mash with back of plunger handle of 3 ml syringe (keep sterile, do not touch). Incubate for 2 min. Add 3× volume of IMDM (12 ml) to stop lysis. Use a 5 ml syringe with 19½ in gauge needle to further dissociate cells and pass through 40 μm cell strainer on top of a 50 ml conical tube to obtain single cell suspension. Rinse dish with additional 5 ml IMDM and filter through cell strainer. Spin collected cells, remove IMDM and lysis buffer, and resuspend in HBSS/2 % FBS. Primary PTEN-deficient leukemic mice usually have a large thymus. Use the rough portion of two glass slides to mash up the thymus in HBSS/2 % FBS and filter through 40 μm cell strainer placed on 50 ml conical tube to generate single cell suspension of thymocytes.

2. Count nucleated cells of BM, SP, and Th using hemocytometer.
3. Prepare to transplant doses of 1000 to 10^6 PTEN deficient hematopoietic cells into secondary recipients. Use syngeneic mice for recipients (C57/B6 background is commonly used), or if murine donor cells are mixed background, use *NOD-SCID-IL2Rγ−/−(NSG)* mice for recipients. Recipient mice may be sublethally irradiated. Note: dose depends on strain background and can range between 180 and 800 Gy.
4. Prepare cells for transplantation in HBSS/2 % FBS buffer and administer 100–200 μl volume per injection.
5. Transplant cells by injecting into the lateral tail veins or by injecting into the retro-orbital sinus. If injecting by retro-orbital sinus an anesthetic or sedative must be used to avoid permanent damage to the eye. An insulin syringe with 28G ½ in needle can be used for injections.
6. Monitor recipient mice after transplantation for development of leukemia. Peripheral blood (PB) may be sampled every 2 weeks and assessed for presence of blast cells using CD45-SSC analysis by flow cytometry (see below). Recipients are sacrificed upon becoming moribund. BM and spleen are collected for pathological, flow cytometric, and genetic analysis.

3.3.3 Limiting Dilution Analysis to Determine LIC Frequency

This technique can be combined with cell surface staining and FACS sorting to identify populations enriched for LIC activity.

1. Transplant range of cell doses (10–10^5 cells) into recipient mice (syngeneic or NSG). Evaluate leukemia development in recipients (PB sampling for CD45-SSC analysis, pathological signs such as weight loss, hunched posture, and unkempt fur).
2. Use limiting dilution analysis to determine the frequency of the LIC based on number of surviving mice (leukemia-free,

event-free) [46]. Various resources are available for this analysis including L-Calc software (Stem Cell Technologies) and logarithmic plots.

3. Use surface staining and sorting to determine whether specific hematopoietic populations (e.g., HSCs, MPP, or lymphoid progenitor) are enriched in LIC activity. Such populations include lymphoid progenitors ($CD3^{+}ckit^{mid}Lin^{-}$), HSC (LSK, $Flk2^{-}$, $CD48^{-}$), or myeloid blasts ($CD11b^{+}$, $CD45^{high}$, $CD4^{-}$).

3.3.4 Serial Transplantation to Examine Self-Renewal Capacity

Leukemia-initiating cells show self-renewal activity. Self-renewal activity of PTEN deficient LICs can be assayed by serial transplantation into secondary, tertiary, and quaternary recipients.

1. BM is harvested from primary PTEN-deficient leukemic mice. Single cell suspensions are prepared in HBSS/2 % FBS.
2. BM from primary PTEN-deficient mice is transplanted by tail vein or retro-orbital injection into secondary recipient mice. These mice are monitored for leukemia development (PB sampling for CD45-SSC analysis, and pathological signs such as weight loss, hunched posture, and unkempt fur).
3. BM from secondary mice that become moribund from leukemia can be harvested and transplanted into tertiary recipients. Dose may be 200–1000 cells.
4. Cell surface markers of putative LICs (such as $CD3^{+}ckit^{mid}Lin^{-}$) can be used to sort for specific populations and test self-renewal capacity upon serial transplantation.

3.4 PTEN and Neural Stem Cells

3.4.1 BrdU Pulse Labeling

1. Dissolve BrdU in sterile solution of warm 7 mM NaOH in 0.9 % NaCl at 10 mg/kg.
2. Inject 100–200 mg/kg IP daily for the appropriate length of time for the experiment.
3. Sacrifice animal via whole animal perfusion and proceed to free-floating immunofluorescence/immunohistochemistry (*see* Subheading 3.4.2).
4. Co-stain tissue with antibodies against BrdU and desired markers relevant to study, to identify active proliferating populations.
5. Count the number of BrdU+ cells at the desired time points and in the regions of interest (*see* **Note 14**).

3.4.2 Free-Floating Immunofluorescence/Immunohistochemistry

1. Whole-animal perfusion fixation for rodents is required for optimal histology [47]. Remove brain and post-fix by placing in 4 % PFA at 4 °C overnight.
2. After post-fix, briefly wash brain wash in 0.1 M TBS. Tissue can be stored in 0.1 M TBS with 0.05 % NaN_3 at 4 °C.
3. Equilibrate tissue in 30 % sucrose in PBS at 4 °C overnight or until tissue loses buoyancy (*see* **Note 15**).

4. Embed tissue in OCT and section on cryostat. Recommended tissue thickness is 30–100 μm.
5. Wash tissue three times in 0.1 M TBS for 10 min/wash on slow shaker.
6. Incubate tissue in 1 N HCl at room temperature or 0.01 M sodium citrate at 60 °C for 20 min for antigen retrieval (*see* **Note 16**).
7. Wash tissue three times in 0.1 M TBS for 10 min/wash on slow shaker.
8. Incubate tissue in blocking solution for 1 h at room temperature. Gently shake.
   (a) Blocking solution: 10 % normal serum in 0.1 M TBS with 0.5 % Triton X-100
9. Incubate tissue in primary antibody, 4 °C overnight. Gently shake. If no 4 °C shaker is available, then incubate while shaking at room temperature for 30 min before placing in 4 °C overnight.
   (a) Primary antibody dilution solution: 0.1 M TBS with 0.5 % Triton X-100.
10. Wash tissue three times in 0.1 M TBS for 10 min/wash.
11. Immunofluorescence: Incubate tissue in secondary antibody for 30 min at room temperature. Protect from light from this step forward.
    (a) Secondary antibody dilution solution: 10 % normal serum in 0.1 M TBS.

For DAB/immunohistochemistry, skip to **step 15**.

12. Wash tissue three times in 0.1 M TBS for 10 min/wash.
13. Incubate tissue in DAPI solution for 5 min.
14. Wash tissue one time in 0.1 M TBS for 10 min/wash.
15. Mount sections on slides by gently floating them onto the glass slide using a paint brush. Adding a little Triton will help get the tissue onto the slide without wrinkling. Let dry overnight and cover slip with antifade medium. Protect from light. End of Immunofluorescence protocol.
16. DAB/immunohistochemistry: Incubate tissue in biotinylated-secondary antibody for 30 min at room temperature. Secondary antibody dilution solution: 10 % normal serum in 0.1 M TBS. After secondary antibody incubation, make ABC solution because it has to sit for 30 min before use.
17. Wash tissue three times in 0.1 M TBS. 10 min/wash.
18. Incubate tissue in ABC solution. ABC solution: 1:100 dilution of A and B in 0.1 M TBS.

19. Wash tissue three times in 0.1 M TBS for 10 min/wash.
20. Incubate in DAB solution for 10–60 s, or until reaction is complete. The timing of the reaction with each antibody varies and must be optimized.
21. Wash tissue three times in 0.1 M TBS for 10 min/wash.
22. Mount sections on slides by gently floating them onto the glass slide using a paint brush. Adding a little Triton will help get the tissue onto the slide without wrinkling. Let dry overnight.
23. Dehydrate slides for 2 min in each solution: 70 % (1×), 95 % (3×), 100 % (3×).
24. Incubate slides in xylene three times for 5 min/incubation.
25. Coverslip with permount and dry.

3.4.3 Isolation of NSCs from Mouse Brain

1. Remove brain and store immediately into ice-cold Hanks' balanced salt solution (HBSS) to preserve. Chill on ice for 1 min.
2. Place brain upside down in Zivic Instruments Mouse Brain Slicer. Cut brain with razor blades to obtain 1 mm slices and place in cold HBSS on ice.
3. Obtain desired slice and submerge in cold HBSS in a petri dish under the dissecting microscope (*see* **Note 14**).
4. Use #11 scalpel and fine forceps to cut out the subventricular, subgranular, and subcallosal zones, and place into 500 μl of HBSS on ice.
5. Centrifuge at 200 × *g* for 5 min.
6. Remove buffer being careful not to disturb the pellet.
7. Resuspend in Accumax enzyme at room temp, 2× the volume of the tissue pellet. Do not put in water bath as enzyme degrades at 37 °C. Break up largest chunks of tissue with pipette and incubate at room temperature for 15 min.
8. Add 10 ml of HBSS and filter through 40–70 μm cell strainers.
9. Centrifuge at 200 × *g* for 5 min to wash off Accumax (*see* **Note 17**).
10. Resupsend in HBSS or the appropriate experimental buffer.

3.4.4 Flow Cytometry

1. Proceed from isolation of NSCs from mouse brain (*see* Subheading 3.3).
2. Count nucleated cells using hemocytomer. Adjust volume to approximately 1×10^7 cells per ml in HBSS/2 % FBS.
3. Perform red blood cell lysis. Resuspend cells in red blood cell lysis buffer at a concentration of 1×10^7 cells per ml. Incubate

at room temperature for 5 min and promptly wash with HBSS/2 % FBS buffer.

4. Resuspend cells in HBSS/2 % FBS at 1 to 10×10^7 cells per ml.
5. Incubate cells with antibodies against surface antigens in dark on ice using up to 1×10^7 cells per test in 100 µl HBSS/2 % FBS. Incubation times with surface staining antibodies are typically 30–60 min depending on the manufacturer's recommendation. Magnetic separation may be used for enrichment or depletion of cells prior to staining and/or sorting for NSCs.
6. Wash cells in HBSS/2 % FBS. Resuspend in HBSS/2 % FBS at concentration desired for flow cytometry or sorting, typically $1–10 \times 10^6$ cells/ml.
7. A viability dye can be added to samples for FACS analysis to identify dead cells. Add 5 µl of 50 µg/ml 7-AAD per 500 µl of sample for final concentration of 0.5 µg/ml 7-AAD. Alternatively, add 5 µl of 10 µg/ml DAPI per 500 µl of sample for a 100 ng/ml final concentration. Vortex to mix. Incubate for 5–15 min and acquire immediately on FACS analyzer as viability dyes are toxic to cells.

3.4.5 Side Population Analysis

1. Thaw Hoechst dye just prior to use. Keep cool until needed.
2. Resuspend cells in NSC media at 10^6 cells/ml and divide into two samples.
3. To the experimental sample, add Hoechst to a final concentration of 5 µg/ml. To the control samples, add 50 µMverapamil.
4. Incubate cells in a circulating 37 °C water bath for exactly 90 min, gently agitating several times.
5. Spin down the cells in a centrifuge at $200–300 \times g$ at 4 °C and remove the supernatant. Resuspend cells at 10^8 cells/ml in ice-cold HBSS+. All further proceedings should be carried out at 4 °C or on ice.
6. Add 2 µg/mlof 7-AAD or PI to the cells and mix 5–10 min before FACS analysis.
7. Analyze Hoechst 33342 efflux on the FACS Aria II using the 375 nm laser.
8. Hoechst 33342 is excited at 350 nm ultraviolet, and the resulting fluorescence is measured at two wavelengths using 424/44 BP and 675 LP filters for detection of Hoechst blue and red, respectively.
9. Sort side population (SP) positive and negative fractions.

3.4.6 CFSE Labeling

1. Resuspend cells in PBS/0.1 % BSA at a final concentration of 5×10^7 cells/ml.
2. Add 2 µl of 5 mM CFSE per milliliter cells (final 10 µM) in a tube that is ≥6× the volume of cells. Incubate 10 min at 37 °C.

3. Quench staining by adding ice-cold RPMI 1640/10 % FBS and incubating 5 min on ice.
4. Wash cells three times in the culture medium (*see* Subheading 3.4.7).
5. Harvest cells and stain with antibodies for immunophenotyping.
6. Set up a flow cytometer for excitation at 488 nm. Collect green fluorescence (CFSE) with a 525 nm band-pass filter. Collect immunophenotyping signals with a 575 nm band-pass filter for orange fluorescence (for PE) or a 675 nm band-pass filter for red fluorescence (for PerCP or PE/Cy5).

3.4.7 In Vitro Neurosphere Culture

1. Proceed from Isolation of NSCs from Mouse Brain (*see* Subheading 3.3).
2. Add 800–900 ml of NSC sphere media to Accumax/tissue and gently pipette to break up the chunks. Using fire-polished glass pipettes of decreasing diameter increases cell survival while breaking up the tissue. Regular plastic pipettes are okay but the fire-polished glass pipettes are better.
3. Filter cell suspension with a 40–70 μm cell filter.
4. Resuspend cells in Neurosphere media at 100,000 cells/ml and incubate at 37 °C with 5 % CO_2 (Note that the Accumax does not need to be rinsed off of the cells because it will break down automatically when the cells are placed back into the incubator at 37 °C).
5. Double the volume of Neurosphere media after 3 or 4 days of incubation.

After 1 week, count the number of spheres with diameter >50 mm. Spheres can be used for differentiation assay or continuously passaged (*see* **Note 18** and Subheading 3.8).

3.4.8 Passaging Neurosphere Culture

1. Remove cells and media from flask (if some cells have adhered take only what is floating).
2. Centrifuge at 200 × *g* for 5 min.
3. Remove media being careful not to disturb the pellet.
4. Add 200 ml Accumax and incubate for 15 min at room temperature.
5. Add 600 ml complete media.
6. Triturate with fire-polished pipette.
7. Add 200 ml of complete media for a total of 1 ml.
8. Count cells and replate as described above.
9. Replate the cells at your desired density in new flasks: high-density (bulk) cultures = 50,000 to 100,000 cells/ml or clonal density 1000 cells/ml.

3.4.9 Differentiation of Intact Neurospheres

1. Place sterile glass cover slips into the wells of a 24-well plate.
2. Coat cover slips with poly-L-lysine (PLL) at room temp for at least 1 h (dilute frozen stock 1:1000).
3. Wash cover slips 2× with water.
4. Final wash of cover slips with PBS or media 1×.
5. Air-dry cover slips with the lid off of the plate inside the sterile hood.
6. Add a 30 μl drop of undiluted mouse laminin onto the center of a cover slip, remove the laminin right before placing the sphere(s) onto the cover slip and move the laminin drop onto the next cover slip.
7. Remove spheres from flask with as little media as possible using a glass pipette and "squeezie-bulb" and place into approx. 10 ml of growth factor-free media to rinse any growth factor off of the sphere.
8. Remove spheres from rinse media and place onto the PLL and laminin-coated cover slip in a very small volume of media (approx. 30 μl). A maximum of 15 per cover slip is recommended.
9. Put plate into the incubator for 30–60 min *before* adding the media to the wells.
10. Very gently and slowly flood each well with 500 μl of differentiation (growth factor-free) media by pipetting it down the side of the well.
11. Return plate(s) to incubator. Differentiate mouse neurospheres for 5–7 days (human cells 7–14 days) before fixing and staining.

3.4.10 Immunocytochemistry of Differentiated Neurospheres

1. Wash wells once with PBS; just add PBS and aspirate, no need to incubate.
2. Add 1 ml 4 % PFA to each well and place on slow shaker for 1 h at RT; to make 4 % PFA, add 1 vial (10 ml) of 32 % PFA to 70 ml cold PBS; can keep this PFA in the fridge for <1 week.
3. Wash tissue three times in 0.1 M TBS. 10 min/wash on slow shaker.
4. Add 250 μl of blocking solution for 1 h at RT on slow shaker.
5. Aspirate block and add primary antibody diluted in fresh block overnight at 4 °C. *Cover plate with foil if using fluorescent antibodies.*
6. Wash primary antibody off with 3× PBS washes at 7 min. each.
7. Add secondary antibody diluted in fresh block and cover entire plate with foil; incubate on slow shaker for 2 h.
8. Wash tissue three times in 0.1 M TBS. 10 min/wash.

9. Incubate tissue in DAPI solution for 5 min.
10. Wash tissue one time in 0.1 M TBS. 10 min/wash.
11. Keep cover slips in last wash and mount onto slides using anti-fade mounting media using fine forceps; aspirate excess mounting media be sure to keep slides covered from light. Let dry overnight before viewing. Store slides at 4 °C after viewing.

3.5 PTEN and Prostate Cancer Stem Cells

3.5.1 Isolation of Prostate Cells from Mouse Prostate Tumors

1. After euthanizing a male mouse with an endogenous prostate tumor (e.g., *Pb-Cre+/−;PtenL/L*(*Pten*-null) mouse), remove most of the urogenital system by holding up the bladder and cutting the connective tissue below the urogenital sinus.
2. Place the urogenital system (containing the prostate, bladder, seminal vesicles, urethra, and ureters) in a 10 cm dish.
3. Pull away pieces of fat still attached to the prostate. Fat should be easy to remove and should not require excessive force to remove.
4. Using two dissecting tweezers, hold onto the bladder and remove the two ureters.
5. With the tweezers, separate the anterior lobes of the prostate from the seminal vesicles by severing the blood vessels connecting them.
6. Remove the seminal vesicles by pulling the seminal vesicles away from the prostate at the base using tweezers.
7. Holding onto the prostate with one pair of tweezers, remove the bladder from the prostate.
8. Remove the urethra by pulling it away from the prostate with tweezers (*see* **Note 19**).
9. Transfer the prostate to a new 10 cm dish containing 5 ml of prostate media.
10. Place 10 cm dish with prostate into a sterilized tissue culture hood.
11. Place prostate into new 10 cm dish and serrate the tissue for 1–2 min with a razor blade.
12. Add 10 ml of collagenase media and serrate for another 2–3 min with a razor blade until prostate chunks are small enough to pass through a 5 ml pipette.
13. Transfer minced prostate tissue into a 15 ml Falcon tube.
14. Seal cap on 15 ml Falcon tube with Parafilm.
15. Place tube on a rotator, and rotate at 37 °C overnight (12–16 h) (*see* **Note 20**).
16. Add DNase I to the prostate tissue in collagenase media, and rotate an additional 15 min at 37 °C (*see* **Note 21**).

17. Wash cells two times in 10 ml of PBS and spin down at 250 × *g* for 5 min at 4 °C.
18. Remove supernatant and add 5 ml of pre-heated Trypsin/0.05 % EDTA to the cell pellet. Mix the cells thoroughly with trypsin using a pipette, and place into a 37 °C incubator for 5 min.
19. Add 5 ml of prostate media to inactivate trypsin and transfer solution to a new 50 ml Falcon tube.
20. Pass solution through an 18-G syringe five times, and pass through a 21-G syringe five times (*see* **Note 22**).
21. Pass solution through a 40 μm nylon mesh filter into a new 50 ml Falcon tube.
22. Wash filter with an additional 10 ml of prostate media.
23. Spin down collected cells at 250 × *g* for 5 min at 4 °C.
24. Remove supernatant, resuspend cells in 1 ml of prostate media, and transfer to an Eppendorf tube.
25. Count cells using a hemocytometer and trypan blue exclusion.

3.5.2 Isolation of Prostate Cells from Human Tumor Specimens

1. Obtain human tissue samples from a human pathology laboratory.
2. Marked regions of prostate tissue are removed with scissors.
3. Transfer prostate tissue to 10 cm dish and place in a sterilized tissue culture hood.
4. Use scissors to cut tissue into small chunks. Add 5 ml of prostate media and continue to mince tissue until chunks are small enough to pass through 5 ml pipette.
5. Transfer prostate media with prostate tissue chunks to 50 ml Falcon tube.
6. Spin down tissue at 500 × *g* for 5 min at 4 °C.
7. Remove supernatant from cell pellet and resuspend in 30 ml of PBS.
8. Repeat spin wash.
9. Resuspend cell pellet in collagenase/dispase solution. (Add 10 ml per gram of tissue.)
10. Seal cap on 50 ml Falcon tube with Parafilm.
11. Place tube on a rotator, and rotate at 37 °C overnight (12–16 h).
12. Add DNase I to the prostate tissue in collagenase media, and rotate for an additional 15 min at 37 °C (*see* **Note 21**).
13. Wash cells two times in 25 ml of PBS and spin down at 500 × *g* for 5 min at 4 °C.

14. Remove supernatant and add 5 ml of pre-heated trypsin/ 0.05 % EDTA to the cell pellet. Mix the cells thoroughly with trypsin/0.05 % EDTA using a pipette, and place into a 37 °C incubator for 5 min.
15. Add 5 ml of prostate media to inactivate trypsin and transfer solution to a new 50 ml Falcon tube.
16. Pass solution through an 18-G syringe five times, and pass through a 21-G syringe five times (*see* **Note 22**).
17. Pass solution through a 100 μm nylon mesh filter into a new 50 ml Falcon tube.
18. Wash filter with an additional 10 ml of prostate media.
19. Spin down collected cells at 500 × *g* for 5 min at 4 °C.
20. Remove supernatant, resuspend cells in 1–3 ml of prostate media.
21. Count cells using a hemocytometer and trypan blue exclusion.

3.5.3 Prostate Cancer Cell Lines

1. Culture *Pten*-null prostate cell lines in 10 cm dishes with DMEM (CaP2, CaP8) or RPMI-1640 (PC3) supplemented with 100 units/ml penicillin, 100 μg/ml streptomycin, 10 % FBS, and 4 mM L-glutamine.
2. Wash cells with 5 ml of PBS.
3. Remove PBS and add 5 ml of pre-heated trypsin/0.05 % EDTA. Place into a 37 °C incubator for 5 min.
4. Add 5 ml of cold DMEM or RPMI-1640 media to inactivate Trypsin.
5. Pipet vigorously to remove remaining attached cells. Transfer to a 15 ml Falcon tube.
6. Spin down cells at 250 × *g* for 5 min at 25 °C.
7. Resuspend cells in 1 ml of DMEM or RPMI-1640 media.

3.6 Enrichment of Prostate Stem Cells

Various cell surface markers, including CD133 [25], CD117 [48], Sca-1 [49], CD49f [49], CD44 [50], Trop2 [32], and CD166 [51], have been demonstrated to select for prostate cells with enriched stem cell activities. Moreover, prostate stem cells can also be isolated by side population analysis [52], aldehyde dehydrogenase (ALDH) activity [53], and through their ability to grow as spheres in Matrigel culture [54] and retain labels such as BrdU over long latencies [55].

3.6.1 Immunomagnetic Separation Based on Cell Surface Marker Expression

Prior to FACS staining, negative selection of specific contaminating cell populations can be carried out using immunomagnetic separation. The Lineage Cell Depletion Kit (mouse or human) can be used to remove hematopoietic cells and erythrocytes from prostate cells. Follow the instructions described in the kit.

1. Positive enrichment of stem cell populations based on stem cell surface marker expression can be used to isolate or prepare stem cell populations prior to FACS staining.
2. The Anti-Sca-1 Microbead Kit (FITC) can be used to enrich for Sca-1$^+$ stem cell fractions and label Sca-1$^+$ cells with FITC. Follow the manufacturer's instructions described in the kit. This kit can be used in combination with the Lineage Cell Depletion Kit.
3. The CD117 Microbead Kit can be used to enrich for CD117$^+$ stem cells fractions. Follow the manufacturer's instructions described in the kit. This kit can be used in combination with the Lineage Cell Depletion Kit.
4. The CD44 Microbead Kit can be used to enrich for CD44$^+$ stem cells fractions. Follow the manufacturer's instructions described in the kit. This kit can be used in combination with the Lineage Cell Depletion Kit.
5. The CD133 Microbead Kit can be used to enrich for CD133$^+$ stem cells fractions. Follow the manufacturer's instructions described in the kit. This kit can be used in combination with the Lineage Cell Depletion Kit.

3.6.2 FACS Analysis of Prostate Stem Cell Marker Expression

1. Label Eppendorf tubes for the following: A. Unstained cells. B. Single antibody compensation controls (e.g., FITC, PE, APC). C. FACS Samples. Aliquot $1–5 \times 10^4$ cells into 200 μl of Prostate Media per tube for unstained and single antibody compensation controls.
2. Divide the remaining cells into new Eppendorf tubes at up to 3×10^6 cells per tube in a total volume of 1 ml of prostate media.
3. For the single antibody compensation controls, add 1 ml of each single antibody used for FACS analysis.
4. For FACS samples, add the following amount of each antibody needed for isolation of specific prostate stem cell populations (*see* **Note 23**):
   (a) Live cell gating (*see* **Note 24**).
      - Propidium iodide (PI):
         - Add 5–10 min prior to running FACS sample through the machine.
         - 2 μg/ml final concentration.
         - Live cells are PI$^-$.
      - 7-AAD.
         - Add 10–15 min prior to running FACS sample through the machine.
         - 2 μg/ml final concentration.
         - Live cells are 7AAD$^-$.

(b) Murine prostate cancer stem cells.

- $Lin^{-}Sca\text{-}1^{+}CD49f^{hi}$ (LSC^{hi}) (*see* **Note 25**).
  - CD45-PE (2 μl).
  - CD31-PE (2 μl).
  - Ter119-PE (2 μl).
  - Sca-1-FITC (4 μl).
  - CD49f-Alexa647 (3 μl).
- LSC^{hi};$CD166^{hi}$ (*see* **Note 25**).
  - CD45-FITC (4 μl).
  - CD31-FITC (4 μl).
  - Ter119-FITC (4 μl).
  - Sca-1-PE-cy7 (2 μl).
  - CD49f-Alexa647 (3 μl).
  - CD166-PE (2 μl).
- $Lin^{-}Trop2^{+}CD49f^{Hi}$ (LTC^{hi}) (*see* **Note 25**).
  - CD45-FITC (4 μl).
  - CD31-FITC (4 μl).
  - Ter119-FITC (4 μl).
  - Trop2-APC (6 μl).
  - CD49f-PE (3 μl).
- $Lin^{-}Sca\text{-}1^{+}CD133^{+}CD44^{+}CD117^{+}$ (*see* **Note 26**).
  - Sca-1-PE-cy7 (2 μl).
  - CD133-PE (2 μl).
  - CD44-APC-Alexa fluor 750 (2 μl).
  - CD117-APC (2 μl).
- $Epcam^{lo}CD24^{lo}$ (stem cells with mesenchymal characteristics).
  - CD45-PE (2 μl).
  - CD31-PE (2 μl).
  - Ter119-PE (2 μl).
  - Epcam-APC-Cy7 (2 μl).
  - CD24-FITC (4 μl).

(c) Human prostate cancer stem cells.

- $Trop2^{+}CD49f^{hi}CD24^{lo/+}$.
  - CD45-APC-eFluor780 (2 μl).
  - Trop2-APC (6 μl).
  - CD49f-PE (3 μl).
  - CD24-FITC (4 μl).

- $CD44^+/CD24^-$.
  - CD44-PE (2 μl).
  - CD24-FITC (4 μl).
- $CD44^+/CD133^+$.
  - CD44-PE (2 μl).
  - CD133-APC (2 μl).
- $CD44^+/\alpha2/\beta1^+$.
  - α2/β1 (10 μg/ml) (*see* **Note 27**).
  - CD44-PE (2 μl).
- $CD44^+/CD133^+/\alpha2/\beta1^+$.
  - α2/β1 (10 μg/ml) (*see* **Note 27**).
  - CD44-PE (2 μl).
  - CD133-APC (2 μl).
- $ALDH^+/CD44^+/\alpha2/\beta1^+$.
  - α2/β1 (10 μg/ml) (*see* **Note 27**).
  - CD44-PE (2 μl).
  - ALDH (*see* **Note 28**).
  - Follow the steps outlined in Subheading 3.6.8 and in the ALDEFLUOR Kit to stain cells with high ALDH activity.

5. After mixing the antibodies with the cells, wrap all tubes in aluminum foil and place on a shaker at 4 °C for 20–30 min.
6. Spin down the cells on a benchtop centrifuge at 500 ×*g* for 2 min at 4 °C.
7. Aspirate the media off, wash in 1 ml of prostate media, and spin down again at 500 ×*g* for 2 min at 4 °C.
8. For primary antibodies that require a secondary antibody conjugated to a fluorophore, add the secondary antibody at a 1:100 concentration, wrap tubes in aluminum foil, and place on a shaker at 4 °C for 15 min.
9. Spin down the cells on a benchtop centrifuge at 500 ×*g* for 2 min at 4 °C.
10. Aspirate the media off, wash in 1 ml of prostate media, and spin down again at 500 ×*g* for 2 min at 4 °C.
11. Aspirate the media off and resuspend in 500 μl of Prostate Media for the compensation tubes and 1 ml for the FACS sample tubes.
12. Transfer cells to FACS tubes with 40 μm filter tops. Add cell solution through the filter into the FACS tube. Add an additional 0.5–1 ml of Prostate Media to wash the filter. 4–5 ml of cells can be combined into the same FACS tube.

13. Centrifuge briefly at 500 × *g* for 10 s at 4 °C to ensure all liquid has passed through the filter and into the tube.
14. Keep all samples on ice covered with aluminum foil.
15. If collection of sorted cells is desired, make FACS collection media prior to running samples through the FACS machine.
16. Add 6 ml of FACS collection media into a 15 ml Falcon tube for each sorted population.
17. Run samples on a FACS Aria II cell sorter.
18. Run individual single-antibody controls and unstained cells on the FACS Aria II to set up correct compensation and voltage for sorting.
19. Set FACS gates as desired.
20. Collect up to four different cell populations simultaneously.
21. After FACS sorting is complete, spin down cells at 300 × *g* for 5 min at 4 °C.
22. Count cells using a hemocytometer.

3.6.3 In Vitro Selection in Matrigel

1. Dilute prostate cells to 2.5×10^5 cells/ml in PrEGM media.
2. Mix 40 μl of cell solution with 60 μl of cold Matrigel (*see* **Note 29**).
3. Pipette the mixture around the rim of a well of a 12-well plate using a P200 micropipetter.
4. Swirl the plate so that the mixture is even around the rim, and place the plate in a 37 °C incubator for 30–45 min to allow the Matrigel to solidify.
5. Add 800 μl of warm PrEGM media to each well. Make sure to add the media to the center of the well so as not to disturb the Matrigel around the rim (*see* **Note 30**).
6. Perform a half media change every 3 days. Aspirate 400 μl of media from the center of the plate, and add 400 μl of fresh, warmed PrEGM to the center of the plate.
7. After 7–10 days of culturing, spheres can be enumerated.

3.6.4 In Vitro Selection with Type I Collagen

1. Culture prostate cells with DMEM (mouse cells) or RPMI-1640 (human cells) supplemented with 100 units/ml penicillin, 100 μg/ml streptomycin, and 4 mM L-glutamine in 10 cm dish coated with type I collagen (52 μg/ml).
2. After 5 min of culture, wash dish twice with PBS.
3. Remove PBS and add 5 ml of preheated trypsin/0.05 % EDTA. Place into a 37 °C incubator for 5 min.
4. Add 5 ml of cold DMEM or RPMI-1640 media to inactivate trypsin.

5. Pipet vigorously to remove remaining attached cells. Transfer to a 15 ml Falcon tube.
6. Spin down cells at 300 × *g* for 5 min at 25 °C.
7. Resuspend cells in 1 ml of DMEM or RPMI-1640 media.

3.6.5 Label-Retaining Cells in Mice (BrdU Pulse-Chase)

1. Inject mice by intraperitoneal (i.p.) injection with BrdU (dissolved to a final concentration of 10 mg/ml in PBS) at 100 mg/kg of body weight.
2. Administer i.p. injections (pulse) every 12 h for a total of 72 h.
3. 4–6 Weeks after the last BrdU injection (chase), euthanize the mice and harvest tumors.
4. Harvest prostate cells as described in Subheading 3.5.1.
5. Cytospin 2–3 × 10^4 cells to coated glass slides at 500 rpm (30 × *g*) for 5 min.
6. Fix cells in 4 % paraformaldehyde (PFA) for 15–20 min.
7. Wash slides three times for 5 min with dH_2O.
8. Permeabilize with 1 % Triton X-100 for 15 min.
9. Wash slides three times for 5 min with PBS.
10. Denature DNA using 3 M hydrochloric acid in 1 % Triton X-100 for 20 min.
11. Wash slides three times for 5 min with PBS.
12. Treat with 0.1 M sodium borate for 15 min.
13. Wash slides three times for 5 min with PBS.
14. Block slides with Background Sniper for 20 min.
15. Incubate with mouse monoclonal anti-BrdU antibody at a 1:500 concentration for 5 h at room temperature or overnight at 4 °C.
16. Wash slides three times with PBS + 0.01 % Tween 20.
17. Incubate with goat anti-mouse IgG conjugated to Alexafluor 594 for 60 min at a 1:1000 concentration.
18. Wash slides three times with PBS + 0.01 % Tween 20.
19. Mount slides with Prolong Gold antifade with DAPI reagent.
20. Count $BrdU^+$ label-retaining cells (LRCs), which represent putative prostate stem cells, using a fluorescence microscope.

3.6.6 Label-Retaining Cells in Cell Culture Systems (BrdU Pulse-Chase)

1. Grow cell lines in normal culture conditions.
2. Pulse cells in media containing 10 μM BrdU for 4–6 days.
3. Culture cells in BrdU-free media (chase) for 8–10 days.
4. At the end of the chase period, trypsinize cells as described in Subheading 3.5.3.
5. For BrdU immunostaining, follow **steps 5–20** as outlined in Subheading 3.6.5.

3.6.7 Side Population Analysis

1. Dilute cells to a concentration of 2×10^6 cells/ml in pre-warmed standard culture media and divid into two separate tubes.
2. Treat one tube with 50 μM verapamil, and leave the other untreated (*see* **Note 31**).
3. Incubate both tubes with 5 μg/ml Hoechst 3342 for 90 min at 37 °C in a shaking water bath. Tubes should be mixed several times during the 90-min incubation period.
4. After 90 min, spin down cells in a centrifuge at 4 °C.
5. Resuspend the cells in ice-cold PBS. All further proceedings should be carried out at 4 °C or on ice.
6. Add 2 μg/ml of 7-AAD or PI to the cells and mix 10 min before FACS analysis.
7. Analyze Hoechst 33342 efflux on the FACS Aria II using the 375 nm laser.
8. Hoechst 33342 is excited at 350 nm ultraviolet, and the resulting fluorescence is measured at two wavelengths using 424/44 BP and 675 LP filters for detection of Hoechst blue and red, respectively.
9. Sort side population (SP) positive and negative fractions.

3.6.8 ALDH Activity

1. The ALDEFLUOR kit is used to isolate prostate cell populations with high ALDH activity.
2. Suspend cells in the ALDEFLUOR assay buffer containing ALDH substrate for 30 min at 37 °C, with or without the ALDH inhibitor diethylaminobenzaldehyde (DEAB), according to the manufacturer's instructions.
3. ALDH converts the substrate to a green fluorescent product that can be detected in the fluorescein isothiocyanate channel.
4. ALDH positive and negative fractions are sorted using a FACS Aria II.

3.7 Functional Assays for Assessment of Prostate Stem Cell Properties

A defining property of cancer stem cells is their self-renewal capacity. A number of experimental systems have been designed over the past decade to prospectively identify prostate stem cells and CSCs and quantitatively measure their self-renewal capacity, including in vitro 2-dimensional colony or 3-dimensional Matrigel sphere forming assays [56–58] and in vivo renal capsule reconstitution assays [59, 60]. Colony- and sphere-forming assays measure the ability of cells to grow clonally in serum-free conditions, a hallmark of stem cells. Asymmetrical cell division is another property of stem cells, demonstrating that cells have the multipotency to give rise to differentiated daughter cells, as well as retain cells with self-renewal capacity during cell division. In vivo prostate regeneration assays, which include subcutaneous, renal capsule, and orthotopic transplantation of cells into immunocompromised mice, measure not only the ability of prostate cancer stem cells to regenerate tumors, but also their ability to differentiate to form prostatic glandular structures and other prostate cell lineages. The in vivo dissemination assay measures the ability of prostate stem cells to survive in circulation and regenerate tumors in distant organs with foreign microenvironments, such as the lungs, liver, and bone, defining a subset of CSCs with metastatic seeding capacity. Although lineage tracing techniques have been recently used to demonstrate the ability to prostate cancer stem cells to give rise to differentiated progeny in vivo, these tools are outside of the scope of this review.

3.7.1 In Vitro Clonogenic Assays: Colony Forming Assay with 3T3 Feeder Layer

1. Plate 3T3 cells in DMEM supplemented with 100 units/ml penicillin, 100 μg/ml streptomycin, 10 % FBS, and 4 mM L-GLUTAMINE in a 6-well dish (5×10^4 cells) or a 60 mm dish (1×10^5 cells).
2. Irradiate cells with 500 rad of γ-irradiation 12 h after plating.
3. Change media and add fresh DMEM supplemented with 100 units/ml penicillin, 100 μg/ml streptomycin, 10 % FBS, and 4 mM L-glutamine 1 h after irradiation, and keep in incubator until cells are ready for plating.
4. Dilute isolated prostate tumor cells to 1.25×10^5 cells/ml in PREGM media.
5. Add 40 μl of cells into each well containing 3 ml of PREGM, and shake plate to evenly disperse cells.
6. Change media every 3 days.
7. Allow colonies to form for 7–14 days.
8. Fix plates with cold acetone for 2 min.
9. Remove acetone, and wash cells with PBS for 5 min.
10. Stain cells with trypan blue for 1 h.
11. Aspirate off trypan blue and wash plate twice with PBS for 5 min (*see* **Note 32**).
12. Enumerate colonies at 10× magnification using a light microscope.

3.7.2 Colony-Forming Assay with Matrigel

1. Pipet 20 μl of cold Matrigel into each well of a 6-well dish and distribute it using a cell scraper.
2. Put plates in a 37 °C incubator for 30 min to allow Matrigel to solidify.
3. Wash with PBS for 5 min.
4. Aspirate off PBS and add 2.5 ml of pre-warmed PREGM media.
5. Follow **steps 5–12** as outlined in Subheading 3.7.1.

3.7.3 Colony-Forming Assay with Soft Agar

1. Place agarose solution in 55 °C water bath to maintain it in a liquid phase.
2. Dilute agarose to a 0.8–1.0 % solution by mixing it with DMEM supplemented with 100 units/ml penicillin, 100 μg/ml streptomycin, 10 % FBS, and 4 mM L-glutamine.
3. Add 0.8–1.0 % agarose solution to a 60 mm dish (bottom layer agarose) and allow to solidify for 30 min at room temperature.
4. Once the agarose layer has solidified, add 1 ml of media to the dish to keep agarose from drying out.
5. Resuspend 3000–5000 FACS sorted cells in 1 ml of DMEM supplemented with 100 units/ml penicillin, 100 μg/ml streptomycin, 10 % FBS, 4 mM L-glutamine, and 0.4 % agarose (top layer agarose), and add to plates. Follow **steps 6–12** as outlined in Subheading 3.7.1.

3.7.4 Matrigel Sphere Formation Assay

1. Follow the steps outlined in Subheading 3.6.3.
2. For passaging of spheres to study self-renewal, or for isolation of spheres for other downstream assays, the Matrigel can subsequently be digested for collection of spheres.
3. Aspirate PrEGM media from each well, and add 1 ml of Dispase. Mix well by pipetting.
4. Incubate plate at 37 °C for 1 h.
5. Collect spheres in a 15 ml Falcon tube and pellet at 200 × *g* for 5 min at 25 °C.
6. Resuspend cell pellet in 1 ml of warm trypsin/0.05 % EDTA, and transfer into a 1.5 ml Eppendorf tube.
7. Incubate the tube at 37 °C for 5 min.
8. Pass spheres through a 27-G needle attached to a 1 ml syringe five times.
9. Spin down cells in benchtop centrifuge at 500 × *g* for 2 min at 25 °C.
10. Aspirate off the supernatant, and resuspend pellet in PrEGM media.
11. Centrifuge cells again in benchtop centrifuge at 500 × *g* for 2 min at 25 °C.
12. Resuspend cells in 500 μl PrEGM media.
13. Filter cells through a 40 μm nylon mesh filter.
14. Centrifuge cells again in benchtop centrifuge at 500 × *g* for 2 min at 25 °C.
15. Resuspend cells in 500 μl PrEGM media and count cells using a hemocytometer.
16. Passage spheres three to five times to assess long-term self-renewal.

3.7.5 Free-Floating Sphere Formation Assay

1. Plate 5000–10,000 cells diluted in 5 ml of sphere media per well of a 6-well ultralow attachment plate.
2. After 7–14 days in culture, enumerate spheres using a light microscope at 10× magnification.
3. To confirm that spheres floating in culture are viable, harvest spheres by collecting the media from each well into a 15 ml conical tube and centrifuging at $200\times g$ for 5 min at 25 °C.
4. Dissociate spheres into single cells using **steps 6–15** as outlined Subheading 3.7.4.
5. Follow instructions as described in the Annexin-V PE Apoptosis Detection Kit to assess cell viability.
6. Viable cells will be Annexin-V-negative and 7-AAD-negative cells.

3.7.6 Asymmetrical Cell Division Using Time-Lapse Microscopy

1. Transfect prostate cells with a Numb-GFP retroviral reporter as previously described [55].
2. FACS sort cells 72 h after transfection to purify GFP^+ cells containing the Numb-GFP reporter.
3. Plate sorted cells at clonal density in a 6-well plate in prostate media (*see* **Note 33**).
4. Place plate on the incubator stage of a Nikon Biostation Time lapse System.
5. Maintain plate at 37 °C, 5 % CO_2, and > 95 % humidity.
6. Collect phase and GFP images continuously with a 20× objective at 30 min–1 h intervals for 24–72 h depending on the rate of cell division.
7. Image a total of 20–30 different fields.
8. Perform analysis using the Nikon NIS-Elements software.
9. During symmetrical division, Numb-GFP expression should be dispersed evenly between two daughter cells after cell division.
10. During asymmetrical division, Numb-GFP expression should localize into one daughter cell but not both after cell division.

3.8 In Vivo Prostate Tumor Regeneration Assays

3.8.1 Subcutaneous Implantation

1. Resuspend 5×10^3 to 1×10^6 prostate cells in 50–100 μl of PREGM media and place into a 1.5 ml Eppendorf tube (*see* **Note 34**).
2. Obtain 6–8-week-old *NOD;SCID; IL2rγ-null* (*NSG*) male mice.
3. Anesthetize mice with a ketamine/xylazine mixture.

4. Once animals are anesthetized, shave the area where cells will be injected.
5. Mix cell suspension with an equal volume of cold Matrigel immediately before injection. Keep remaining cell suspension/Matrigel mixture on ice (*see* **Note 35**).
6. Lift up the back of the skin like a tent with forceps, inject the cell suspension/Matrigel mixture, and slowly pull out the needle. Hold the skin with forceps for 60 s to make sure Matrigel has solidified.
7. Check mice three times per week for 6–8 weeks for the appearance of subcutaneous masses, and measure the size of tumor masses with a caliper. Tumor volume is calculated with the formula $(\text{length} \times \text{width}^2)/2$.
8. Sacrifice animals when tumor burden exceeds a mass >2.0 cm in one dimension.
9. At this point harvested tumor masses can either by fixed in 10 % buffered formalin and processed for histology, or serially passaged into new recipient *NSG* mice to assess long-term self-renewal capacity.
10. For secondary transplants, isolate the transplanted cells from primary grafts by FACS sorting prostate tumor cells using the same markers that were originally used for their isolation.
11. Repeat **steps 9–25** as described in Subheading 3.5.1 in order to dissociate tumor masses into single suspensions prior to FACS sorting.
12. Repeat **steps 1–14** in Subheading 3.7.4 to serially passage prostate tumor cells.
13. Stain formalin-fixed, paraffin-embedded prostate tissue sections with antibodies against PTEN, P-AKT, and P-S6 to distinguish transplanted *Pten* null cells from endogenous cell types (*see* **Note 36**).

3.8.2 Renal Capsule Graft Assay

1. Prepare urogenital sinus mesenchyme (UGSM) as previously described [60].
2. Mix prostate stem cells in PREGM media (5×10^3 to 1×10^5) with UGSM cells ($1–2 \times 10^5$) in a 1.5 ml Eppendorf tube (*see* **Note 37**).
3. Spin down in a benchtop centrifuge at $500 \times g$ for 2 min at room temperature.
4. Aspirate off media and resuspend cell pellet in 40 to 50 μl of type I collagen mixture (*see* **Note 38**).
5. Pipette cell/collagen mixture into a 6 cm tissue culture dish.
6. Place the plate in a 37 °C incubator for 15–30 min to allow collagen to solidify.

7. Add pre-warmed UGSM media to the plate, and place it back in the incubator until surgery.
8. Obtain 6–8-week-old *NOD;SCID; IL2rγ-null* (*NSG*) male mice.
9. Anesthetize mice with a ketamine/xylazine mixture.
10. Once the mouse has been anesthetized, place the mouse in the prone position and shave a 3 cm × 3 cm area in the mid-back region.
11. Sterilize the area by swapping it with alcohol and Betadine swabs three times.
12. Lift up the back of the skin with forceps and cut along the midline with scissors to create an incision approximately 1 in. in length.
13. Place the mouse on its side, and the location of kidney should be easily identified.
14. Make a small incision in the peritoneum that is slightly longer than the length of the kidney itself.
15. Once the kidney has been located, use the small fat pad at the top of the kidney to pull it to the surface and expose the kidney outside of the skin.
16. Use forceps to gently pinch and lift the capsule of the kidney from the parenchyma of the kidney so that a 2–4 mm incision is made in the capsule.
17. After lifting the edge of the capsule, push the collagen graft underneath the membrane.
18. Allow the kidney to slip back into the mouse.
19. Suture the peritoneum together with absorbable sutures.
20. Use metal wound clips to staple the skin closed.
21. Remove staples 7–10 days after surgery.
22. Harvest grafts 6–12 weeks later.
23. Weigh the tissue mass, fix in 10 % buffered formalin, and process for histology.
24. Stain formalin-fixed, paraffin-embedded prostate tissue sections with antibodies against PTEN, P-AKT, and P-S6 to identify transplanted cells with *Pten* loss from other endogenous cell types.

3.8.3 Orthotopic Transplantation

1. Aliquot $2–5 \times 10^3$ prostate tumor cells into 5 μl of PBS.
2. Immediately before surgery, mix 5 μl of cell mixture with 5 μl of Matrigel, and keep mixture on ice.
3. Obtain 6–8-week-old *NOD;SCID; IL2rγ-null* (NSG) male mice.
4. Anesthetize mice with a ketamine/xylazine mixture.

5. Once the mouse has been anesthetized, place the mouse in the supine position, and shave the area between the penis and the rib cage.
6. Sterilize the area by swapping it with alcohol and Betadine swabs three times.
7. Using forceps to hold the skin, make a 1 in. incision directly above the penis
8. Push fat to the side, and gently tug on the bladder with blunt forceps to pull the entire urogenital system from the peritoneum.
9. Identify the two anterior lobes of the prostate, which are found directly underneath the seminal vesicles.
10. Load a 10 μl Hamilton syringe with 10 μl of the cell/Matrigel mixture.
11. Holding the distal end of one anterior lobe with blunt forceps, place the Hamilton syringe through the distal end of the anterior lobe, with the tip of the needle coming into close proximity to the urethra.
12. Inject 5 μl of the cell/Matrigel mixture slowly into the first anterior lobe.
13. Hold the needle in the anterior lobe for 60 s to make sure Matrigel has settled.
14. Inject the remaining 5 μl of the cell/Matrigel mixture into the second anterior lobe.
15. 4–8 weeks post-transplantation, harvest prostates for immunohistochemical analysis.
16. Stain formalin-fixed, paraffin-embedded prostate tissue sections with antibodies against PTEN, P-AKT, and P-S6 to identify transplanted cells with *Pten* loss from endogenous prostate cells (*see* **Note 36**).

3.8.4 *In Vivo* Dissemination Assay

1. Resuspend prostate cells to a concentration of 5×10^4–5×10^6 cells/ml in PBS.
2. Inject 200 μl of the cell suspension into the tail veins of 6–8-week-old NSG mice (*see* **Note 39**).
3. Sacrifice mice 4–8 weeks after transplantation.
4. Inflate the lungs by injecting 10 ml of PBS into a trachea of mice postmortem using a 10 cc syringe attached to an 18G needle.
5. Harvest the lungs and remove the heart with scissors.
6. Fix lungs for 24 h in 10 % buffered formalin at room temperature.
7. Remove formalin, wash three times with 70 % ethanol, and process tissue for histological examination.

8. For assessment of micro- and macrometastases in the lungs, stain formalin-fixed, paraffin-embedded tissue sections with antibodies against pan-cytokeratin, the androgen receptor (AR), and Ki67. While macrometastases should express all three markers, small micrometastases will express low levels of Ki67 (*see* **Note 40**).

4 Notes

1. Make medium fresh for usage up to 1 month. Medium should be warmed at 37 °C before use. Avoid frequent and prolonged warming since L-glutamine degrades at higher temperature.
2. Antibodies are conjugated to fluorophores and may be purchased from Biolegend, eBiosciences, and BD Biosciences. Multicolor flow cytometry is used and design of multicolor staining panels is critical. Antigen abundance, fluorophore properties, and cytometer configuration must be considered. Online resources (Biolegend, eBiosciences, and BD Biosciences websites) and programs FlowJo (TreeStar) and Fluorish provide guidance.
3. BD instruments are commonly used and include BD LSRFortessa, BD FACSCanto, or BD LSRII for analysis, and BD FACSAria II or III for sorting. Multiple lasers and detectors allow for complex panel design using many conjugates.
4. MethoCult™ M3434 and MyeloCult™ M5300 are stored at −20 °C for up to 1 year or at 4 °C for up to 1 month. To avoid freeze-thaw cycles, methylcellulose media should be aliquoted into 3 or 4 ml aliquots for performing assay in duplicate or triplicate. The addition of antibiotics is not required if aseptic techniques are used.
5. *Pten* deficient mESCs reach high cell density faster than *Wt* mESCs.
6. As an alternative, real-time PCR can be applied using WT and *Pten* deficient (KO1)-specific primers to determine the representation of each cell types.
7. For biochemical analysis, ES cells are passaged twice without MEF feeder cells at a density of 10^7 cells/10 cm plate. Two days later, these cells are further passaged without feeders at 1:6 ratio.
8. As an alternative for quick feeder removal, ES culture is trypsinized and replated on 0.2 % gelatin-coated plate. After incubating for 40 min in incubator, carefully transfer the cell suspension (unattached cells, composed mainly of ESCs) into a new gelatin-coated plate and continue culture.

9. Lindgren et al. [5] discovered that the loss of *Pten* causes the emergence of a subpopulation (~5–10 %) of aggressive and teratoma-initiating embryonic carcinoma cells during differentiation in vitro. These cells are double positive for SSEA-1 and c-kit staining.
10. BM from a WT mouse yields ~50×10^6 white blood cells. Work fast, keep cells cold, use pre-chilled buffers. Filter cells through 40 μm filters to obtain single-cell suspension.
11. Refer to manufacturer data sheets for suggested amount of antibody to use per test (may range from 0.01 to 0.25 μg). It is recommended to titrate antibodies to determine optimal concentration. Incubation times with surface staining antibodies are typically 30–60 min depending on the manufacturer's recommendation.
12. Magnetic separation may be used for enrichment or depletion of cells. Enrichment or depletion of populations is often used for sorting LT-HSCs. For example, c-kit enrichment or lineage cell depletion may be performed using magnetic beads prior to sorting for LT-HSCs for Lin^- $Sca1^+c\text{-}kit^+CD150^+CD48^-Flk2^-$.
13. In vivo BrdU labeling may also be achieved by administering BrdU to mice in their drinking water (1 mg/ml). The BrdU solution should be freshly made and changed daily. BrdU labeling can occur for various time periods depending on the study. However, prolonged feeding of BrdU can have toxic effects for the animal and should be avoided.
14. The subventricular and subgranular zones continuously self-renew and serve as internal positive controls. The subventricular zone is a particularly good positive control as it is the largest germinal layer in the brain. Its progeny travel along the rostral migratory stream and differentiate in the olfactory bulb, a process that takes 2 weeks. Therefore, one can also analyze the rostral migratory stream to observe SVZ progeny, which will be BrdU positive after 1 week of BrdU administration.
15. Tissue can be processed for paraffin embedding by placing in 70 % EtOH instead of sucrose. However, paraffin sections are limited to a thickness of 10 μM and rendering whole-brain analysis challenging. The protocol for free-floating staining can be used for paraffin sections.
16. Antigen retrieval methods can vary among antigens and antibodies. It is recommended to follow the antibody manufacturer's guidelines and optimize conditions for each assay.
17. Accumax is inactivated at 37 °C. Therefore, when culturing NSCs directly from the brain, washing is not necessary. After

accumax digestion, resuspend cells in pre-warmed NSC media and place into the incubator.

18. After initial incubation, spheres must be passaged every 4–7 days to minimize death of the cells residing at the center of the sphere. Spheres that appear bright throughout and are less than 150 μm are healthy but larger spheres have a necrotic core that appears dark in the center when looking at them under the microscope. Therefore, to minimize cell death and to maximize proliferation mouse spheres should be broken up into a single cell suspension frequently.
19. For analysis and harvesting of prostate stem cells localized within the proximal region of the prostate and anterior portion of the urethra, do not remove the urethra from the prostate.
20. For analysis and harvesting of prostate stems cells from the basal layer and the proximal region, it is optimal to treat prostate tissue with collagenase overnight to fully dissociate all epithelial cell subtypes. However, if this is not desired, 2–3 h of incubation is sufficient. Collagenase incubation time should also be optimized based on the amount of tissue that is to be digested.
21. DNase I treatment will increase the total cellular yield. However, if the cell population of interest is in high abundance, this step can be skipped.
22. Repeat for a higher cellular yield.
23. Antibody concentrations can vary based on cell concentration, cell type, and the fluorophore used. It is best to optimize FACS antibody concentrations for each particular assay.
24. Usually only one of these stains is chosen to discern live cells by FACS.
25. If the Lineage Cell Depletion Kit was used prior to FACS staining, the CD45, CD31, and Ter119 FACS antibodies do not need to be used.
26. Lineage depletion was carried out using the Lineage Cell Depletion Kit prior to FACS staining.
27. The α2/β1 is not directly conjugated to a fluorophore. Stain with α2/β1 and the appropriate fluorophore-conjugated secondary antibody before staining with the other FACS antibodies. Stain with the α2/β1 antibody for 30 min on ice, followed by staining with either APC or FITC-conjugated IgG at a 1:100 concentration for 15 min on ice. Wash cells three times with PBS before staining with additional FACS antibodies.
28. This step should be done last, after staining with α2/β1 and CD44-PE.

29. Matrigel stored at −20 °C should be thawed at 4 °C for 3–4 h prior to use. Matrigel should be kept on ice before mixing with cell solution. Be careful not to pipet air into the Matrigel/cell solution, as this well create large air bubbles that are difficult to remove.
30. Media must be warmed to 37 °C in a water bath before adding it to the wells. If the media is not warm enough, the Matrigel could dissolve or lose its integrity.
31. As verapamil is an ABC pump inhibitor, the sample treated with verapamil should have a greatly reduced side population (SP^+).
32. Alternatively, 0.2 % crystal violet, 500 μg/ml nitrobluetetrazolium (NBT), or Giemsa can also be used to stain colonies.
33. For human prostate cells, DMEM should be replaced with RPMI-1640 media.
34. To assess the proportion of cells are tumor-initiating cells (TICs) in a given population, limiting dilution analysis can be carried out, in which a progressively lower number of cells can be injected into mice (i.e., 1×10^3, 1×10^2, 10, 1 cell) to determine what percentage of cells have TIC qualities.
35. The total volume injected subcutaneously into mice should be no more than 200 μl.
36. To distinguish transplanted prostate cells from endogenous host cells, cells can be transduced with lentiviral vectors containing GFP or RFP. In addition, if tracking of disseminated prostate cells from the primary injection site is desired, cells can be transduced with a lentiviral vector containing both RFP and Luciferase (Lenti-Luc-RFP), the latter which can be used for intravital bioluminescent imaging (BLI).
37. Set up a graft control containing UGSM alone (3.5×10^5 cells) to exclude the possibility of tissue formation from contaminating urogenital sinus epithelial cells. In order to make sure that tissue grows from donor cells, GFP-expressing prostate cells isolated from GFP transgenic mice (C57BL/6-TgN from Jackson Laboratory) can also be used as the implantation material.
38. Be careful not to introduce air bubbles into mixture.
39. Another method to study the dissemination of tumor cells from the blood stream is to transplant cells through intracardiac injection. This method is particularly useful if studying metastasis to the bone.
40. In order to assess the kinetics of metastasis formation, as well as track the dissemination of tumor cells into other organ sites (e.g., liver, bone), prostate cells can be transduced with a lentiviral vector containing luciferase (Lenti-Luc) prior to transplantation to trace cells through intravital bioluminescent imaging (BLI).

References

1. Sun H, Lesche R, Li DM, Liliental J, Zhang H, Gao J, Gavrilova N, Mueller B, Liu X, Wu H (1999) PTEN modulates cell cycle progression and cell survival by regulating phosphatidylinositol 3,4,5,-trisphosphate and Akt/protein kinase B signaling pathway. Proc Natl Acad Sci U S A 96(11):6199–6204
2. Stiles B, Gilman V, Khanzenzon N, Lesche R, Li A, Qiao R, Liu X, Wu H (2002) Essential role of AKT-1/protein kinase B alpha in PTEN-controlled tumorigenesis. Mol Cell Biol 22(11):3842–3851
3. Freeman DJ, Li AG, Wei G, Li HH, Kertesz N, Lesche R, Whale AD, Martinez-Diaz H, Rozengurt N, Cardiff RD, Liu X, Wu H (2003) PTEN tumor suppressor regulates p53 protein levels and activity through phosphatase-dependent and -independent mechanisms. Cancer Cell 3(2):117–130
4. Di Cristofano A, Pesce B, Cordon-Cardo C, Pandolfi PP (1998) Pten is essential for embryonic development and tumour suppression. Nat Genet 19(4):348–355. doi:10.1038/1235
5. Lindgren AG, Natsuhara K, Tian E, Vincent JJ, Li X, Jiao J, Wu H, Banerjee U, Clark AT (2011) Loss of Pten causes tumor initiation following differentiation of murine pluripotent stem cells due to failed repression of Nanog. PLoS One 6(1), e16478. doi:10.1371/journal.pone.0016478
6. Morrison SJ, Uchida N, Weissman IL (1995) The biology of hematopoietic stem cells. Annu Rev Cell Dev Biol 11:35–71. doi:10.1146/annurev.cb.11.110195.000343
7. Yilmaz OH, Valdez R, Theisen BK, Guo W, Ferguson DO, Wu H, Morrison SJ (2006) Pten dependence distinguishes haematopoietic stem cells from leukaemia-initiating cells. Nature 441(7092):475–482
8. Zhang J, Grindley JC, Yin T, Jayasinghe S, He XC, Ross JT, Haug JS, Rupp D, Porter-Westpfahl KS, Wiedemann LM, Wu H, Li L (2006) PTEN maintains haematopoietic stem cells and acts in lineage choice and leukaemia prevention. Nature 441(7092):518–522
9. Guo W, Lasky JL, Chang CJ, Mosessian S, Lewis X, Xiao Y, Yeh JE, Chen JY, Iruela-Arispe ML, Varella-Garcia M, Wu H (2008) Multi-genetic events collaboratively contribute to Pten-null leukaemia stem-cell formation. Nature 453(7194):529–533. doi:10.1038/nature06933
10. Tesio M, Oser GM, Baccelli I, Blanco-Bose W, Wu H, Gothert JR, Kogan SC, Trumpp A (2013) Pten loss in the bone marrow leads to G-CSF-mediated HSC mobilization. J Exp Med 210(11):2337–2349. doi:10.1084/jem.20122768
11. Lesche R, Groszer M, Gao J, Wang Y, Messing A, Sun H, Liu X, Wu H (2002) Cre/loxP-mediated inactivation of the murine Pten tumor suppressor gene. Genesis 32(2):148–149
12. Magee JA, Ikenoue T, Nakada D, Lee JY, Guan KL, Morrison SJ (2012) Temporal changes in PTEN and mTORC2 regulation of hematopoietic stem cell self-renewal and leukemia suppression. Cell Stem Cell 11(3):415–428. doi:10.1016/j.stem.2012.05.026
13. Lee JY, Nakada D, Yilmaz OH, Tothova Z, Joseph NM, Lim MS, Gilliland DG, Morrison SJ (2010) mTOR activation induces tumor suppressors that inhibit leukemogenesis and deplete hematopoietic stem cells after Pten deletion. Cell Stem Cell 7(5):593–605. doi:10.1016/j.stem.2010.09.015
14. Guo W, Schubbert S, Chen JY, Valamehr B, Mosessian S, Shi H, Dang NH, Garcia C, Theodoro MF, Varella-Garcia M, Wu H (2011) Suppression of leukemia development caused by PTEN loss. Proc Natl Acad Sci U S A 108(4):1409–1414. doi:10.1073/pnas.1006937108
15. Schubbert S, Cardenas A, Chen H, Garcia C, Guo W, Bradner J, Wu H (2014) Targeting the MYC and PI3K pathways eliminates leukemia-initiating cells in T-cell acute lymphoblastic leukemia. Cancer Res 74(23):7048–7059. doi:10.1158/0008-5472.CAN-14-1470
16. Kriegstein A, Alvarez-Buylla A (2009) The glial nature of embryonic and adult neural stem cells. Annu Rev Neurosci 32:149–184. doi:10.1146/annurev.neuro.051508.135600
17. Galvez-Contreras AY, Quinones-Hinojosa A, Gonzalez-Perez O (2013) The role of EGFR and ErbB family related proteins in the oligodendrocyte specification in germinal niches of the adult mammalian brain. Front Cell Neurosci 7:258. doi:10.3389/fncel.2013.00258
18. Gregorian C, Nakashima J, Le Belle J, Ohab J, Kim R, Liu A, Smith KB, Groszer M, Garcia AD, Sofroniew MV, Carmichael ST, Kornblum HI, Liu X, Wu H (2009) Pten deletion in adult neural stem/progenitor cells enhances constitutive neurogenesis. J Neurosci 29(6):1874–1886. doi:10.1523/JNEUROSCI.3095-08.2009
19. Groszer M, Erickson R, Scripture-Adams DD, Dougherty JD, Le Belle J, Zack JA, Geschwind

DH, Liu X, Kornblum HI, Wu H (2006) PTEN negatively regulates neural stem cell self-renewal by modulating G0-G1 cell cycle entry. Proc Natl Acad Sci U S A 103(1): 111–116
20. Groszer M, Erickson R, Scripture-Adams DD, Lesche R, Trumpp A, Zack JA, Kornblum HI, Liu X, Wu H (2001) Negative regulation of neural stem/progenitor cell proliferation by the Pten tumor suppressor gene in vivo. Science 294(5549):2186–2189
21. Yue Q, Groszer M, Gil JS, Berk AJ, Messing A, Wu H, Liu X (2005) PTEN deletion in Bergmann glia leads to premature differentiation and affects laminar organization. Development 132(14):3281–3291. doi:10.1242/dev.01891
22. English HF, Santen RJ, Isaacs JT (1987) Response of glandular versus basal rat ventral prostatic epithelial cells to androgen withdrawal and replacement. Prostate 11(3):229–242
23. Bonkhoff H, Remberger K (1996) Differentiation pathways and histogenetic aspects of normal and abnormal prostatic growth: a stem cell model. Prostate 28(2):98–106. doi:10.1002/(SICI)1097-0045(199602)28:2<98::AID-PROS4>3.0.CO;2-J
24. Tsujimura A, Koikawa Y, Salm S, Takao T, Coetzee S, Moscatelli D, Shapiro E, Lepor H, Sun TT, Wilson EL (2002) Proximal location of mouse prostate epithelial stem cells: a model of prostatic homeostasis. J Cell Biol 157(7):1257–1265. doi:10.1083/jcb.200202067
25. Richardson GD, Robson CN, Lang SH, Neal DE, Maitland NJ, Collins AT (2004) CD133, a novel marker for human prostatic epithelial stem cells. J Cell Sci 117(Pt 16):3539–3545. doi:10.1242/jcs.01222
26. Chaffer CL, Weinberg RA (2011) A perspective on cancer cell metastasis. Science 331(6024):1559–1564. doi:10.1126/science.1203543
27. Taylor BS, Schultz N, Hieronymus H, Gopalan A, Xiao Y, Carver BS, Arora VK, Kaushik P, Cerami E, Reva B, Antipin Y, Mitsiades N, Landers T, Dolgalev I, Major JE, Wilson M, Socci ND, Lash AE, Heguy A, Eastham JA, Scher HI, Reuter VE, Scardino PT, Sander C, Sawyers CL, Gerald WL (2010) Integrative genomic profiling of human prostate cancer. Cancer Cell 18(1):11–22. doi:10.1016/j.ccr.2010.05.026
28. Wang S, Gao J, Lei Q, Rozengurt N, Pritchard C, Jiao J, Thomas GV, Li G, Roy-Burman P, Nelson PS, Liu X, Wu H (2003) Prostate-specific deletion of the murine Pten tumor suppressor gene leads to metastatic prostate cancer. Cancer Cell 4(3):209–221
29. Mulholland DJ, Xin L, Morim A, Lawson D, Witte O, Wu H (2009) Lin-Sca-1+CD49fhigh stem/progenitors are tumor-initiating cells in the Pten-null prostate cancer model. Cancer Res 69(22):8555–8562. doi:10.1158/0008-5472.CAN-08-4673
30. Goldstein AS, Huang J, Guo C, Garraway IP, Witte ON (2010) Identification of a cell of origin for human prostate cancer. Science 329(5991):568–571. doi:10.1126/science.1189992
31. Mulholland DJ, Kobayashi N, Ruscetti M, Zhi A, Tran LM, Huang J, Gleave M, Wu H (2012) Pten loss and RAS/MAPK activation cooperate to promote EMT and metastasis initiated from prostate cancer stem/progenitor cells. Cancer Res 72(7):1878–1889. doi:10.1158/0008-5472.CAN-11-3132
32. Goldstein AS, Lawson DA, Cheng D, Sun W, Garraway IP, Witte ON (2008) Trop2 identifies a subpopulation of murine and human prostate basal cells with stem cell characteristics. Proc Natl Acad Sci U S A 105(52):20882–20887. doi:10.1073/pnas.0811411106
33. Choi SY, Gout PW, Collins CC, Wang Y (2012) Epithelial immune cell-like transition (EIT): a proposed transdifferentiation process underlying immune-suppressive activity of epithelial cancers. Differentiation 83(5):293–298. doi:10.1016/j.diff.2012.02.005
34. Knoechel B, Roderick JE, Williamson KE, Zhu J, Lohr JG, Cotton MJ, Gillespie SM, Fernandez D, Ku M, Wang H, Piccioni F, Silver SJ, Jain M, Pearson D, Kluk MJ, Ott CJ, Shultz LD, Brehm MA, Greiner DL, Gutierrez A, Stegmaier K, Kung AL, Root DE, Bradner JE, Aster JC, Kelliher MA, Bernstein BE (2014) An epigenetic mechanism of resistance to targeted therapy in T cell acute lymphoblastic leukemia. Nat Genet. doi:10.1038/ng.2913
35. Valamehr B, Jonas SJ, Polleux J, Qiao R, Guo S, Gschweng EH, Stiles B, Kam K, Luo TJ, Witte ON, Liu X, Dunn B, Wu H (2008) Hydrophobic surfaces for enhanced differentiation of embryonic stem cell-derived embryoid bodies. Proc Natl Acad Sci U S A 105(38):14459–14464. doi:10.1073/pnas.0807235105
36. Morrison SJ, Lagasse E, Weissman IL (1994) Demonstration that Thy(lo) subsets of mouse bone marrow that express high levels of lineage markers are not significant hematopoietic progenitors. Blood 83(12):3480–3490
37. Spangrude GJ, Heimfeld S, Weissman IL (1988) Purification and characterization of mouse hematopoietic stem cells. Science 241(4861):58–62

38. Christensen JL, Weissman IL (2001) Flk-2 is a marker in hematopoietic stem cell differentiation: a simple method to isolate long-term stem cells. Proc Natl Acad Sci U S A 98(25):14541–14546. doi:10.1073/pnas.261562798
39. Kiel MJ, Yilmaz OH, Iwashita T, Yilmaz OH, Terhorst C, Morrison SJ (2005) SLAM family receptors distinguish hematopoietic stem and progenitor cells and reveal endothelial niches for stem cells. Cell 121(7):1109–1121. doi:10.1016/j.cell.2005.05.026
40. Yilmaz OH, Kiel MJ, Morrison SJ (2006) SLAM family markers are conserved among hematopoietic stem cells from old and reconstituted mice and markedly increase their purity. Blood 107(3):924–930. doi:10.1182/blood-2005-05-2140
41. Lemieux ME, Eaves CJ (1996) Identification of properties that can distinguish primitive populations of stromal-cell-responsive lympho-myeloid cells from cells that are stromal-cell-responsive but lymphoid-restricted and cells that have lympho-myeloid potential but are also capable of competitively repopulating myeloablated recipients. Blood 88(5):1639–1648
42. Song MS, Salmena L, Pandolfi PP (2012) The functions and regulation of the PTEN tumour suppressor. Nat Rev Mol Cell Biol 13(5):283–296. doi:10.1038/nrm3330
43. Shen H, Boyer M, Cheng T (2008) Flow cytometry-based cell cycle measurement of mouse hematopoietic stem and progenitor cells. Methods Mol Biol 430:77–86. doi:10.1007/978-1-59745-182-6_5
44. Szilvassy SJ, Humphries RK, Lansdorp PM, Eaves AC, Eaves CJ (1990) Quantitative assay for totipotent reconstituting hematopoietic stem cells by a competitive repopulation strategy. Proc Natl Acad Sci U S A 87(22):8736–8740
45. Till JE, Mc CE (1961) A direct measurement of the radiation sensitivity of normal mouse bone marrow cells. Radiat Res 14:213–222
46. Smith LG, Weissman IL, Heimfeld S (1991) Clonal analysis of hematopoietic stem-cell differentiation in vivo. Proc Natl Acad Sci U S A 88(7):2788–2792
47. Gage GJ, Kipke DR, Shain W (2012) Whole animal perfusion fixation for rodents. J Vis Exp JoVE (65). doi:10.3791/3564
48. Leong KG, Wang BE, Johnson L, Gao WQ (2008) Generation of a prostate from a single adult stem cell. Nature 456(7223):804–808. doi:10.1038/nature07427
49. Xin L, Lawson DA, Witte ON (2005) The Sca-1 cell surface marker enriches for a prostate-regenerating cell subpopulation that can initiate prostate tumorigenesis. Proc Natl Acad Sci U S A 102(19):6942–6947. doi:10.1073/pnas.0502320102
50. Patrawala L, Calhoun T, Schneider-Broussard R, Li H, Bhatia B, Tang S, Reilly JG, Chandra D, Zhou J, Claypool K, Coghlan L, Tang DG (2006) Highly purified CD44+ prostate cancer cells from xenograft human tumors are enriched in tumorigenic and metastatic progenitor cells. Oncogene 25(12):1696–1708. doi:10.1038/sj.onc.1209327
51. Jiao J, Hindoyan A, Wang S, Tran LM, Goldstein AS, Lawson D, Chen D, Li Y, Guo C, Zhang B, Fazli L, Gleave M, Witte ON, Garraway IP, Wu H (2012) Identification of CD166 as a surface marker for enriching prostate stem/progenitor and cancer initiating cells. PLoS One 7(8):e42564. doi:10.1371/journal.pone.0042564
52. Chen Y, Zhao J, Luo Y, Wang Y, Wei N, Jiang Y (2012) Isolation and identification of cancer stem-like cells from side population of human prostate cancer cells. J Huazhong Univ Sci Technolog Med Sci 32(5):697–703. doi:10.1007/s11596-012-1020-8
53. Burger PE, Gupta R, Xiong X, Ontiveros CS, Salm SN, Moscatelli D, Wilson EL (2009) High aldehyde dehydrogenase activity: a novel functional marker of murine prostate stem/progenitor cells. Stem Cells 27(9):2220–2228. doi:10.1002/stem.135
54. Lukacs RU, Goldstein AS, Lawson DA, Cheng D, Witte ON (2010) Isolation, cultivation and characterization of adult murine prostate stem cells. Nat Protoc 5(4):702–713. doi:10.1038/nprot.2010.11
55. Qin J, Liu X, Laffin B, Chen X, Choy G, Jeter CR, Calhoun-Davis T, Li H, Palapattu GS, Pang S, Lin K, Huang J, Ivanov I, Li W, Suraneni MV, Tang DG (2012) The PSA(-/lo) prostate cancer cell population harbors self-renewing long-term tumor-propagating cells that resist castration. Cell Stem Cell 10(5):556–569. doi:10.1016/j.stem.2012.03.009
56. Lawson DA, Xin L, Lukacs RU, Cheng D, Witte ON (2007) Isolation and functional characterization of murine prostate stem cells. Proc Natl Acad Sci U S A 104(1):181–186. doi:10.1073/pnas.0609684104
57. Xin L, Lukacs RU, Lawson DA, Cheng D, Witte ON (2007) Self-renewal and multilineage differentiation in vitro from murine prostate stem cells. Stem Cells 25(11):2760–2769. doi:10.1634/stemcells.2007-0355
58. Lukacs RU, Lawson DA, Xin L, Zong Y, Garraway I, Goldstein AS, Memarzadeh S, Witte ON (2008) Epithelial stem cells of the prostate and their role in cancer progression.

Cold Spring Harb Symp Quant Biol 73:491–502. doi:10.1101/sqb.2008.73.012

59. Cunha GR, Lung B (1978) The possible influence of temporal factors in androgenic responsiveness of urogenital tissue recombinants from wild-type and androgen-insensitive (Tfm) mice. J Exp Zool 205(2):181–193. doi:10.1002/jez.1402050203

60. Xin L, Ide H, Kim Y, Dubey P, Witte ON (2003) In vivo regeneration of murine prostate from dissociated cell populations of postnatal epithelia and urogenital sinus mesenchyme. Proc Natl Acad Sci U S A 100(Suppl 1):11896–11903. doi:10.1073/pnas.1734139100

Part VII

Methods to Study *PTEN* Using Animal Models

Chapter 16

Modeling Cancer-Associated Mutations of PTEN in Mice

Antonella Papa* and Pier Paolo Pandolfi*

Abstract

Manipulation of mammalian genome and generation of genetically engineered mouse models (GEMMs) has revolutionized the scientific approach to address biological questions. To date, a number of gene-targeting strategies have been devised and are available to investigators for the generation of genetically modified mouse lines. Nevertheless, irrespective of the methodological approach selected, there remain critical molecular steps that need to be performed and put in place in order to obtain controlled and well-characterized new animal models. Here we provide technical details for the (1) handling and maintenance of mouse embryonic stem (ES) cells; (2) analysis of genomic DNA by Southern Blot; and (3) sequencing and PCR analysis of recombined genomic DNA. These experimental steps have been undertaken for the generation of new mouse models harboring cancer-associated PTEN mutations.

Key words PTEN, Point mutations, Knock-in mice, Embryonic stem cells, Southern blot analysis, DNA sequencing

1 Introduction

Gene targeting of ES cells and generation of derived mouse models have significantly contributed to the functional characterization of thousands of genes and their role in the pathogenesis of human diseases [1]. As knowledge increases, so does the number of new techniques and approaches available to scientists to advance the meaning of their research and expedite the reach of results and conclusions. For instance, a new revolution in the functional genetics world is currently underway owing to the development of the CRISPR/Cas9 system for gene targeting [2]. The CRISPR/Cas9 system is an "old" bacterial defense system recently adapted for the manipulation of mammalian genomes and, although still in its early age, has already proved to be highly efficient to permanently target and modify mammalian gene loci, with single or multiple hits targeted at once [3, 4]. Beyond the scope of this chapter, we

*Both are equal contribution

Leonardo Salmena and Vuk Stambolic (eds.), *PTEN: Methods and Protocols*, Methods in Molecular Biology, vol. 1388, DOI 10.1007/978-1-4939-3299-3_16,

describe molecular steps of some laboratory procedures whose relevance and reliability remain essential for the monitoring and characterization of genomic DNA targeted for short sequences deletion, insertion, or for the introduction of point mutations.

In sum, careful handling of ES cells will help preserve their stability and guarantee successful germ line transmission to the progeny; strategic DNA digestions, reliable Southern blot analysis, and an effective sequencing approach demonstrate correct and successful targeting of genomic loci [5].

We report our experience with the successful generation of new knock-in mouse models for the tumor suppressor Pten and focus on practical molecular steps that apply to many investigators, regardless of the gene targeting strategy employed.

2 Materials

2.1 ES Cell Culture

A dedicated tissue culture hood should be selected to exclusively perform ES cells work and avoid cross contamination from different types of cell lines. Medium vessels, plates, multi-wells, and tips should be opened in the dedicated hood. All solutions should be sterile-filtered.

1. ES cells medium: KNOCKOUT(TM) D-MEM and 100× MEM Non-Essential Amino Acids Solution (Life Technologies), Hyclone Fetal Bovine Serum (FBS) heat inactivated (15 % final concentration), 2 mM l-Glutamine, 100× Penicillin and Streptomycin solution, 55 μM of 2-mercaptoethanol (Gibco), and 500 μl of Leukemia Inhibitory Factor (LIF) (ESGRO 10^7 U/mL).
2. pH 7.4 Phosphate buffer saline.
3. 0.05 % Trypsin–EDTA.
4. DR4 Mouse Embryonic Fibroblasts (MEFs) (ATCC) and medium: High-glucose DMEM, 10 % FBS, 2 mM l-glutamine, 100× MEM non-essential amino acid solution, and 100× penicillin and streptomycin solution.
5. Mitomycin C (Sigma), store at 4 °C.
6. 2 % Gelatin, stock solution: weigh 2 g of gelatin and mix to a total volume of 100 mL of H_2O; autoclave to sterilize and store at 4 °C. Before use, dilute 1:10 in sterile-filtered H_2O.
7. 2× ES cell freezing medium: 60 % ES cell medium, 20 % dimethyl sulfoxide (DMSO), 20 % FBS.
8. Centrifugations: all cells have been centrifuged using a Swinging Bucket Rotor (Sorvall, 7500 6441), with 19.2 cm radius (max).

2.2 Genomic DNA Extraction, Digestion, and Southern Blot

1. ES cell lysis buffer: 100 mM Tris–HCl, pH 8.5 5 mM EDTA, 0.2 % SDS, 200 mM sodium chloride (NaCl), Proteinase K, at 50 μg/mL final concentration.
2. Isopropanol.
3. Ethanol 70 %.
4. Tris–EDTA (TE) buffer.
5. Distilled H_2O (dH_2O).
6. Restriction enzymes and 10× Digestion Buffers (New England Bio-Labs).
7. DNA-grade agarose and 10 mg/mL ethidium bromide.
8. TAE Buffer, 50×: 2 M Tris base, 1 M acetic acid glacial, 50 mM EDTA. Add 500 mL of dH_2O to a 1-L graduated cylinder then weigh 242 g of Tris base, pipet 57.1 mL of glacial acetic acid and 100 mL of 0.5 M EDTA. Bring to 1 L final volume with dH_2O. Dilute 1:50 in dH_2O before use.
9. DNA loading dye with bromophenol blue (6× loading dye).
10. Tanks for DNA gel electrophoresis, trays, and combs.
11. Paper towels and Kimwipes.
12. Large plastic container (*see* Fig. 1).
13. Whatman 3MM paper: prepare two smaller sheets of the same size of the DNA gel; and two bigger sheets, 6 cm longer in length than the tray used to set up the DNA transfer.
14. Positively charged nylon membrane, prepare one sheet of membrane of the same size of the DNA gel.
15. Depurination buffer: 0.25 M hydrochloric acid (HCl). Add 2.15 mL of 37 % HCl (12 M) to 97.84 mL of dH_2O. Use glass cylinder and glass pipettes.
16. Denaturing buffer: 1.5 M sodium chloride (NaCl) and 0.5 M sodium hydroxide (NaOH).

Fill a big solid plastic tank with 8 L of dH_2O and weigh 876.6 g of NaCl and 200 g of NaOH. Stir until solution is dissolved then adjust to 10 L final volume with dH_2O.

17. Neutralizing buffer: 1.5 M of NaCl, 0.5 M of Tris base. Fill a big solid plastic tank with 8 L of dH_2O and weigh 876.6 g of NaCl, 605.7 g of Tris base, and add and 276 mL of 37 % HCl. Stir until solution is dissolved then adjust to 10 L final volume with dH_2O.
18. 20× SSC solution: 3 M of NaCl and 300 mM of trisodium citrate ($Na_3C_6H_5O_7$). Add 500 mL of dH_2O to a 1-L graduated cylinder then weigh 175.3 g of NaCl and 88.2 g of $Na_3C_6H_5O_7$. Stir until solution is dissolved then bring to 1 L with dH_2O and adjust pH to 7.0 with HCl. Dilute 1:1 with dH_2O just before use.

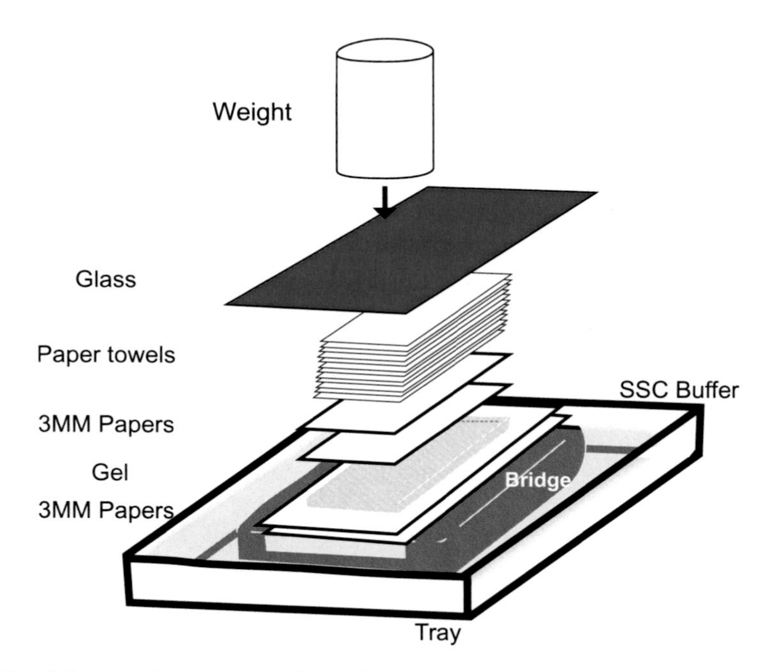

Fig. 1 Gel membrane assembly for Southern blot transfer. Upon denaturation, fragments of digested DNA are transferred to a positively charged nylon membrane by setting up a "sandwich" transfer. DNA gel and membrane are placed on a supporting bridge, in between 3MM paper sheets and compacted with weight on top. Sucking of the SSC buffer from the tray by paper towels piled on top allows DNA migration to the membrane by capillary action

19. UV cross-linker, Stratalinker.
20. Quickhyb hybridization solution (Agilent Technologies).
21. Prime-it II random primer labeling kit (Agilent Technologies).
22. Alpha-32P-dCTP, 50 μCi per membrane.
23. 100 μg/mL of sonicated salmon sperm DNA.
24. Washing solution I: 100 mL of 20× SSC solution and 5 mL of 20 % SDS in 1 L of dH_2O.
25. Washing solution II: 10 mL of 20× SSC solution and 5 mL of 20 % SDS in 1 L of dH_2O.
26. Stripping solution: 5 mL of 20× SSC solution and 25 mL of 20 % SDS in 1 L of dH_2O.

2.3 Genomic DNA Amplification and Sequencing

To verify that homologous recombination occurred in the desired genomic locus, genomic DNA from ES cells is extracted and amplified by PCR. A set of primers should be designed to surround the desired nucleotide region and generate a PCR product no longer than 200 bps.

1. HotStarTaq Master Mix Kit (1000 U), (Qiagen).
2. Set of primers.
3. DNase-RNase-free H_2O.
4. Filter tips and sterile 0.2 mL Eppendorf tubes.

3 Methods

3.1 ES Cell Culture and Maintenance

Wild-type and transgenic V6.5 ES cells (129-C57Bl/6 hybrid ES cell line) are cultured on 22 mm wells (12 well plates) and on a feeder layer of DR4 MEFs at 37 °C, 5 % CO_2 (*see* **Note 1**).

3.1.1 Preparation of Feeder Layer for ES Cell Culture

DR4-MEFs neomycin, hygromycin, puromycin, and 6-thioguanine-resistant are commercially available and are grown according to the Todaro and Green protocol [6]. Primary MEFs are maintained in MEF medium and must be mitotically inactivated before use (*see* **Note 2**). To this end:

1. Grow one 10 cm dish of DR4 MEFs to confluence.
2. Pipet 5 mL of MEF medium into a 15 mL tube and add sterile-filtered mitomycin C to obtain a 10 μg/mL final concentration.
3. Vortex and mix well.
4. Aspirate off the medium from the 10 cm dish of DR4 MEFs and add 5 mL of mitomycin C solution for 2–3 h, at 37 °C (*see* **Note 3**).
5. At the end of the treatment, aspirate the medium, wash twice with 5 mL of PBS and add 2 mL of trypsin–EDTA.
6. Incubate for 2 min at 37 °C then add 5 mL of full medium to inactivate the trypsin.
7. Disgregate cells to single-cell solution by pipetting ten times up and down.
8. Transfer the cell suspension to a 15 mL tube and centrifuge for 5 min at $308.8 \times g$.
9. Aspirate the medium and resuspend the pellet in 12 mL of MEFs medium.
10. Plate 1 mL of cell suspension into each well of the 12-well plate previously coated with 0.2 % gelatin*.

    *Beforehand: prepare 0.2 % sterile solution of gelatin in sterile-filtered H_2O. Pipet 500 μl of 0.2 % gelatin in each well of a 12-well plate and leave for 5 min at room temperature (RT). Aspirate excess solution and leave the plate to air dry. Seed MEFs for ES cell culture.
11. Leave overnight (o/n) to adhere.

3.1.2 Thawing and Passaging of ES Cells

1. Remove frozen vial of ES cells from liquid nitrogen (LN2) tank and quickly transfer to a 37 °C water bath. A fast thawing will preserve viability and quality of ES cells.
2. Transfer ES cells into a 15 mL tube, gently add 5 mL of ES cells medium and resuspend by pipetting up and down, 4–5 times.
3. Centrifuge at 137.3 (g) for 5 min; aspirate off the supernatant and resuspend ES cells pellet with 3 mL of freshly prepared ES cells medium.
4. Plate 1 mL each into three wells of a 12-well plate with freshly seeded feeder cells (*see* **Note 4**).
5. Leave to attach o/n. After 24–48 h check under the microscope the appearance of ES cells colonies.
   Maintain ES cells at medium confluence at all time and split 1:4 when at 80 % of capacity. Refresh pre-warmed medium every other day, or every day if reaching confluence.
   When confluent:

1. Pre-warm PBS, trypsin–EDTA, and ES cell medium in a 37 °C heating bath for 20 min (*see* **Note 5**).
2. Spray bottles with ethanol 70 %, wipe and dry with paper towels; transfer to dedicated tissue culture hoods.
3. Take ES cells plate and place it in the hood.
4. Gently lift the lid, tilt the plate and aspirate the medium without touching the base.
5. Pipet 500 μl of PBS into each well without disturbing the colonies and gently swirl the plate to rinse the wells.
6. Aspirate off the PBS and pipet 200 μl of trypsin–EDTA into each well. Transfer the plate into the ES cells incubator and leave at 37 °C for 2 min.
7. Place the plate in the hood and disgregate the colonies to single cells by pipetting 5–10 times up and down with a p200 pipet.
8. Add 500 μl of ES cell medium to inactivate the trypsin and transfer the suspension into a 15 mL tube.
9. Rinse the well with 2 mL of ES cell medium to harvest the leftover cells and combine altogether.
10. Centrifuge at 137.3 (g) for 5 min. Transfer the tube into the ES cell hood, open the lid and gently aspirate off the supernatant (*see* **Note 6**).
11. Resuspend the cell pellet with 4 mL of ES cells medium and dispense 1 mL each into four well of the 12-well plate with pre-seeded feeder cells.
12. Transfer the plate into the ES cell incubator and leave to attach.

3.1.3 Electroporation, Picking, and Freezing of ES Cells

To establish new genetically modified ES cells, design and clone a targeting vector containing long DNA sequences with high homology to the genomic locus to be targeted. Whether the strategy is designed for knockouts, knock-ins, or for the generation of conditional alleles, as general rule high complementarity between the cloned arms and the genomic locus of interest favors the homologous recombination between endogenous and exogenous DNA (*see* **Notes** 7 and **8**).

Once ready, the targeting vector is amplified, purified and resuspended to 1 μg/μl in sterile dH_2O. Use between 25 and 50 μg of purified DNA per electroporation (*see* **Note 9**).

To electroporate ES cells, grow and expand wild-type ES cells up to a 10 cm dish. When at 80 % of confluence, cells are ready to be electroporated.

To this end:

1. Feed ES cells with pre-warmed fresh media 4 h before electroporation.
2. Trypsinize, count and centrifuge the ES cells as described above.
3. Resuspend 1.1×10^7 ES cells in 0.9 mL of sterile PBS and pipet into a 15 mL sterile tube.
4. Gently add the digested and purified DNA and flick the tube 5–6 times to mix.
5. Leave the DNA-ES cells mix at room temperature for 5 min.
6. Set electroporator with the following parameters: 230 V and 500 μF (*see* **Note 10**).
7. Pipet the DNA-ES cells mix into a sterile electroporation cuvette without touching the metal plates.
8. Place the cuvette in the electroporation holder with the metal plates in contact with the electrodes.
9. Push the start and pulse the DNA-ES cell mix. Foam should appear when electricity runs through the mix.
10. Leave the electroporated mix at room temperature for 5 min.
11. Move the cuvette to the ES cell hood and pipet the electroporated mix into a clean 15 mL tube.
12. Add 11 mL of fresh ES cell medium and dispense 4 mL each into 10 cm dishes with pre-seeded feeder cells. Add 6 mL of fresh medium and leave to adhere.
13. After 48 h check under the microscope for ES cell colonies.
14. Start antibiotics selection at the required concentration (*see* **Note 11**).
15. Add fresh ES cell medium every other day.

16. ES cells will start to die after 4–5 days of selection; medium can then be changed every day.
17. After 8–10 days of selection, resistant ES cells colonies will be easily detectable and ready to be picked.
18. Pick ES colonies with sharp edges and bright appearance. Flat and large clones should be left behind.

ES colonies are picked in sterile condition, in the ES cells hood:

1. Pipet 40 μl of pre-warmed trypsin–EDTA in each well of a V-bottom 96-well plate and place it in the ES hood.
2. Take the 10 cm dishes with selected ES colonies, aspirate the medium and wash with 10 mL of PBS.
3. Add 6 mL of fresh PBS.
4. Spray with plenty of 70 % ethanol, wipe and place a microscope under the ES hood.
5. To pick colonies, set a p20 pipet to 20 μl. Point with the tip to a single ES colony, slide the tip under the colony and gently aspirate to suck the colony into the pipet.
6. Transfer the colony into one well of the 96 well with trypsin previously prepared.
7. Repeat the process to pick all colonies that look bright and sharp.
8. When half 96-well plate is filled with picked ES cells, transfer the plate to the ES cell incubator and leave for 2 min at 37 °C.
9. Add 100 μl of ES cell medium in each well and pipet up and down to break the colonies. The uses of multichannel pipette will easy the process.
10. Transfer each ES colony into one well of a 24 well with pre-seeded feeder cells. Label every well with progressive numbers to identify the position of each individual clone.
11. Leave to adhere for 48 h and then check under the microscope for ES colonies.
12. Grow to confluence refreshing the medium every other day and maintaining the antibiotic selection.
13. When confluent, trypsinize and freeze one-half of the well for backup.

To this end:

14. Wash confluent ES cells with PBS.
15. Add 500 μl of trypsin–EDTA and incubate for 2 min at 37 °C, then add 1.5 mL of ES cell medium to inactivate the trypsin. Pipet ES cell suspension up and down to disgregate colonies.

16. Prepare cryogenic vials: carefully label each vial with progressive numbers and according to the numeration used to label the 24 wells. Add 1 mL of 2× ES cell freezing medium into each vial.
17. Transfer 1 mL of ES cell suspension into the cryogenic vial with the corresponding number. Pipet up and down to mix with the 2× ES cell freezing medium.
18. Close vials and leave for 2–3 days in a −80 °C incubator, then transfer to LN2 tank for long-term storage.
19. Add 1 mL of fresh ES cell medium to the second half of ES cell suspension; leave in the same well and grow to confluence.

When confluent, proceed to genomic DNA extraction, digestion, and Southern blot.

Once positive ES cells clones have been identified, expand and freeze down a few vials for backup.

For screening purpose, 24-well plates allow you to recover 10 μg of genomic DNA per well, enough for 2–3 Southern blot experiments. However, few 22 mm wells of the same ES cell colony can be combined in order to extract larger amount of DNA.

3.2 Genomic DNA Extraction, Digestion, and Southern Blot

This protocol applies to 22 mm multi-well plates. If larger wells are in use, scale up solution volumes and follow the same procedures.

Grow ES cells to confluence and then prepare for DNA extraction.

3.2.1 Extraction of Genomic DNA

1. Gently aspirate off the ES cell medium.
2. Wash with 1 mL of PBS.
3. Pipet 500 μl of ES cell lysis buffer per well with freshly added Proteinase K.
4. Incubate at 55 °C o/n (*see* **Note 12**).

3.2.2 DNA Purification

DNA purification can be performed in non-sterile conditions:

1. Collect your samples and transfer to the bench.
2. Pipet lysed samples into 1.5 mL Eppendorf tubes (*see* **Note 13**).
3. Add 1 mL of isopropanol per sample.
4. Mix vigorously and shake up and down until precipitated DNA becomes apparent.
5. Centrifuge at 16,000 × *g* for 5 min (benchtop centrifuge). A white pellet should be visible at the bottom of the Eppendorf tube.
6. Gently pipet off the supernatant without disturbing the pellet and wash with 500 μl of ethanol 70 %.
7. Centrifuge at 16,000 × *g* for 5 min.

8. Pipet off as much supernatant as possible without disturbing the pellet.
9. Leave to air-dry for 5–10 min on the bench (*see* **Note 14**).
10. Add 60 μl of TE buffer.
11. Leave at RT o/n to allow genomic DNA to rehydrate and solubilize.
12. The day after, pipet genomic DNA up and down to assure complete solubilisation; do not vortex (*see* **Note 15**).
13. Store at 4 °C (*see* **Note 16**).

3.2.3 Enzymatic Digestion of Genomic DNA

Enzymatic digestions are carried out in small volumes (45 μl final volume).

1. Pipet 16 μl of genomic DNA into an Eppendorf tube.
2. Add 20 μl of dH2O and mix well by pipetting up and down.
3. Add 4.5 μl of 10× digestion buffer and mix well by pipetting up and down.
4. Add 4.5 μl of restriction enzyme and mix well by pipetting up and down (*see* **Note 17**).
5. Digest o/n at 37 °C, or at required temperature as specified by the manufacturer (*see* **Note 18**).

In case of double digestions, adjust the final volume reaction to 50 μl.

1. Pipet 20 μl of genomic DNA into an Eppendorf tube.
2. Add 20 μl of dH_2O and mix well by pipetting up and down.
3. Add 5 μl of digestion buffer and mix well by pipetting up and down.
4. Add 5 μl of 10× restriction enzyme and mix well by pipetting up and down (*see* **Note 19**).
5. Digest o/n at 37 °C, or at required temperature as specified by the manufacturer.

At the end of the digestion, add plenty of loading dye and prepare for DNA gel run.

3.3 Southern Blotting

For optimal separation of the bands, run the gel o/n at 20 V (*see* **Note 20**).

3.3.1 Gel Run and DNA Separation

1. Prepare a 0.8 % agarose gel with 1× TAE buffer (*see* **Notes 21** and **22**).
2. Leave to solidify (40 min) and transfer into the running tank.
3. Fill electrophoresis tank with fresh 1× TAE buffer in dH_2O.
4. Cover the gel with TAE buffer and gently remove the comb.

5. Load DNA ladder marker.
6. Load digested genomic DNA.
7. Run o/n at 20 V.
8. When the bromophenol blue band of the loading dye has reached the bottom of the gel (corresponding to 370 bps band), assess the quality of the digested DNA under UV light.
9. Transfer the gel to a solid plastic container.
10. Place the gel on top of the UV transilluminator (*see* **Note 23**).
11. Position a fluorescent ruler alongside the gel, lined up with the top of the wells (*see* **Note 24**).
12. Cover the gel with protective shield and activate UV light.
13. Quickly expose the gel and print out few pictures for records.
14. Carefully move the gel back to the plastic container and rinse with dH2O (*see* **Note 25**).
15. Submerge the gel in denaturing buffer and gently rock for 30 min at RT (discoloring of the loading bands is expected).
16. Discard the denaturing buffer and quickly rinse the gel with dH_2O.
17. Submerge the gel in neutralizing buffer and gently rock for 1 h at RT.
18. Discard the neutralizing buffer and quickly rinse the gel with dH_2O.
19. Equilibrate the gel in 1× SSC buffer.

3.3.2 DNA Transfer

For DNA transfer:

1. Prepare 2 L of 10× SSC buffer in dH_2O.
2. Place a large plastic tray on the bench and fill it with SSC buffer (select appropriate size tray to contain the gel).
3. Place a support bridge in the tray (*see* Fig. 1).
4. Pipet 10 mL of 1× SSC buffer over the bridge and position one of the two bigger Whatman 3 MM paper right on top.
5. With a roller, or a 2 mL pipette, distend the paper over the bridge and allow the SSC solution to be absorbed.
6. Pipet few mL of SSC buffer over the paper and remove air bubbles between the paper and the bridge (do not brake or fold the paper)
7. Submerge the overhanging edges of the paper in the SSC buffer filling the tray and allow soaking up.
8. Repeat the procedure with the second sheet of paper.
9. Place the gel, upside down, atop the paper sheets.
10. Roll the pipette over the gel to carefully remove air bubbles.

11. Place the dry nylon membrane on top of the gel and eliminate air bubbles as above (*see* **Note 26**).
12. Complete the sandwich by assembling the last 2 dry sheets of Whatman 3MM papers on top of the nylon, without disturbing the layers underneath.
13. Surround the edges of the gel with plastic wrap and seal to avoid evaporation of the SSC buffer from the tray.
14. Stack a few paper towels on top of the two Whatman 3MM papers to favor uptake of SSC buffer.
15. Place a solid flat object, a glass plate or a book, on top of the paper towels.
16. Place a 500 g–1 Kg weight to compact the sandwich and increase absorption (*see* Fig. 1).
17. Allow to transfer o/n.

The morning after:

1. Remove weight and paper towels.
2. Lift and flip the gel, place it on the bench by lying on the nylon and Whatman 3 MM papers.
3. Mark the wells with a pencil.
4. Remove and appropriately dispose the gel.
5. Transfer the nylon (DNA side facing up) to a UV light cross-linker and set at 120 mJ. This step will permanently fix the DNA fragments to the membrane.
6. Briefly rinse the membrane in dH2O and proceed with the DNA probing.

3.3.3 DNA Probing

For probing of DNA:

1. Warm up the hybridizing buffer for 20 min at 65 °C.
2. Select a glass cylinder with rubber-lined screw caps to prevent leaking of buffers.
3. Transfer the membrane into the glass cylinder (*see* **Note 27**).
4. For half liter volume cylinders, pipet 15 mL of heated hybridizing buffer and leave to roll at 65 °C for 1 h. For larger cylinder, increase the volume of the buffer to cover the whole membrane.

In the meantime, prepare and purify the radioactive probe:

1. Labelling: Prime-it II random primer labeling kit, 35 min.
2. Purification: Illustra ProbeQuant G-50 μ Columns, 5 min.

Accurately follow the manufacturer's instructions. Always wear personal protective equipment (PPE) and work safely behind shields.

When the labeled probe is ready:

1. Boil radioactive probe and sonicated salmon sperm DNA at 95 °C for 5 min.
2. After a quick spin at 16,000 × *g*, pipet the radioactive probe and 10 μl of sonicated salmon sperm DNA in the glass cylinder and gently swirl to evenly mix with the hybridizing buffer.
3. Hybridize the membrane by gently rolling for 2 h at 65 °C.

At the end of the incubation, prepare to expose the radioactive membrane.

1. Remove the glass cylinder from the 65 °C incubator and safely place it behind the protective shield. Use heat proof gloves as the bottle will be very hot.
2. Properly dispose the radioactive solution into dedicated bin.
3. Add 20 mL of washing solution I and gently roll for 10 min at 65 °C.
4. Dispose the solution into dedicated bin.
5. Add 20 mL of washing solution II and gently roll for 5–10 min at 65 °C.
6. Dispose the solution into dedicated bin.

To expose the membrane safely handle radioactive membrane and materials with PPE and protective shields.

1. With tweezers, lift one corner and pull the membrane out of the glass cylinder.
2. Briefly lay the membrane on paper towels to dry excess washing solution.
3. Place the hybridized membrane in the developing cassette.
4. Cover with saran wrap.
5. With soft paper towels gently distend the saran wrap on top of the membrane, remove air bubbles and wipe off excess buffer.
6. Transfer to darkroom and set up the exposure.
7. Place two western blot films on top of the membrane and expose o/n (*see* **Note 28**).

The day after, develop both films and rapidly close the cassette.

1. Place the films on the bench.
2. Place a ruler alongside the film and sign wells and DNA ladder by using the pictures from the DNA gel run.
3. Estimate the size of the DNA bands.

If re-probing is needed, strip the nylon membrane:

1. Rinse the nylon membrane with dH_2O
2. Place the membrane in a plastic tray and fill with stripping buffer.
3. Gently rock for 15 min at 60 °C.
4. Properly dispose the buffer in dedicated bin.
5. Rinse with dH_2O
6. Transfer membrane into a glass cylinder and incubate with hybridizing buffer for 1 h.

Proceed with the probing as described above.

3.4 Genomic DNA Amplification and Sequencing

Isolate genomic DNA from targeted ES cells as described above.

1. Quantify and dilute genomic DNA to 100 ng/μl in DNase/RNase-free H_2O.
2. Set up PCR reactions following the manufacturer's instructions and kit.
3. Use 2 μl of genomic DNA in 25 μl final volume reaction.
4. At the end of the PCR cycles, run 10 μl of reaction on a 2 % agarose gel.
5. Place the gel in the transilluminator and confirm the presence of amplified bands.
6. Set aside 5 μl of reaction for sequencing, using the same set of primers.

ES cell clones positive to Southern blot and sequencing analyses can then be characterized for karyotyping. ES cell clones with normal karyotype can be injected into recipient blastocysts (*see* Fig. 2). Two independent ES cell clones per mutation should be injected in order to generate and establish two distinct mouse cohorts for biological replicates.

4 Notes

1. The V6.5 ES cell line was originally generated in the laboratory of Pr. Rudolph Jaenisch. Wild-type V6.5 ES cells were used to generate two knock-in mouse models for the tumor suppressor Pten (PtenC124S/+ and PtenG129E/+ mice) [7]. Briefly, wild-type V6.5 ES cells were electroporated to introduce targeting vectors carrying point mutations towards the *exon5* of the *Pten* genomic locus. Homologous recombination led to the introduction of Pten point mutations as well as the insertion of the neomycin resistance cassette. Following neomycin selection (300 μg/mL of neomycin refreshed every

Work-flow with Mouse Embryonic Stem (ES) cells:

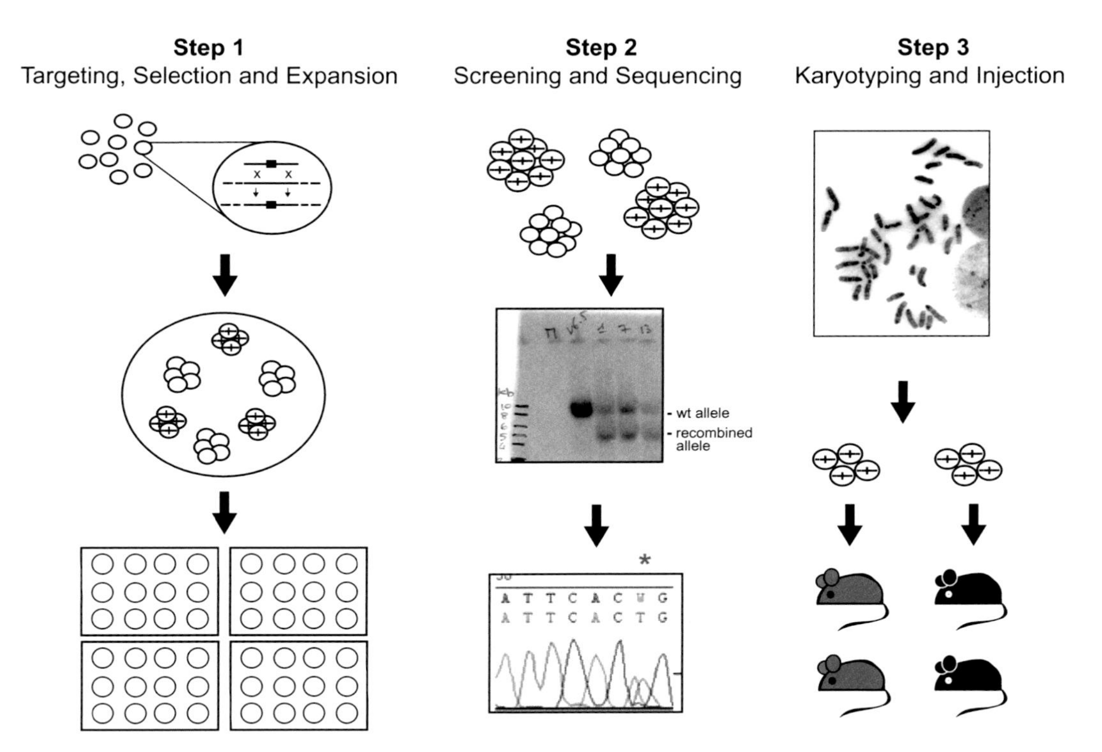

Fig. 2 Workflow showing experimental steps undertaken for the generation of new knock-in mice for Pten. *Step 1:* ES cells were electroporated with purified targeting vector and selected to isolate recombined clones. *Step 2:* ES cell clones were screened by Southern blot analyses and further sequenced for confirmation. *Step 3:* positive ES cell clones with normal karyotype were injected into recipient mouse blastocyst to generate chimeric mice

day for 8–10 days), resistant ES cell clones were picked, expanded, and tested for homologous recombination by Southern blot analysis. Wild-type V6.5 ES cells were used as control samples.

It is recommended to maintain ES cells at low number passages to prevent loss of stability and differentiation. As few as 2–3 wells of 12-well plates at 80 % of confluence were used to prepare for injection.

2. Alternatively, MEFs can be inactivated by γ-irradiation; or purchased already inactivated.
3. One 10 cm dish of early passage DR4 MEFs (P2–P3) contains on average $5–6 \times 10^6$ cells, sufficient to cover a whole 12-well plate without over-seeding.
4. ES cells are generally frozen at 5×10^6 cells/mL. 1 mL of frozen cell suspension should be plated into three wells of a 12-well plate. 5–10 % cell death is generally observed.

5. It is recommended not to over-heat ES cells medium in order to preserve stability of components, serum and LIF.
6. When removing the medium by using vacuum systems, gently aspirate the supernatant without dipping the tip. Leave a thick layer of medium to cover the cell pellet in order to avoid accidental sucking off the cells. Alternatively, use a p1000 pipette to pipet off the medium.
7. To facilitate the selection of recombined ES cell colonies, selectable markers are used for positive and negative selection. Specifically for the *Pten* knock-in mice, murine *Pten* exon 5 mapping chromosome 19 was targeted with a targeting vector expressing the neomycin cassette (positive selection) and the herpes simplex virus thymidine kinase (tk) cassette (negative selection). In addition, a unique restriction enzyme must be included in the targeting strategy, in order to allow linearization of the targeted vector before electroporation.
8. Large size targeting vectors tend to recombine during the transformation and amplification in bacteria. To prevent this, use bacterial strains carrying mutation/deletion of the *rec*A, B, and C genes in order to reduce recombination ability.
9. The targeting vector is digested with single cut in a unique site outside the recombination sequences. Overnight digestion is followed by ethanol precipitation and resuspension of the vector with 50 μl of sterile-filtered H_2O.
10. Electroporation parameters have been optimized for the Biorad GenePulser electroporator. If a different apparatus is used, it is recommended to assess the electroporation efficiency by testing voltage and charge parameters using reporter genes.
11. Antibiotic selection is maintained to prevent contamination during the picking phase. Once positive clones are selected and expanded, the antibiotics can be released.
12. Cover the multi-well plate with two layers of wet paper towels and wrap with plastic (Saran wrap) to prevent desiccation of samples.
13. Lysed genomic DNA may form viscous clumps difficult to pipet. To ease this step, transfer the samples to eppendorf tubes as soon as out of the incubator, when still warm and easier to work with. Use p1000 pipette and cut 3–4 mm off the tip in order to allow access of thick genomic clumps.
14. Do not overdry genomic DNA as this may cause loss of integrity. However, no drops of ethanol should be left in the tube as this may hamper enzymatic digestions.
15. Pipet the samples up and down with a p200. If the DNA is not fully dissolved, incubate at 55 °C for 1 h then repeat the pipetting to solubilize clumps.

16. Sample freezing is allowed for long-term storage; however avoid multiple thawing and freezing cycles as this would degrade the DNA.
17. If possible, high-concentration restriction enzymes are recommended.
18. For best digestion efficiency, perform o/n digestion in incubators rather than heating blocks. Incubators allow an even distribution of the heat and prevent evaporation of the sample to the lid of the Eppendorf.
19. Generally, equal amount of each restriction enzyme, in terms of enzymatic units (U/μl) should be used unless otherwise required (i.e., in case of uneven digestion efficiency between the selected enzymes). Do not exceed 10 % reaction volume with restriction enzymes as this may reduce the overall digestion efficiency.
20. If the predicted bands are less than 10 Kbs in length, and if the distance between the expected bands is big enough (at least 5 Kbs), shorter runs of 4–5 h at 80 V can be performed.
21. 0.8 % agarose gels are very delicate and easy to break. Handle with care and use containers to transport as they may slip off the tray.
22. Select the size of the comb considering the volume of the digestion reaction plus the volume of the loading dye.
23. Gently slide the gel from the tray to the transilluminator, leftover TAE buffer helps to slip off the gel over the glass. For big-size DNA gel, lift from a bottom corner and hold with the whole hand. Gently place the gel on the transilluminator glass.
24. The positioning of the fluorescent ruler helps to mark the DNA ladder on the film/blot after the exposure time, and gives an estimate of the size of the expected bands.
25. If high-molecular-weight bands are expected (above 15 kb), add a depurination step by gently rocking the gel in 0.25 M HCl solution at RT for exactly 10–20 min. Rinse with dH2O and proceed to denaturation.
26. For nylons bigger than 10 cm^2, briefly rehydrate with dH_20 and soak into 1× SSC buffer before transfer.
27. Select a glass cylinder big enough to contain the membrane. Lift the membrane and roll one edge over the other to form a closed tube. Insert the rolled nylon membrane in the glass cylinder and let adhere to the wall. Pipet few mL of dH_20 in the cylinder and allow the membrane to unwind by rolling the cylinder. Do not let edges to fold or overlap.
28. To reduce background noise, incubate cassette at –80 °C o/n. The morning after, remove the frozen cassette and leave to warm up for 20 min.

References

1. Capecchi MR (1989) Altering the genome by homologous recombination. Science 244(4910):1288–1292
2. Hsu PD, Lander ES, Zhang F (2014) Development and applications of CRISPR-Cas9 for genome engineering. Cell 157(6): 1262–1278
3. Cong L et al (2013) Multiplex genome engineering using CRISPR/Cas systems. Science 339(6121):819–823
4. Mali P et al (2013) RNA-guided human genome engineering via Cas9. Science 339(6121):823–826
5. Southern EM (1975) Detection of specific sequences among DNA fragments separated by gel electrophoresis. J Mol Biol 98(3): 503–517
6. Todaro GJ, Green H (1963) Quantitative studies of the growth of mouse embryo cells in culture and their development into established lines. J Cell Biol 17:299–313
7. Papa A et al (2014) Cancer-associated PTEN mutants act in a dominant-negative manner to suppress PTEN protein function. Cell 157(3):595–610

Chapter 17

C. elegans Methods to Study PTEN

Shanqing Zheng and Ian D. Chin-Sang

Abstract

C. elegans encodes a PTEN homolog called DAF-18 and human PTEN can functionally replace DAF-18. Thus *C. elegans* provides a valuable model organism to study PTEN. This chapter provides methods to study DAF-18/PTEN function in *C. elegans*. We provide methods to genotype *daf-18/Pten* mutants, visualize and quantify DAF-18/PTEN in *C. elegans*, as well as to study physiological and developmental processes that will provide molecular insight on DAF-18/PTEN function.

Key words PTEN, Tumor suppressor, DAF-18, *C. elegans*, Molecular biology

1 Introduction

PTEN (*p*hosphatase and *ten*sin homolog deleted on chromosome 10), also called MMAC1 or TEP1, was first discovered in 1997, and is recognized as the second most (after P53) frequently mutated tumor suppressor in human cancer, and germline mutations in PTEN can lead to *PTEN* hamartoma tumor syndrome (PHTS) [1–6]. While significant advances have been made since its discovery we have yet to fully understand all of PTEN's functions. Using model organisms such as *C. elegans* will help to further understand the role of PTEN on a genetic and molecular basis.

C. elegans has been used as a model organism to study the genetics and molecular biology of many genes [7–9]. *C. elegans* encodes one PTEN homolog called *daf-18* and was named after a mutant phenotype: abnormal *DA*uer *F*ormation (DAF) [10–13]. Like human PTEN, DAF-18 contains a phosphatase domain, a C2 domain, and a PDZ binding domain. Missense mutations in human PTEN that alter PTEN enzymatic activity such as phosphatase inactive C124S, lipid phosphatase inactive G129E [14], and protein phosphatase trapping D92A [15] all have the corresponding protein variants in the *C. elegans* DAF-18 protein: C169S, G174E and D137A respectively [16–18]. PTEN and DAF-18 have evolutionarily conserved functions as human PTEN can functionally

Leonardo Salmena and Vuk Stambolic (eds.), *PTEN: Methods and Protocols*, Methods in Molecular Biology, vol. 1388, DOI 10.1007/978-1-4939-3299-3_17,

replace DAF-18 and rescues the dauer and longevity defects in *daf-18* mutants [18]. Furthermore, when human PTEN is expressed in *C. elegans* it is phosphorylated on regulatory serine and threonine amino acids suggesting the kinases responsible for PTEN phosphorylation at these residues are present in *C. elegans* [18]. This chapter provides various methods to study the function of DAF-18/PTEN in *C. elegans*. DAF-18/PTEN has numerous developmental and physiological roles such as oocyte maturation, axon guidance, vulva formation, germ cell proliferation, and axon regeneration but in this chapter we focuses on its best understood roles: L1 arrest, adult longevity, and dauer formation. For methods on the other roles please see Ref. 19.

2 Materials

2.1 General C. elegans Maintenance

1. Nematode growth media (NGM) plates: NaCl 3 g, agar 20 g, peptone 2.5 g in 975 mL dH_2O. Sterilize by autoclaving. Cool down to 55 °C add sterile solutions of 1 mL $CaCl_2$ (1 M), 1 mL cholesterol (5 mg/mL in ethanol), 1 mL $MgSO_4$ (1 M) and 25 mL 1 M KPO buffer (108.3 g KH_2PO_4, 35.6 g K_2HPO_4 dissolved in 1 L dH_2O). Pour 10 mL NGM per plate into 6 cm diameter petri dishes.
2. M9 buffer: Na_2HPO_4 5.8 g, KH_2PO_4 3.0 g, NaCl 0.5 g, NH_4Cl 1.0 g, dH_2O to 1 L. Sterilize by autoclaving.
3. *E. coli* (OP50 strain): *E. coli* OP50 is a uracil auxotroph whose growth is limited on NGM plates can be obtained from the CGC (https://www.cbs.umn.edu/research/resources/cgc), and should be cultured on a streak plate of LB agar: 10 g Bacto-tryptone, 5 g Bacto-yeast, 5 g NaCL, 15 g agar, H_2O to 1 L, pH 7.5. Using a single colony from the streak plate, aseptically inoculate LB Broth: 10 g Bacto-tryptone, 5 g Bacto-yeast, 5 g NaCl, H_2O to 1 L, pH to 7.0 using 1 M NaOH. Allow inoculated cultures to grow overnight at 37 °C.

2.2 L1 Arrest Assay

1. Bleach solution: 10 % bleach, 1 M NaOH.
2. Adult life-span assay: 1 M 5-fluoro-2-deoxyuridine (FUdR): FUdR 2.5 g in 10 mL dH_2O. Store at −20 °C.

2.3 Dauer Assay

1 % sodium dodecyl sulfate (SDS): SDS 1 g, dH_2O to 100 mL, sterilize by autoclaving.

2.4 daf-18/Pten PCR Genotyping

Lysis buffer: 1× PCR buffer (see below for 10× PCR buffer).

Proteinase K: 20 mg/mL.

10× PCR buffer: 100 mM Tris, 500 mM KCl, 15 mM $MgCl_2$ pH 8.3.

dNTP mix: 25 mM each dNTP.

Oligos: oIC641:5′AACTTACCACTGGCAACGAATGAATAC, oIC642: 5′AATCGTGGAGATAAGGCTGCTAAATC (5–10 μM).

Thermostable DNA polymerase: e.g., Tag, Pfu, Q5 5 U/μL.

Liquid nitrogen.

Agarose.

Ethidium bromide (EtBr) 10 mg/mL.

2.5 Antibody Staining

Primary and secondary antibodies:

40× Borate buffer stock: H_3BO_3 6.2 g (1 M), NaOH 2.0 g (0.5 M), add dH_2O to 100 mL store at room temperature. Dilute to 1× using dH_2O.

10× PBS (for 1 L): NaCl 80 g, KCl 2.0 g, Na_2HPO_4:$7H_2O$ 11.5 g, KH_2PO_4 2.0 g, sodium azide (toxic) 2 g, Dissolve in 900 mL dH_2O, adjust to pH = 7.2 with 10 M NaOH, top off to 1 L with dH_2O. Autoclave and store at room temperature.

Antibody buffer A: 1× PBS, 1 % bovine serum albumin (BSA), 1.5 % Triton-100, 1 mM EDTA.

Antibody buffer B: 1× PBS, 0.1 % BSA, 0.5 % Triton X-100, 1 mM EDTA.

2× Fixation solution: 160 mM KCl, 40 mM NaCl, 20 mM Na_2EGTA, 10 mM spermidine HC, 30 mM Na Pipes, pH = 7.4, 50 % methanol. Just prior to use add βME to 2 %.

Tris-Triton buffer: 100 mM Tris–HCL pH = 7.4, 1 % Triton X-100, 1 mM EDTA.

10 % Paraformaldehyde: Make fresh each time. Dissolve 1 g paraformaldehyde in 10 mL dH_2O and add 50 μL 10 M NaOH dissolve in 65 °C water bath.

β-Mercaptoethanol (βME).

1 M Dithiothreitol (DTT): Dissolve 1.54 g in 10 mL dH_2O. Store at −20 °C in 1 mL aliquots.

3 % Hydrogen peroxide (H_2O_2).

4′,6′-Diamidino-2-phenylindole hydrochloride (DAPI) 1 mg/mL in 1× PBS.

Prolong® Diamond (Life Technologies).

2.6 Protein Preparation for Western Blots

Primary and secondary antibodies:

Protein gel sample buffer: 2 mL Tris (1 M, pH = 6.8), 4.6 mL glycerol (50 %), 1.6 mL SDS (10 %), 0.4 mL bromophenol blue (0.5 %), 0.4 mL βME.

2.7 General Equipment

Micro-centrifuge, vortex/mixer, smaller microfuge rocker or Nutator, 60 cm petri dish, slides, cover slips, platinum wire pick, compound microscope, dissecting microscope, heat block, 20 °C incubator, 25 °C incubator, water bath, thermocycler, gel electrophoresis apparatus, Pasteur pipettes, Pipetman.

3 Methods

3.1 L1 Arrest Assay

When *C. elegans* hatch in the absence of food they enter a developmental arrest known as L1 arrest ([20, 21], Fig. 1). The insulin/IGF-1 signaling pathway (IIS) is a major signaling pathway controlling *C. elegans* L1 arrest. DAF-18 like human PTEN antagonizes the IIS pathway and *daf-18/pten* mutants are short lived in the L1 arrest. DAF-18/PTEN has functions independent of the terminal IIS FOXO transcription factor, DAF-16, as *daf-16* null mutants live longer than *daf-18* in L1 arrest ([22], Fig. 2). Thus, the *daf-18*L1 arrest phenotype may help identify DAF-18/PTEN pathways independent of DAF-16/FOXO [21, 22]. *C. elegans* eggs are more resistant to bleach when compared to the adults. When adults are treated with a bleach solution they disintegrate allowing the eggs to be released and if they hatch in the absence of food they will arrest at the L1 stage. Below is a protocol for L1 arrest longevity assay:

1. Wash gravid adults from 1 to 3 6 cm NGM plates with 1 ml M9 buffer and transfer to a 1.5 mL centrifuge tube.
2. Centrifuge at 1000 ×*g* for 1 min to pellet worms.
3. Aspirate the M9 and resuspend the worm pellet in 1 mL of bleach solution and agitate for 1 min.
4. Centrifuge at 1000 ×*g* for 1 min, aspirate, and resuspend again in another 1 mL bleach solution. Agitate/mix until the worm bodies are no longer visible leaving only the eggs (less than 3 min, *see* **Note 1**).
5. Centrifuge embryos at 3000 ×*g* for 1 min, aspirate off the bleach solution being careful not to disturb the egg pellet.
6. Perform a minimum of 3 × 1 mL washes in M9, each time spinning at 3000 ×*g* for 1 min and aspirating to the pellet.
7. After the final wash, resuspend eggs in M9 buffer (1–10 mL) to less than ten eggs per μL and place on a rocker/shaker at 20 °C for 15 h to let eggs hatch (*see* **Note 2**).
8. Each day or every other day, remove an aliquot containing a minimum of 30 L1 arrested worms and plate on nematode growth media (NGM) agar plates. Do this in triplicate for each sample.
9. Score % survival by counting the number of worms that are alive (moving) and dividing that number by the total number of worms in the aliquot (*see* **Note 3**). Each time point on the survival curves should represent the average of at least three independent trials.

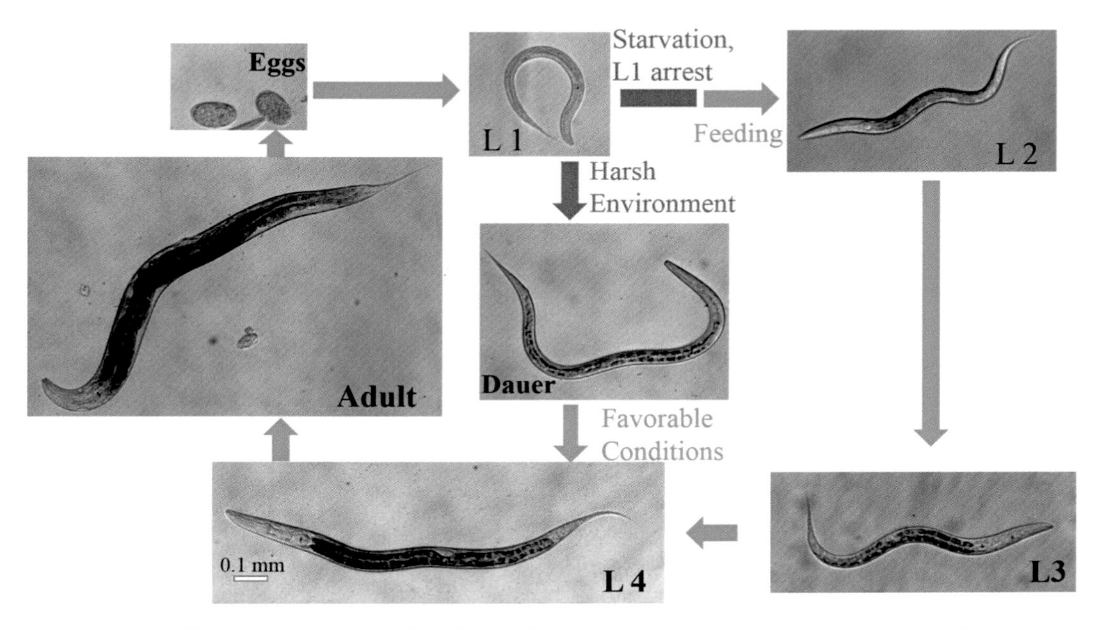

Fig. 1 *C. elegans* life stages. Food is required for larval development, as hatching in the absence of food causes L1 arrest. L1 arrested larvae can survive for weeks until food becomes available. Dauer formation is an alternative developmental program (indicated by *red arrows*) and is induced by harsh environments such as crowding, low food and high temperature. Dauer larvae can survive up to 4 months until favorable conditions return. *daf-18/Pten* mutants are defective for both L1 arrest and dauer larvae formation

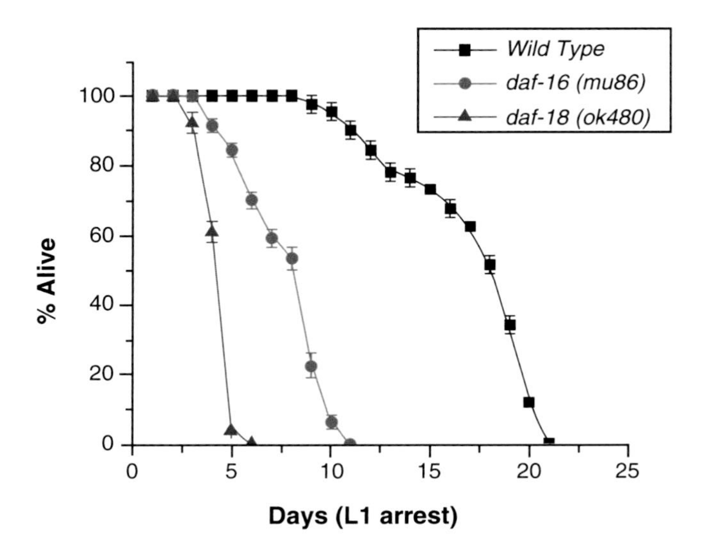

Fig. 2 L1 arrest assay. *daf-18(ok480)* mutant worms do not live past 5 days in L1 arrest. In contrast wild-type *C. elegans* can survive over 20 days in L1 arrest. The terminal regulator in the insulin like signaling pathway is the FOXO transcription DAF-16. *daf-16(mu86)* mutants survive L1 arrest longer than *daf-18/Pten* mutants suggesting that DAF-18/PTEN has roles independent of DAF-16/FOXO signaling

10. To compare the survival rates between strains, the L1 arrests are carried out in triplicate with at least 100 L1s and the mean survival rate calculated with the Kaplan-Meier method, that is the fraction of living animals over a time course [23]. The significance of difference in overall survival rate is performed using the log-rank test. There are several commercial software programs available to plot the Kaplan–Meier curves and statistics such as Statistical Package for the Social Sciences (SPSS) or Statistical Analysis Software (SAS) package [24] and free online software such as OASIS (Online Application for the Survival Analysis) plots survival curves and performs many statistical tests on the data (http://sbi.postech.ac.kr/oasis/surv) [25]. For an example L1 arrest experiment *see* Fig. 2.

3.2 Adult Life-Span Assay

daf-18/Pten mutants are short lived and conversely insulin receptor mutants *daf-2/IR* are long lived. Adult lifespan assay involves the examination of a population of adult worms until the last worm in the population dies. The data is presented as a survival curve and provides population measurements such as the mean, median and maximal lifespan. The following method is a typical lifespan experiment:

1. Day 0: Transfer 100 L4 larval stage worms to 5 plates (20 per plate) and put in the 20 °C incubator (*see* **Note 4**).
2. Day 2: Transfer live adults to fresh plates being careful not to transfer any eggs or larvae. A worm is counted as dead if it fails to move after gentle prodding (*see* **Note 5**).
3. Days 4, 6, and 8: Transfer live adults every other day after Day 2, until no more eggs are being laid (*see* **Note 6**).
4. Every other day make note of how many worms are alive until the last worm is dead.
5. As in the L1 arrest assay the data are plotted as a Kaplan-Meier curves and the log-rank test is used to calculate statistical survival distributions between the different groups.

3.3 Dauer Assay

Dauer (enduring) larva is a second time point where *C. elegans* can arrest its development (Fig. 1). Worms will undergo dauer diapause at the second molt when they are in harsh environmental conditions, such as, crowding, absence of food, or high temperature [26]. DAF-18/PTEN negatively regulates the IIS to control dauer formation in *C. elegans* [10]. *daf-18/pten* mutants are dauer defective, in contrast to either the insulin receptor, *daf-2*, or PI3K, *age-1* mutants which cause animals to form dauer larvae even in the presence of food. *daf-18/pten* mutants can suppress these phenotypes. Studying the effects of DAF-18/PTEN on dauer diapause will reveal key molecular insights into developmental plasticity and identify new players in the DAF-18/PTEN pathways. Dauer larvae are thinner and have a specialized cuticle. Their mouths are closed by an internal plug and they do not pump their pharynx [26]. The dauer-specific cuticle and the lack of pharyngeal pumping makes them resistant to many chemicals, including 1 % SDS [26]. Thus SDS resistance provides a convenient method in assaying for dauer larvae.

3.3.1 Assay for Dauer Defective Mutants (e.g., daf-18 Mutants, See Note 7)

1. Pick 3–5 adult animals to freshly seeded plates. These plates were allowed to grow until starvation at 20 °C, ~4 days.
2. Transfer a 1 × 1 cm chunk of the agar (with starved worms on it) to a fresh NGM plate seeded with OP50 *E. coli* and make note which plate it came from. You will use this plate to recover dauer-defective strains.
3. To test for dauer formation on the starved plate, add 1 mL 1 % SDS.
4. Let sit for 15 min and all larvae except dauer larvae will be killed.
5. Plates in which all animals are killed by SDS are considered dauer defective. Use the plate from **step 2** to recover the dauer defective (Daf-d) mutation (e.g., *daf-18/pten*).

3.4 Genotyping daf-18/Pten Alleles

3.4.1 Single-Worm PCR

PTEN homozygous knockouts are lethal in *Drosophila*, mouse, and zebrafish [27–31]. In contrast, *C. elegans* has an unparalleled advantage, as the homozygous *daf-18/PTEN* null mutants, despite a shorter life-span, are viable and fertile. Single adult worm lysis followed by PCR can determine *daf-18/pten* insertion/deletion alleles. Below we provide a single-worm PCR method to determine *wild type* and *daf-18*(*ok480*) alleles (*see* **Note 8** for determining other *daf-18* alleles). The RB712 *daf-18* (*ok480*) knockout mutant was identified by the *C. elegans* knockout consortium (Oklahoma Medical Research Foundation) and is available through the *Caenorhabditis* Genetics Centre (CGC). The *daf-18* (*ok480*) allele is a 956 bp deletion in exon 4 and 5 and likely to be a null allele [16]. The *daf-18*(*ok480*) deletion allele can be determined by PCR using primers oIC641 and oIC642 which amplifies a 1703 bp region in *wild-type* animals and a 747 bp region in *daf-18*(*ok480*) mutants. The single-worm PCR protocol was modified from [32].

1. Make fresh 100 μL of PCR lysis buffer by adding proteinase K to 1× PCR buffer (95 μL 1× PCR buffer + 5 μL 20 mg/mL proteinase K). Scale up as needed.
2. Place 10 μL of PCR lysis buffer in the inside of the lid of a 200 μL PCR microcentrifuge tube.
3. Pick a single worm into lysis buffer (*see* **Note 9** about isolating worms after genotyping).
4. Immediately spin down to bottom of tube by spinning in PicoFuge for 3 s or just flick the tube and the drop will move down from the lid.

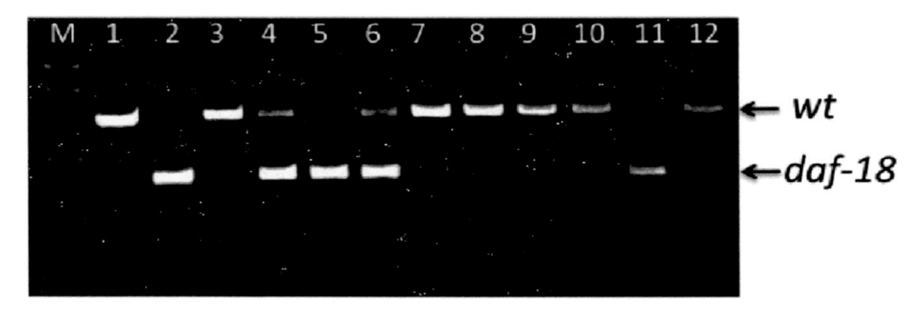

Fig. 3 Single-worm PCR *daf-18*/Pten genotyping. 12 single worms were picked to test for the *daf-18(ok480)* deletion by PCR. Wild-type *daf-18* produces a 1703 bp PCR product while *daf-18*(ok480) produces a 747 bp PCR product. Worms 1, 3, 7, 8, 9, and 10 are homozygous wild type, worm 2 is homozygous for *daf-18*(ok480), and worms 4, 5, 6, 11 are heterozygous

5. Drop into liquid nitrogen in leave for at least 10 min (you can leave all your tubes in liquid nitrogen until you have finished picking all your worms you want to genotype).
6. To perform the worm lysis use the PCR thermocycler and program it to heat to 65 °C for 60–90 min followed by inactivation of proteinase K by heating to 95 °C for 15 min (*see* **Note 10**).
7. Perform PCR (50 μL reaction) by adding 45 μL of PCR "master mix" (1× PCR buffer, 0.5 μM each of oIC641 and oIC642, 0.2 mM/each dNTPs, 50 U/mL thermostable DNA polymerase) to each tube.
8. Run PCR reaction with the following cycling conditions: 95 °C 45 s, 53 °C 45 s, 72 °C 2 min, for 30–35 cycles, followed by a 10 min extension at 72 °C and hold at 4 °C when complete.
9. Run a 10 μL aliquot on a 1.2 % agarose Gel with ethidium bromide (EtBr) to a final concentration of approximately 0.4 μg/mL (4 μlL of lab stock (10 mg/mL) solution per 100 mL gel). *See* Fig. 3 for an example of 12 single worms genotyped.

3.5 Visualizing and Quantifying DAF-18/PTEN

Endogenous DAF-18 levels are low; however overexpression lines (e.g., *quIs18* [16]) and tagged lines (e.g., DAF-18-GFP) allow the visualization and quantification of DAF-18/PTEN protein in various genetic backgrounds. Below we provide a protocol DAF-18/PTEN in situ immunolocalization for mixed-staged and embryo preparations. DAF-18/PTEN Antibody staining was modified from Michael Koelle (http://medicine.yale.edu/lab/koelle) and [33].

3.5.1 Mixed-Stage Antibody Staining

1. Wash mixed staged worms off 1–4 6 cm plates with 1.5 mL M9 buffer, and transfer to a microcentrifuge tube and spin them at 1000 ×*g* for 1 min to pellet worms. Resuspend with 1 mL dH_2O and spin 1000 ×*g* 30 s to wash out the bacteria.
2. Aspirate the supernatant leaving 100 μL and place on ice.

3. Add 200 μL cold 2× fixation solution (with 2 % ßME added fresh), and 100 μL fresh 10 % paraformaldehyde (final concentration is 2.5 %). Mix well, and freeze in liquid nitrogen for at least 10 min (*see* **Note 11**).
4. Thaw in a 37–42 °C water bath for 1–2 min making sure it does not thaw completely.
5. Incubate on ice with occasional agitation for 30 min. Wash the worms twice 1 mL Tris–Triton buffer and resuspend in 1 mL 1 % ßME diluted in Tris–Triton buffer for 2 h at 37 °C on the rotator.
6. Wash in 1 mL 1× borate buffer.
7. Incubate in 1 mL 10 mM DTT diluted in 1× borate buffer for 15 min at room temperature.
8. Wash in 1 mL 1× borate buffer.
9. Incubate in 0.3 % H_2O_2 diluted in 1× borate buffer 15 min at room temperature on rotator.
10. Spin at 3000 × *g* aspirate and wash 1 min in 1 mL 1× borate buffer.
11. Incubate 15 min in 1 mL antibody buffer B (*see* **Note 12**).
12. Primary antibody incubation: Spin at 3000 × *g* for 1 min and remove as much liquid as possible, and add 1 mL of primary antibody diluted in antibody buffer A (*see* **Note 13** for antibodies and dilutions). Mix by pipetting up/down. Incubate at room temperature overnight (agitate occasionally if you can).
13. Wash the worms four times 1 mL antibody buffer B for 25 min each on a rotator at room temperature.
14. Aspirate and add 500 μL secondary antibody (e.g., FITC anti-rabbit) diluted in antibody buffer A (*see* **Note 13** for antibodies and dilutions). At this step you may add 1 μL of 1 mg/mL DAPI to stain for nuclei).
15. Wash the worms with 1 mL antibody buffer B four times for 25 min between each wash on a rotator at room temperature and aspirate leaving the worms in about 50 μL.
16. The stained worms can be stored for months at 4 °C in the dark.
17. Mounting: place 3 μL worm suspension on a microscope slide. We use Prolong® Diamond (Life Technologies) anifade mounting media. Add a drop of Prolong Diamond anti-fade to a slide with worms, mix, and gently drop cover slip over sample. Use a kimwipe to wick out excess liquid. Allow to cure for 24 h at room temperature in the dark, then the slides are ready for imaging. *See* Fig. 4 for example of anti-DAF-18 antibody stainings.

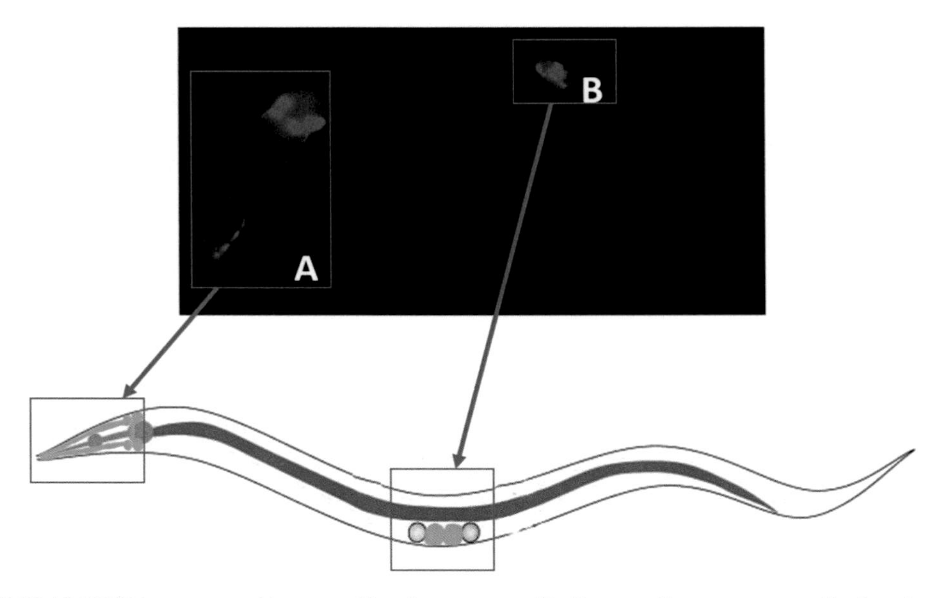

Fig. 4 DAF-18/PTEN is expressed in several head neurons and in the germline precursor cells. A cartoon worm is shown below and the boxed regions depict where anti-DAF-18 immunofluorescence is observed. (**a**) DAF-18 antibody recognizes head neurons in transgenic worms that carry multiple copies of the *daf-18* gene *(quIs18)*. (**b**) Endogenous levels of DAF-18 can be detected in the two germline precursor cells

3.5.2 Embryo Antibody Staining

1. Wash 3–4 6 cm NGM plates of worms off with 1–2 mL M9. Use microfuge tubes to spin 1000 × *g* worms. Wash 2× with 1 mL M9 buffer. Pool worms into a microcentrifuge tube.
2. Add 1 mL of Bleach solution to worm pellet. Vortex occasionally and let sit for 3 min.
3. Spin embryos at 3000 × *g* for 30 s. Aspirate off bleach solution and add 1 mL of new bleach solution. Vortex again and let sit for 3 min. You can leave longer than in embryo preparations for L1 assay (above) but try not to exceed greater than 10 min in the bleach solution as the embryos tend to get damaged.
4. Wash 3 × 1 mL M9 buffer spinning at 3000 × *g* for 1 min and aspirating between each wash.
5. After the final spin, aspirate off all but 100 μL of the M9 buffer. Be careful not to lose the embryos.
6. Add 200 μL 2× fixation solution (with 2 % BME, 2 μL). Mix by inversion.
7. Add 100 μL 10 % paraformaldehyde solution (made fresh). Mix and freeze immediately in liquid nitrogen for 10 min. At this stages embryos may be kept at −80 °C for several weeks until ready to use.
8. Thaw on ice for at least 20 min.

9. Wash 1× with Tris–Triton buffer for 2 min.
10. Wash 2× with antibody buffer A for 10 min each wash.
11. Add 500 μL primary antibody diluted in antibody buffer A (*see* **Note 13** for dilutions).
12. Let sit for 4 h to overnight at room temperature.
13. Wash 4× with antibody buffer B 1 mL each with 10 min between each wash.
14. Add 500 μL of secondary antibody (e.g., FITC anti-rabbit) diluted in antibody buffer A antibody (*see* **Note 13** for antibodies and dilutions).
15. Let sit for at least 2 h at room temperature in the dark. At this step you may add 1 μL of 1 mg/mL DAPI to stain for nuclei).
16. Wash 3× with 1 mL antibody buffer B. Aspirate down to 50 μL.
17. Mount as in the mixed-stage preparation above.

3.6 Protein Preparation for Western Blotting

Visualizing DAF-18/PTEN protein levels on Western blots provides a quick way to see how DAF-18/PTEN protein is regulated in various genetic backgrounds [16, 34]. Human PTEN can also be expressed in *C. elegans* and since several antibodies are commercially available this "humanized" worm strain provides us with a valuable tool to assess human PTEN functions and its regulation in *C. elegans* (Fig. 5).

1. Grow 3–4 6 cm NGM plates of each strain you want to test to the point where they have just cleared the *E. coli* OP50. Do not used starved plates.
2. Wash off worms with several mL of M9 buffer and collect in a 15 mL conical tube and spin at 1000 ×*g* to obtain worm pellet

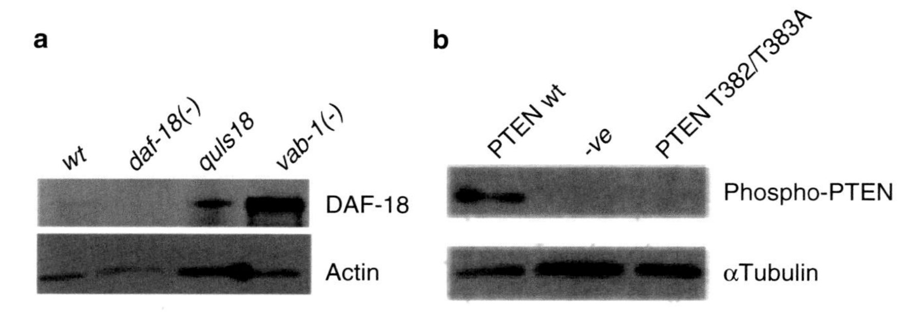

Fig. 5 DAF-18/PTEN Western blots. (**a**) Endogenous levels of DAF-18 (wt) are low and absent in the *daf-18*(−) null allele. Over expression of *daf-18 (quIs18)* or a mutation in the Eph Receptor increases the DAF-18/PTEN levels. (**b**) Human PTEN functions in *C. elegans*. Phospho-specific PTEN antibodies recognize wild type PTEN wt but the signal is abolished in the untransformed worm (*−ve*) or the PTEN T382A T383A mutant. Loading controls (actin or tubulin) are shown below [16, 18]

(about 100 μL). Transfer the worm pellet to a microcentrifuge tube using a glass Pasteur pipette.

3. Wash the worms two times with 1 mL M9 aspirating down to the pellet each time.
4. Aspirate as much M9 without disturbing the worm pellet. Freeze worms in liquid nitrogen or put in the −80 °C freezer. At this point you can leave them in the −80 °C to and prepare other worm protein lysates.
5. Add 500 μL Protein gel sample buffer + 100 μL 1 M DTT to the worm pellet and boil the sample for 10 min. Centrifuge at full speed for 10 min. Transfer supernatant to new microcentrifuge tube.
6. Load 10–30 μL on a standard 10 % SDS PAGE gel followed by Western blotting.
7. Probe blots with anti-DAF-18/PTEN antibodies and secondary HRP-anti-rabbit antibodies (*see* **Note 14** for antibodies and dilutions).

4 Notes

1. It is crucial you do not leave the embryos in bleach solution too long as this will decrease the survival of the L1s.
2. Since the viability of *C. elegans* during L1 arrest depends on the density of worms, the eggs are diluted to a final concentration of no more than ten eggs/μL [20].
3. Score immediately after plating the L1s in the M9 droplet so that the worms remain suspended in liquid. This causes them to thrash and facilitates the identification of living worms. Generally worms that do not thrash within 10 s and do not respond to touch stimuli from a worm pick qualify as dead. Because each time point represents the average percentage survival from an aliquot it is possible that at a later time point the survival rate is higher, e.g., day 3, 94 % and day 6, 96 %. In this case survival curves can be smooth to a non-increasing function. For the example above, the survival at day 3 would be corrected to 96 % as the difference is likely due to pipetting and counting errors. The smoothing does not significantly change the mean survival rates but allows for application of log-rank statistics [24].
4. About 100 young adults should be tested in each longevity assay and each strain should be tested in triplicate.
5. Transferring adult worms can be labor intensive and may also lead to inadvertent injury of worms. To avoid this, 5-fluoro-2-deoxyuridine (FUdR) 120 μM can be added to the NGM plates. FUdR inhibits DNA synthesis and prevents the hatch-

ing of eggs and larval growth. However, there have been reports that show FUdR can extend worm lifespan [35]. Alternatively, genetic mutants such as *fem-1(hc17)* exhibit conditional sterility, thus it eliminates the need for daily transfers of worms away from their progeny [16, 36].

6. If any plates show fungal contamination, transfer the worms onto a fresh plate and move to new plates every few hours for a day. This can help eliminate the fungal spores and the worms can be reintroduced into the experiment. If fungal or bacterial contamination is too persistent discard the whole plate and make note of how many animals were censored. You may have to censor other animals, such as animals that died from injury or premature death due to internal hatching.
7. SDS is not always needed to determine dauer larvae as you can visually score for dauer larvae which are distinguished by a radially constricted body, darkly pigmented intestine, dauer alae, and a constricted pharynx (Fig. 1).
8. Another common *daf-18/Pten* allele *daf-18* (*e1375*) is a partial loss of function allele and has a 30 base pair insertion combined with a deletion of two nucleotides within the fourth exon [11, 10]. Single-worm PCR can also detect the *daf-18* (*e1375*) allele with the primers 5′ GTTCGGCAATCTGG CATTCCAT and 5′CTTCGAGATGTACTGCGTTGC. The mutant PCR product is 28 bp larger (298 bp) than the wild-type PCR product (270 bp) and can be separated on a 3 % agarose gel [10]. The *daf-18*(*nr2037*) deletion allele can be determined by single-worm PCR using the primers 5′CTATTGAAGGAGGACTAACACAGGC and 5′GCCAAC GAAGTGCTAAATCGAC. A diagnostic 0.6-kb or 1.6-kb fragment is indicative of the presence of the *nr2037* or wild-type allele, respectively [12]. The *daf-18*(*nr2037*) is not available through the CGC.
9. From a population of mixed genotypes you can pick individual worms to a plate and let it lay eggs for 1 day. The single adult can then be PCR tested for its genotype. Based on the PCR result homozygous animals can be recovered from their progeny. Genotypes should also be confirmed by expected phenotypes (e.g. dauer defective).
10. If you do not want to use the worm DNA right away at this stage you can store it in a −80 °C freezer.
11. At this stage you may place the frozen worm pellet in a −80 °C freezer indefinitely at this point.
12. At this point the worms are stable and can be stored in antibody buffer B at 4 °C for up to a month.
13. This protocol works for several antibodies. We have used rabbit anti-DAF-18 (1:25 dilution), chicken and mouse anti-GFP

(1: 200 dilution) [16, 34]. For fluorescent secondary antibodies such as FITC conjugated goat anti-rabbit IgG we first reconstitute in Antibody buffer A at 1 mg/mL stock (store in 1 mL aliquots at −80 °C) and use at 1: 200–500 dilution.

14. Antibodies for Western blots include rabbit anti-DAF-18 (1:250 dilution), chicken and mouse anti-GFP (1: 200 dilution), anti-human PTEN monoclonal (D5G7), (D4.3), or (6H2.1), and anti-phospho-PTEN (Ser380/Thr382/383) (Cell Signaling) and are used at a dilution of 1:1000. Secondary antibodies such as goat anti-rabbit conjugated to horseradish peroxidase (HRP) are used at 1: 5–10,000 dilution.

References

1. Li J, Yen C, Liaw D, Podsypanina K, Bose S, Wang SI, Puc J, Miliaresis C, Rodgers L, McCombie R, Bigner SH, Giovanella BC, Ittmann M, Tycko B, Hibshoosh H, Wigler MH, Parsons R (1997) PTEN, a putative protein tyrosine phosphatase gene mutated in human brain, breast, and prostate cancer. Science 275(5308):1943–1947
2. Steck PA, Pershouse MA, Jasser SA, Yung WK, Lin H, Ligon AH, Langford LA, Baumgard ML, Hattier T, Davis T, Frye C, Hu R, Swedlund B, Teng DH, Tavtigian SV (1997) Identification of a candidate tumour suppressor gene, MMAC1, at chromosome 10q23.3 that is mutated in multiple advanced cancers. Nat Genet 15(4):356–362
3. Li DM, Sun H (1997) TEP1, encoded by a candidate tumor suppressor locus, is a novel protein tyrosine phosphatase regulated by transforming growth factor beta. Cancer Res 57(11):2124–2129
4. Liaw D, Marsh DJ, Li J, Dahia PL, Wang SI, Zheng Z, Bose S, Call KM, Tsou HC, Peacocke M, Eng C, Parsons R (1997) Germline mutations of the PTEN gene in Cowden disease, an inherited breast and thyroid cancer syndrome. Nat Genet 16(1):64–67. doi:10.1038/ng0597-64
5. Waite KA, Eng C (2002) Protean PTEN: form and function. Am J Hum Genet 70(4):829–844. doi:10.1086/340026
6. Zhou XP, Woodford-Richens K, Lehtonen R, Kurose K, Aldred M, Hampel H, Launonen V, Virta S, Pilarski R, Salovaara R, Bodmer WF, Conrad BA, Dunlop M, Hodgson SV, Iwama T, Jarvinen H, Kellokumpu I, Kim JC, Leggett B, Markie D, Mecklin JP, Neale K, Phillips R, Piris J, Rozen P, Houlston RS, Aaltonen LA, Tomlinson IP, Eng C (2001) Germline mutations in BMPR1A/ALK3 cause a subset of cases of juvenile polyposis syndrome and of Cowden and Bannayan-Riley-Ruvalcaba syndromes. Am J Hum Genet 69(4):704–711. doi:10.1086/323703
7. Brenner S (1974) The genetics of *Caenorhabditis elegans*. Genetics 77:71–94
8. Sulston JE, Schierenberg E, White JG, Thomson JN (1983) The embryonic cell lineage of the nematode *Caenorhabditis elegans*. Dev Biol 100:64–119
9. Sengupta P, Samuel AD (2009) Caenorhabditis elegans: a model system for systems neuroscience. Curr Opin Neurobiol 19(6):637–643. doi:10.1016/j.conb.2009.09.009, S0959-4388(09)00133-0 [pii]
10. Ogg S, Ruvkun G (1998) The C. elegans PTEN homolog, DAF-18, acts in the insulin receptor-like metabolic signaling pathway. Mol Cell 2(6):887–893
11. Gil EB, Malone Link E, Liu LX, Johnson CD, Lees JA (1999) Regulation of the insulin-like developmental pathway of Caenorhabditis elegans by a homolog of the PTEN tumor suppressor gene. Proc Natl Acad Sci U S A 96(6):2925–2930
12. Mihaylova VT, Borland CZ, Manjarrez L, Stern MJ, Sun H (1999) The PTEN tumor suppressor homolog in Caenorhabditis elegans regulates longevity and dauer formation in an insulin receptor-like signaling pathway. Proc Natl Acad Sci U S A 96(13):7427–7432
13. Rouault JP, Kuwabara PE, Sinilnikova OM, Duret L, Thierry-Mieg D, Billaud M (1999) Regulation of dauer larva development in Caenorhabditis elegans by daf-18, a homologue of the tumour suppressor PTEN. Curr Biol 9(6):329–332
14. Myers MP, Pass I, Batty IH, Van der Kaay J, Stolarov JP, Hemmings BA, Wigler MH, Downes CP, Tonks NK (1998) The lipid

phosphatase activity of PTEN is critical for its tumor supressor function. Proc Natl Acad Sci U S A 95(23):13513–13518

15. Tamura M, Gu J, Danen EH, Takino T, Miyamoto S, Yamada KM (1999) PTEN interactions with focal adhesion kinase and suppression of the extracellular matrix-dependent phosphatidylinositol 3-kinase/Akt cell survival pathway. J Biol Chem 274(29):20693–20703
16. Brisbin S, Liu J, Boudreau J, Peng J, Evangelista M, Chin-Sang I (2009) A role for C. elegans Eph RTK signaling in PTEN regulation. Dev Cell 17(4):459–469. doi:10.1016/j.devcel.2009.08.009, S1534-5807(09)00346-3 [pii]
17. Nakdimon I, Walser M, Frohli E, Hajnal A (2012) PTEN negatively regulates MAPK signaling during Caenorhabditis elegans vulval development. PLoS Genet 8(8):e1002881. doi:10.1371/journal.pgen.1002881
18. Solari F, Bourbon-Piffaut A, Masse I, Payrastre B, Chan AM, Billaud M (2005) The human tumour suppressor PTEN regulates longevity and dauer formation in Caenorhabditis elegans. Oncogene 24(1):20–27
19. Liu J, Chin-Sang ID (2015) C. elegans as a model to study PTEN's regulation and function. Methods 77–78:180–190. doi:10.1016/j.ymeth.2014.12.009
20. Artyukhin AB, Schroeder FC, Avery L (2013) Density dependence in Caenorhabditis larval starvation. Sci Rep 3:2777. doi:10.1038/srep02777
21. Baugh LR (2013) To grow or not to grow: nutritional control of development during Caenorhabditis elegans L1 arrest. Genetics 194(3):539–555. doi:10.1534/genetics.113.150847, genetics.113.150847 [pii]
22. Fukuyama M, Sakuma K, Park R, Kasuga H, Nagaya R, Atsumi Y, Shimomura Y, Takahashi S, Kajiho H, Rougvie A, Kontani K, Katada T (2012) C. elegans AMPKs promote survival and arrest germline development during nutrient stress. Biol Open 1(10):929–936. doi:10.1242/bio.2012836
23. Rich JT, Neely JG, Paniello RC, Voelker CC, Nussenbaum B, Wang EW (2010) A practical guide to understanding Kaplan-Meier curves. Otolaryngol Head Neck Surg 143(3):331–336. doi:10.1016/j.otohns.2010.05.007
24. Lee BH, Ashrafi K (2008) A TRPV channel modulates C. elegans neurosecretion, larval starvation survival, and adult lifespan. PLoS Genet 4(10):e1000213. doi:10.1371/journal.pgen.1000213
25. Yang JS, Nam HJ, Seo M, Han SK, Choi Y, Nam HG, Lee SJ, Kim S (2011) OASIS: online application for the survival analysis of lifespan assays performed in aging research. PLoS One 6(8):e23525. doi:10.1371/journal.pone.0023525
26. Cassada RC, Russell RL (1975) The dauerlarva, a post-embryonic developmental variant of the nematode Caenorhabditis elegans. Dev Biol 46(2):326–342
27. Huang H, Potter CJ, Tao W, Li DM, Brogiolo W, Hafen E, Sun H, Xu T (1999) PTEN affects cell size, cell proliferation and apoptosis during Drosophila eye development. Development 126(23):5365–5372
28. Stambolic V, Suzuki A, de la Pompa JL, Brothers GM, Mirtsos C, Sasaki T, Ruland J, Penninger JM, Siderovski DP, Mak TW (1998) Negative regulation of PKB/Akt-dependent cell survival by the tumor suppressor PTEN. Cell 95(1):29–39
29. Suzuki A, de la Pompa JL, Stambolic V, Elia AJ, Sasaki T, del Barco BI, Ho A, Wakeham A, Itie A, Khoo W, Fukumoto M, Mak TW (1998) High cancer susceptibility and embryonic lethality associated with mutation of the PTEN tumor suppressor gene in mice. Curr Biol 8(21):1169–1178
30. Di Cristofano A, Pesce B, Cordon-Cardo C, Pandolfi PP (1998) Pten is essential for embryonic development and tumour suppression. Nat Genet 19(4):348–355
31. Faucherre A, Taylor GS, Overvoorde J, Dixon JE, Hertog J (2008) Zebrafish pten genes have overlapping and non-redundant functions in tumorigenesis and embryonic development. Oncogene 27(8):1079–1086. doi:10.1038/sj.onc.1210730, 1210730 [pii]
32. Williams BD, Schrank B, Huynh C, Shownkeen R, Waterston RH (1992) A genetic mapping system in Caenorhabditis elegans based on polymorphic sequence-tagged sites. Genetics 131(3):609–624
33. Finney M, Ruvkun G (1990) The unc-86 gene product couples cell lineage and cell identity in C. elegans. Cell 63(5):895–905
34. Liu J, Visser-Grieve S, Boudreau J, Yeung B, Lo S, Chamberlain G, Yu F, Sun T, Papanicolaou T, Lam A, Yang X, Chin-Sang I (2014) Insulin activates the insulin receptor to downregulate the PTEN tumour suppressor. Oncogene 33(29):3878–3885. doi:10.1038/onc.2013.347
35. Aitlhadj L, Sturzenbaum SR (2010) The use of FUdR can cause prolonged longevity in mutant nematodes. Mech Ageing Dev 131(5):364–365. doi:10.1016/j.mad.2010.03.002
36. Glenn CF, Chow DK, David L, Cooke CA, Gami MS, Iser WB, Hanselman KB, Goldberg IG, Wolkow CA (2004) Behavioral deficits during early stages of aging in Caenorhabditis elegans result from locomotory deficits possibly linked to muscle frailty. J Gerontol A Biol Sci Med Sci 59(12):1251–1260

INDEX

Leonardo Salmena and Vuk Stambolic (eds.), *PTEN: Methods and Protocols*, Methods in Molecular Biology, vol. 1388, DOI 10.1007/978-1-4939-3299-3, © Springer Science+Business Media New York 2016

P

R

S

T

U